Quality Assurance and Reliability Engineering

Quality Assurance and Reliability Engineering

Edited by **Michelle Vine**

C LANRYE
INTERNATIONAL

New Jersey

Published by Clanrye International,
55 Van Reypen Street,
Jersey City, NJ 07306, USA
www.clanryeinternational.com

Quality Assurance and Reliability Engineering
Edited by Michelle Vine

International Standard Book Number: 978-1-63240-433-6 (Hardback)

Contents

Preface

Quality and reliability engineering covers quality control studies and issues about basic and advanced materials. To understand the concept of quality control, one must learn about the development of techniques of statistical models and cost models. The major emphasis of quality control and quality science is on the methodologies, used for the betterment of the output of the production unit. This branch of engineering finds its applications and its relevance to all the other branches; because without excellent quality and reliability, any structure or procedure can have potential problems. Some of the specific branches which are covered under quality and reliability engineering include mechanical, electrical and systems engineering.

This field of engineering, studies about concepts, mathematical approaches, and applications in production related to quality and reliability engineering. To understand this book, the reader should have knowledge of statistics, statistical quality control, theories of probability, and quality models. This book has been compiled from material contributed by scholars in the field of quality control.

Instead of organizing the book into a pre-formatted table of contents with chapters, sections and then asking the authors to submit their respective chapters based on this frame, the authors were encouraged by the publisher to submit their chapters based on their area of expertise.

Each chapter includes an extensive review of literature as well as current techniques and philosophies in this field. We have been fortunate to have an outstanding group of quality control specialists from all over the world contributing to this publication. However, the very fact that problems have been illustrated from different perspectives, only adds further value to the book. I would also like to express my gratitude to the team at the publishing house for their constant support at every stage of the publication process.

Editor

A Nonparametric Shewhart-Type Quality Control Chart for Monitoring Broad Changes in a Process Distribution

Saad T. Bakir

College of Business Administration, Alabama State University, P.O. Box 271, Montgomery, AL 36101, USA

Correspondence should be addressed to Saad T. Bakir, bakir00@yahoo.com

Academic Editor: Xiaohu Li

This paper develops a distribution-free (or nonparametric) Shewhart-type statistical quality control chart for detecting a broad change in the probability distribution of a process. The proposed chart is designed for grouped observations, and it requires the availability of a reference (or training) sample of observations taken when the process was operating in-control. The charting statistic is a modified version of the two-sample Kolmogorov-Smirnov test statistic that allows the exact calculation of the conditional average run length using the binomial distribution. Unlike the traditional distribution-based control charts (such as the Shewhart X-Bar), the proposed chart maintains the same control limits and the in-control average run length over the class of all (symmetric or asymmetric) continuous probability distributions. The proposed chart aims at monitoring a broad, rather than a one-parameter, change in a process distribution. Simulation studies show that the chart is more robust against increased skewness and/or outliers in the process output. Further, the proposed chart is shown to be more efficient than the Shewhart X-Bar chart when the underlying process distribution has tails heavier than those of the normal distribution.

1. Introduction

Most traditional statistical quality control charts assume that the monitored process has a prespecified known probability distribution (usually normal for continuous measurements). Consequently, the chart properties (control limits, false alarm rate, and the in-control average run length) would be in error if the process distribution were missspecified. To remedy this, a number of distribution-free (or nonparametric) schemes that maintain the same chart properties over a class of distributions have been proposed in the literature. For an overview of nonparametric control charts, see Chakraborti et al. [1, 2].

Another problem is that traditional control charts aim at monitoring a change in one parameter (usually a location or scale) of a process distribution. Realistically, however, when a special cause influences a process, it may cause a shift in more than one parameter (location, scale, skewness, etc.) of the process distribution. To remedy this, we need control charts designed to monitor a broad rather than a one-parameter change in a process distribution. To our knowledge, Bakir [3] was first to suggest such charts based on the two-sample

Kolmogorov-Smirnov and the Cramer-von Mises statistics. Zou and Tsung [4] proposed a nonparametric likelihood ratio chart for monitoring broad changes in a process distribution. Ross and Adams [5] developed nonparametric charts based on the two-sample Kolmogorov-Smirnov and the Cramer-von Mises statistics. Their charts, however, are designed for individual observations whereas the chart proposed in this paper is designed for grouped observations.

In this paper, we propose a nonparametric Shewhart-type control chart for monitoring a broad change in a process probability distribution. To develop the chart, we assume the availability of a training (or reference) sample taken when the process was operating in statistical control. The idea of assuming a training sample was first used by Park and Reynolds [6] to develop distribution-free charts based on the Orban and Wolfe [7] placement statistic. Later, Hackl and Ledolter [8], Willemain and Runger [9], and Chakraborti et al. [10] proposed nonparametric charts assuming the availability of a reference sample. Our proposed chart works by taking a random sample (test sample) from the process output at each monitoring stage. The charting statistic is a modified version of the two-sample Kolmogorov-Smirnov test statistic where the difference of the reference and test empirical distribution functions is maximized only over the training sample values. Such modification allows exact calculation of the conditional average run length of the proposed chart using the binomial distribution. Unlike the traditional distribution-based Shewhart X-Bar (Shew-XB) chart, the proposed chart maintains the same control limits and the same in-control average run length (ARL$_0$) over the class of all (symmetric or asymmetric) continuous distributions. The Shew-XB and its average run length (ARL) will be discussed in Section 5. Given the training sample, the exact conditional ARL of the proposed nonparametric chart is computed using the binomial probability distribution. The unconditional ARL can then be computed approximately using simulations. A preliminary simulation study shows that the proposed nonparametric chart is more efficient (has smaller out-of-control ARL) than the Shew-XB chart under distributions with tails heavier than those of the normal distribution. If the process distribution is actually normal, then the Shew-XB chart is more efficient, as expected. The simulation study also indicates that the proposed chart is more robust against increased skewness and/or outliers in the process output.

The rest of the paper is organized as follows: Section 2 presents notational preliminaries. Section 3 develops the proposed nonparametric control chart, and Section 4 develops its ARL. Section 5 discusses the Shew-XB chart and its ARL. Section 6 investigates the effects of skewness and outliers on the two charts and presents efficiency comparisons.

2. Preliminaries

We assume the availability of a training random sample, $\vec{X}_0 = (X_{01}, X_{02}, \ldots, X_{0m})$ of size $m > 1$ observations taken when the process was operating in-control. The in-control process distribution is assumed to have a continuous cumulative distribution function (CDF), F_0. Let $S_0(z)$ denote the empirical distribution function (EDF) of the training sample, as defined by

$$S_0(z) = \begin{cases} 0 & \text{if } z < X_{0(1)}, \\ \dfrac{i}{m} & \text{if } X_{0(j)} \le z < X_{t(j+1)} \text{ for } j = 1, 2, \ldots, m-1, \\ 1 & \text{if } z \ge X_{0(m)}. \end{cases}$$

$$(1)$$

Here, $X_{0(j)}$ is the jth order statistic of the training sample, $\vec{X}_0$. Then at each sampling instance t, $t = 1, 2, \ldots$, we obtain one test sample $\vec{Y}_t = (Y_{t1}, Y_{t2}, \ldots, Y_{tn})$ of size $n > 1$ from the process output, which is assumed to have a continuous CDF, F_y. Let $S_t(z)$ be the empirical distribution function of the test sample, $\vec{Y}_t$, given by

$$S_t(z) = \begin{cases} 0 & \text{if } z < Y_{t(1)}, \\ \dfrac{i}{n} & \text{if } Y_{t(i)} \le z < Y_{t(i+1)} \text{ for } i = 1, 2, \ldots, n-1, \\ 1 & \text{if } z \ge Y_{t(n)}. \end{cases}$$

$$(2)$$

Here, $Y_{t(i)}$ is the ith order statistic of the test sample, $\vec{Y}_t$.

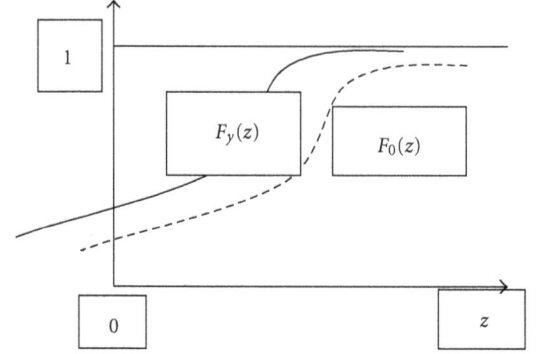

FIGURE 1: $H_a^- : F_y(z) > F_0(z)$.

In practice, we may need to detect one of the following three situations.

Situation 1. Detect whether or not the process tends to produce stochastically smaller observations than the observations of the in-control state. In the terminology of statistical hypothesis testing, we are interested in testing the following null and alternative hypotheses:

$$\begin{aligned} H_0^- &: F_y(z) \le F_0(z) \quad \text{for all } -\infty < z < \infty, \\ H_a^- &: F_y(z) \ge F_0(z) \quad \text{for all } z, \\ &\quad F_y(z) > F_0(z) \quad \text{for at least one } z. \end{aligned}$$

$$(3)$$

Figure 1 depicts Situation 1 graphically and it shows that the process CDF, F_y, has shifted to the left of the in-control CDF F_0.

Situation 2. Detect whether or not the process tends to produce stochastically larger observations than the observations of the in-control state. That is, we are testing the following null and alternative hypotheses:

$$\begin{aligned} H_0^+ &: F_y(z) \ge F_0(z) \quad \text{for all } -\infty < z < \infty, \\ H_a^+ &: F_y(z) \le F_0(z) \quad \text{for all } z, \\ &\quad F_y(z) < F_0(z) \quad \text{for at least one } z. \end{aligned}$$

$$(4)$$

Figure 2 depicts Situation 1 graphically and it shows that the process CDF F_y has shifted to the right of the in-control CDF F_0.

Situation 3. Detect whether or not the process tends to produce smaller and/or larger observations than the in-control state. That is, we are testing the following null and alternative hypotheses:

$$\begin{aligned} H_0 &: F_y(z) = F_0(z) \quad \text{for all } -\infty < z < \infty, \\ H_a &: F_y(z) \ne F_0(z) \quad \text{for at least one } z. \end{aligned}$$

$$(5)$$

Figure 3 depicts Situation 3 graphically.

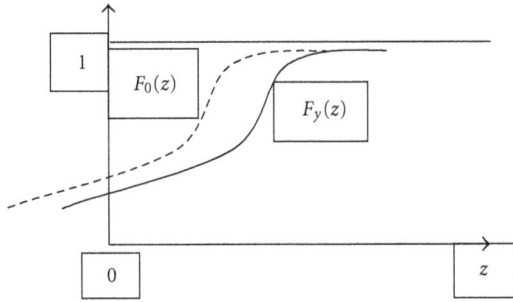

FIGURE 2: $H_a^+ : F_y(z) < F_0(z)$.

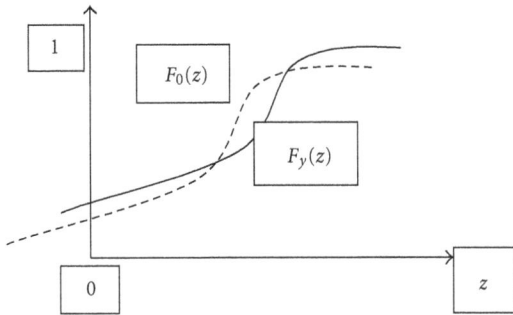

FIGURE 3: $H_a : F_y(z) \neq F_0(z)$.

3. The Proposed Nonparametric Control Chart

In this section, we develop the steps for constructing a distribution-free control chart of the Shewhart type, that is, based on a modified version of the two-sample Kolmogorov-Smirnov statistic. The proposed chart, hereafter, is abbreviated to Shew-KS chart.

Step 1: control characteristic. The characteristic to be controlled (monitored) is the process theoretical probability distribution represented by the CDF, F_y. The purpose is to detect whether or not F_y has shifted away from the process in-control CDF, F_0.

Step 2: sampling plan. Obtain a training sample $\vec{X}_0 = (X_{01}, X_{02}, \ldots, X_{0m})$ of size $m > 1$ when the process was operating in-control. Then obtain a test sample $\vec{Y}_t = (Y_{t1}, Y_{t2}, \ldots, Y_{tn})$ of size $n > 1$ from the process output at each sampling instance t, $t = 1, 2, \ldots$.

Step 3: assumptions. Observations on the process output are independent. The test samples are drawn from unknown continuous distribution with CDF, F_y. The process in-control underlying distribution is assumed continuous with unknown CDF, F_0.

Step 4: pivot statistics. Calculate $S_0(z)$, the EDF of the training sample $\vec{X}_0$. Then at each sampling instance t, $t = 1, 2, \ldots$, calculate the EDF, $S_t(z)$, of the test sample

$\vec{Y}_t = (Y_{t1}, Y_{t2}, \ldots, Y_{tn})$. The pivot statistic for Situation 1 (the lower-sided Shew-KS chart) is

$$\psi_t^- = \min_{z=x_{0j}}[S_0(z) - S_t(z)]. \tag{6}$$

Note that ψ_t^- tends to be negative when the process produces observations smaller than the in-control state, see Figure 1.

The pivot statistic for Situation 2 (the upper-sided Shew-KS chart) is

$$\psi_t^+ = \max_{z=x_{0j}}[S_0(z) - S_t(z)]. \tag{7}$$

Note that ψ_t^+ tends to be positive when the process produces larger observations, see Figure 2.

The pivot statistic for Situation 3 (the two-sided Shew-KS chart) is

$$\psi_t = \max_{z=x_{0j}}|S_0(z) - S_t(z)|. \tag{8}$$

Note 1. The pivot statistics in (6), (7), and (8) will assume only integer values if each is multiplied by the constant mn.

Note 2. The pivot statistics are modified versions of the traditional two-sample Kolmogorov-Smirnov statistic ([11], pp 456–462) where maximization is taken only over the training sample observations, $\vec{X}_0 = (X_{01}, X_{02}, \ldots, X_{0m})$.

Step 5: control sequence (or charting statistics). The control sequences for the lower-sided, the upper-sided, and the two-sided Shew-KS charts, respectively, are

$$\{\psi_t^-, t = 1, 2, \ldots\}, \qquad \{\psi_t^+, t = 1, 2, \ldots\},$$
$$\{\psi_t, t = 1, 2, \ldots\}. \tag{9}$$

Step 6: control limits. For simplification, we consider one upper-sided control limit, L, and let the lower-sided control limit be $-L$. Because the Shew-KS chart is distribution free, the control limit, L, is a constant (design parameter) that depends only on m, n, and the desired in-control ARL$_0$ of the chart. This control limit, however, does not depend on the functional form of the in-control process distribution.

Step 7: signaling rules. The two-sided Shew-KS signals if $\psi_t \geq L$. The lower-sided and upper-sided control charts signal, respectively, if $\psi_t^- \leq -L$, and $\psi_t^+ \geq L$.

Illustration. Let $N(\theta, \sigma^2)$ denote a normal probability distribution with mean θ and variance σ^2. As an illustration of the proposed Shew-KS chart, we generated 20 observations from the standard normal distribution $N(0,1)$ to represent the in-control reference X-sample. Four test Y-samples, each of size 10, were generated. The first two samples, Y1 and Y2, have a $N(0,1)$ distribution. The third and fourth samples, Y3 and Y4, have an $N(2,1)$ and an $N(3,4)$ distributions, respectively. Table 1 depicts the generated samples and the required calculations for the two-sided charting statistic. The resulting Shew-KS chart, shown in Figure 4, gives an out-of-control signal at the third sample when the process mean shifted from zero to two.

TABLE 1: Charting statistic ψ_t for simulated date: reference X: $N(0,1)$, test samples $Y1$, and $Y2$: $N(0,1)$, $Y3$: $N(2,1)$, and $Y4$: $N(3,4)$.

	X: $N(0,1)$	S_0	$Y1$: $N(0,1)$	$\#Y1 \leq X$	$S_1(x)$	$\lvert S_0(x) - S_1(x) \rvert$	$Y2$: $N(0,1)$	$\#Y2 \leq X$	$S_2(x)$	$\lvert S_0(x) - S_2(x) \rvert$	$Y3$: $N(2,1)$	$\#Y3 \leq X$	$S_3(x)$	$\lvert S_0(x) - S_3(x) \rvert$	$Y4$: $N(3,2)$	$\#Y4 \leq X$	$S_4(x)$	$\lvert S_0(x) - S_4(x) \rvert$
1	−1.551	0.05	−1.657	1	0.1	0.05	−2.04	1	0.1	0.05	−0.952	0	0	0.05	−1.401	0	0	0.05
2	−1.074	0.1	−1.223	2	0.2	0.1	−1.41	2	0.2	0.1	−0.562	0	0	0.1	0.963	1	0.1	0
3	−0.738	0.15	−1.067	4	0.4	0.25	−0.5	2	0.2	0.05	0.49	1	0.1	0.05	1.701	1	0.1	0.05
4	−0.722	0.2	−0.891	4	0.4	0.2	−0.4	2	0.2	0	0.5125	1	0.1	0.1	1.896	1	0.1	0.1
5	−0.683	0.25	−0.427	4	0.4	0.15	−0.23	2	0.2	0.05	0.8878	1	0.1	0.15	2.502	1	0.1	0.15
6	−0.198	0.3	−0.04	5	0.5	0.2	0.088	5	0.5	0.2	1.0848	2	0.2	0.1	2.665	1	0.1	0.2
7	−0.085	0.35	0.224	5	0.5	0.15	0.93	5	0.5	0.15	1.2253	2	0.2	0.15	2.681	1	0.1	0.25
8	−0.069	0.4	0.283	5	0.5	0.1	0.174	5	0.5	0.1	1.4254	2	0.2	0.2	4.544	1	0.1	0.3
9	−0.065	0.45	0.906	5	0.5	0.05	0.686	5	0.5	0.05	1.565	2	0.2	0.25	4.623	1	0.1	0.35
10	−0.05	0.5	1.3	5	0.5	0	1.865	5	0.5	0	2.5184	2	0.2	0.3	5.213	1	0.1	0.4
11	0.101	0.55		6	0.6	0.05		7	0.7	0.15		2	0.2	0.35		1	0.1	0.45
12	0.111	0.6		6	0.6	0		7	0.7	0.1		2	0.2	0.4		1	0.1	0.5
13	0.3953	0.65		8	0.8	0.15		8	0.8	0.15		2	0.2	0.45		1	0.1	0.55
14	0.434	0.7		8	0.8	0.1		8	0.8	0.1		2	0.2	0.5		1	0.1	0.6
15	0.5174	0.75		8	0.8	0.05		8	0.8	0.05		4	0.4	0.35		1	0.1	0.65
16	0.5454	0.8		8	0.8	0		8	0.8	0		4	0.4	0.4		1	0.1	0.7
17	1.0735	0.85		9	0.9	0.05		9	0.9	0.05		5	0.5	0.35		2	0.2	0.65
18	1.3542	0.9			1	0.1		9	0.9	0		7	0.7	0.2		2	0.2	0.7
19	1.9396	0.95			1	0.05		10	1	0.05		9	0.9	0.05		4	0.4	0.55
20	2.1318	1			1	0		10	1	0		9	0.9	0.1		4	0.4	0.6
						$\psi_1 =$ 0.25				$\psi_2 =$ 0.2				$\psi_3 =$ 0.5				$\psi_4 =$ 0.7

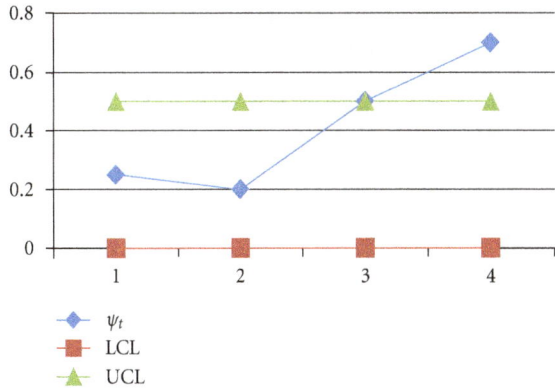

FIGURE 4: Shew-KS chart.

4. Calculating the ARL of the Shew-KS Chart

Values of the ARL are needed for the implementation and the performance evaluation of control charts. The implementation of a control chart requires values of the control limits that lead to some desired values of the in-control ARL_0. When the successive charting statistics of a Shewhart type control chart are independent, the run length distribution is geometric and the ARL = 1/Pr (signal). Unfortunately, this property of independence does not hold for the proposed Shew-KS chart because the successive charting statistics, ψ_t, $t = 1, 2, \ldots$, all depend on the same training sample $\vec{X}_0$. In this section, we develop a method for calculating the ARL of the chart by first conditioning on the training sample $\vec{X}_0$, a method used by Chakraborti [12, 13] and by Vermaat et al. [14].

Recall that the proposed two-sided Shew-KS chart signals at the first sampling instance t, $t = 1, 2, \ldots$ for which $\max_{z=x_{0j}} |S_0(z) - S_t(z)| \geq L$, where $L > 0$. Suppose that the maximum occurs at a value, say, $z = z_{\max}$. Thus, a signal occurs if

$$|S_0(z_{\max}) - S_t(z_{\max})| \geq L. \tag{10}$$

Equivalently, a signal occurs if

$$S_0(z_{\max}) - S_t(z_{\max}) \leq -L, \tag{11}$$

or

$$S_0(z_{\max}) - S_t(z_{\max}) \geq L. \tag{12}$$

It is seen that (11) represents the branch of the chart that detects if the process output, $\vec{Y}_t$, is stochastically smaller than the in-control output, $\vec{X}_0$, see Figure 1. Similarly, (12) detects if the process output is stochastically larger than the in-control output, see Figure 2.

We will work first on the lower branch, (11), of the chart. After rearranging terms and multiplying the inequality by the test sample size, n, (11) becomes

$$nS_t(z_{\max}) \geq n[L + S_0(z_{\max})]. \tag{13}$$

Note that because maximization in (6)–(8) is defined over the X values only, $z_{\max}$ becomes fixed when we condition on $\vec{X}_0$. Consequently, given the training sample, $\vec{X}_0$, $nS_t(z_{\max})$ becomes a binomial random variable, $B_{n,\pi}$, with number of trials $=n$ and probability of success $\pi = F_y(z_{\max})$. Given $\vec{X}_0$, the exact conditional probability of a signal and the exact conditional ARL of the lower branch of the chart, respectively, are

$$\text{cond}P^-_{\text{Shew-KS}} = \Pr\left(B_{n,\pi} \geq n[S_0(z_{\max}) + L] \,\middle|\, \vec{X}_0\right),$$

$$\text{cond}ARL^-_{\text{Shew-KS}} = \frac{1}{\text{cond}P^-_{\text{Shew-KS}}}. \tag{14}$$

Upon taking expectations over the training sample $\vec{X}_0$, the unconditional probability of a signal and the unconditional ARL for the lower branch of the Shew-KS chart are

$$P^-_{\text{Shew-KS}} = E(\text{cond}P^-_{\text{Shew-KS}}),$$

$$ARL^-_{\text{Shew-KS}} = E(\text{cond}ARL^-_{\text{Shew-KS}}). \tag{15}$$

Similarly, (12) can be transformed to show that the signal conditional probability and the conditional ARL of the upper branch of the chart, respectively, are

$$\text{cond}P^+_{\text{Shew-KS}} = \Pr\left(B_{n,\pi} \leq n[S_0(z_{\max}) - L] \,\middle|\, \vec{X}_0\right),$$

$$\text{cond}ARL^+_{\text{Shew-KS}} = \frac{1}{\text{cond}P^+_{\text{Shew-KS}}}. \tag{16}$$

The two-sided Shew-KS chart signals if either one of the lower or the upper branch signals. Therefore, given the training sample $\vec{X}_0$, the conditional probability of a signal and the conditional ARL of the two-sided chart, respectively, are

$$\text{cond}P_{\text{Shew-KS}} = \text{cond}P^-_{\text{Shew-KS}} + \text{cond}P^+_{\text{Shew-KS}}$$

$$\text{cond}ARL_{\text{Shew-KS}} = \frac{1}{\text{cond}P_{\text{Shew-KS}}}. \tag{17}$$

Theoretically, the unconditional probability of a signal and the unconditional ARL of the one-sided and two-sided charts are the expectations, over the training sample, of the respective conditional expressions.

Unfortunately, the required unconditional expectations over the training sample, $\vec{X}_0$, cannot be expressed directly into a closed form. In this paper we use a large number of simulations, $M = 1$ million runs, to estimate these unconditional expectations. At each simulation run, a training sample $\vec{X}_0$ and a test sample $\vec{Y}_t$ are generated where the conditional probability of a signal and the conditional ARL are calculated according to their exact formulas. The International Mathematical and Statistical Library (IMSL) is used to generate pseudo random variables (assuming a certain probability distribution) for the training and test samples, calculate the empirical CDFs, identify $z = z_{\max}$, calculate the exact binomial probabilities, and finally calculate the exact conditional probabilities of a signal and the ARLs. Then we average out these conditional values over the number of simulations to get estimates of the required unconditional expectations. For example, the estimated values for the unconditional probability of a signal and the unconditional ARL for the two-sided Shew-KS chart, respectively, are

$$\hat{P} = \frac{1}{M} \sum_{r=1}^{M} \text{cond} P_r,$$

$$\hat{\text{ARL}} = \frac{1}{M} \sum_{r=1}^{M} \text{cond} \text{ARL}_r,$$

(18)

where r is the simulation run number. Similar calculations are applied to estimate the unconditional expectations of the one-sided charts. The above methods for calculating the signal probability and the ARL can play an important role in the design and implementation of the proposed Shew-KS chart because they allow for calculating control limits that correspond to certain desired values of the in-control ARL for various values of m and n.

5. Calculating the ARL of the Traditional Shew-XB Chart

In this section, we outline an efficient method for evaluating the ARL of the traditional Shew-XB chart in order to compare it to the proposed Shew-KS chart.

The traditional Shew-XB control chart is based on charting the sequence of means $\overline{Y}_t = (1/n) \sum_{j=1}^{n} Y_{tj}$ of the test samples $\vec{Y}_t = (Y_{t1}, Y_{t2}, \dots, Y_{tn})$, $t = 1, 2, \dots$. The control limits are calculated using the sample mean $\overline{X}_0 = (1/m) \sum_{j=1}^{m} X_{0j}$ and the sample standard deviation $S_0 = \sqrt{\sum_{j=1}^{m} (X_{0j} - \overline{X}_0)^2 / (m-1)}$ of the in-control training sample $\vec{X}_0 = (X_{01}, X_{02}, \dots, X_{0m})$. The two-sided Shew-XB chart gives an out-of-control signal at the first sampling instance, t, for which $\overline{Y}_t \leq \overline{X}_0 - kS_0/\sqrt{n}$ or $\overline{Y}_t \geq \overline{X}_0 + kS_0/\sqrt{n}$, where k is a constant chosen (usually equals 3) to achieve a desired in-control ARL. One-sided Shew-XB charts can be obtained by employing one of the signaling rules.

Because the successive signaling events (e.g., $\overline{Y}_t \leq \overline{X}_0 - kS_0/\sqrt{n}$) of the Shew-XB control chart all use the same control limits as estimated from the same training sample, they are no longer independent. Therefore, we cannot use the geometric distribution argument that the ARL = 1/Pr (signal). Jensen et al. [15] presented a literature review on the effects of parameter estimation on control charts properties. Chakraborti [12, 13] used conditional expectation arguments to derive exact formulas for the run length distribution and the ARL of the Shew-XB chart when the in-control mean and/or the variance are estimated. However, almost all studies regarding the effects of parameter estimation on control charts properties assume that the underlying process distribution is normal. This dogmatic restriction to the normal distribution is not appropriate to the distribution-free world of nonparametric statistics where we need to compare the performance of the competing charts under distributions other than the normal.

In this section, we use a conditional expectation argument and simulations to obtain reasonable estimates for the values of the unconditional ARL of the Shew-XB chart under several underlying process distributions.

Given the training sample, $\vec{X}_0$, the exact conditional probabilities of signals for the lower and the upper branches of the Shew-XB chart, respectively, are

$$\text{cond} P_{\text{Shew-XB}}^{-} = F_{\overline{Y}}\left(\overline{X}_0 - \frac{kS_0}{\sqrt{n}} \,\Big|\, \vec{X}_0 \right),$$

$$\text{cond} P_{\text{Shew-XB}}^{+} = 1 - F_{\overline{Y}}\left(\overline{X}_0 + \frac{kS_0}{\sqrt{n}} \,\Big|\, \vec{X}_0 \right),$$

(19)

where $F_{\overline{Y}}$ is the theoretical CDF of the sample mean of the test sample, $\vec{Y}_t$. For example, if the test sample has a normal distribution with mean μ_y and variance σ_y^2, then the conditional probability of a signal of the lower and the upper branches of the Shew-XB charts, respectively, are

$$\text{cond} P_{\text{Shew-XB}}^{-} = \Phi\left(\frac{\overline{X}_0 - \mu_y - k(S_0/\sqrt{n})}{\sigma_y/\sqrt{n}} \,\Big|\, \vec{X}_0 \right),$$

$$\text{cond} P_{\text{Shew-XB}}^{+} = 1 - \Phi\left(\frac{\overline{X}_0 - \mu_y + k(S_0/\sqrt{n})}{\sigma_y/\sqrt{n}} \,\Big|\, \vec{X}_0 \right),$$

(20)

where Φ is the CDF of the standard normal distribution. The conditional probability of a signal for the two-sided chart is

$$\text{cond} P_{\text{Shew-XB}} = \text{cond} P_{\text{Shew-XB}}^{-} + \text{cond} P_{\text{Shew-XB}}^{+}. \quad (21)$$

The exact CDFs of the sample mean are known for many populations beside the normal. We state some results concerning the CDF of the mean of a sample of size n drawn from gamma, Cauchy and Laplace distributions.

(1) For a 3-parameter gamma (shape = α, scale = β, location = θ) distribution with probability density function (PDF)

$$f_y(z) = \frac{(z-\theta)^{\alpha-1}\left[e^{-(z-\theta)/\beta} \right]}{\beta^{\alpha}\Gamma(\alpha)}. \quad (22)$$

The mean of a random sample of size n also has a gamma (shape = $n\alpha$, scale = β/n, location = θ) distribution.

(2) For a Cauchy distribution (scale = β, location = θ), the mean of a random sample of size n has the same Cauchy distribution with PDF

$$f_{\overline{y}}(z) = f_y(z) = \frac{\beta}{\pi\left[\beta^2 + (z-\theta)^2 \right]}, \quad (23)$$

for $-\infty < z < \infty$, $-\infty < \theta < \infty$, $\beta > 0$.

(3) For a Laplace distribution (scale = β, location = θ) with PDF

$$f_y(z) = \frac{1}{\beta} e^{-|z-\theta|/\beta}, \quad (24)$$

for $-\infty < z < \infty$, $-\infty < \theta < \infty$, $\beta > 0$, the distribution of the sample mean is a little bit complicated. Using basic results in Johnson et al. ([16], pp 167), we can express the CDF of the sample mean (when $z > 0$ and $\theta = 0$) as

$$F_{\bar{y}}(z) = 1 - \sum_{j=0}^{n-1} 2^{j+1-2n} \binom{2n-j-2}{n-1} \Pr\left(G_{j+1} > z\right), \quad (25)$$

where G_{j+1} is a 2-parameter gamma (shape $= j + 1$, scale $= \beta/n$) random variable.

The conditional ARLs for the lower, upper, and two-sided Shew-XB chart, respectively, are condARL$^{-}_{\text{Shew-XB}} = 1/\text{condP}^{-}_{\text{Shew-XB}}$, condARL$^{+}_{\text{Shew-XB}} = 1/\text{condP}^{+}_{\text{Shew-XB}}$, and condARL$_{\text{Shew-XB}} = 1/\text{condP}_{\text{Shew-XB}}$.

The unconditional probabilities of signals and the unconditional ARLs for the Shew-XB chart are obtained by taking expectations of their respective conditional values over the training sample $\vec{X_0}$. In practice, we use large number of simulations to estimate these unconditional quantities as described in Section 4, (18).

6. Effects of Skewness, Outliers, and Efficiency Comparisons

In this section, we conduct simulation studies to investigate the sensitivity against skewness, outliers, and the efficiency of both the traditional Shew-XB and the proposed Shew-KS control charts.

6.1. Effects of Skewness. We now examine the effect of skewness on the in-control ARLs of the Shew-XB and the Shew-KS charts. The control limits for the two charts are adjusted so that both charts have the same in-control ARL of 170 under the standard normal distribution. The sample sizes of the training and test sample, respectively, are $m = 39$ and $n = 10$. We used IMSL to generate pseudo random numbers from the three-parameter gamma distribution in (22). We varied the shape parameter α to obtain extremely skewed to almost symmetric distributions. To have a gamma distribution with mean $= 0$ and variance $= 1$, the scale and location parameters are chosen as $\beta = \alpha^{-1/2}$ and $\theta = -\sqrt{\alpha}$. The ARLs of both charts are calculated by first getting the exact conditional ARL and then using one million simulation runs to get the unconditional ARL by averaging the conditional ARL. Table 2 shows that ARL$_0$ of the Shew-KS is chart not affected at all by the skewness of the distribution. The Shew-XB chart, however, changes dramatically as we move from extremely skewed to symmetric distributions. The ARL$_0$ of the Shew-XB chart becomes close to the normal theory value of 170 only when the shape parameter of the gamma distribution is at least 16. Table 2 depicts two anomalies in the ARL of the Shew-XB chart when the shape parameter $\alpha = 2$ or 3. For explanation of these anomalies, refer to Vermaat et al. ([14], pp. 343).

6.2. Effects of Outliers. There are situations in which the in-control process output is contaminated by few outliers; for

TABLE 2: In-control ARLs (in groups of size 10) of the Two-sided Shew-XB and the Shew-KS charts for a process with a gamma (shape α, scale β, location θ) Distribution. Training Sample Size $m = 39$ and Test Sample Size $n = 10$.

			Shew-XB	Shew-KS
Shape α	Scale $\beta = \alpha^{-1/2}$	Location $\theta = -\sqrt{\alpha}$	$k = 2.75$	$CL = 0.541$
1	1	-1	1099.2	165
2	0.7071	-1.4142	2125.4	167.3
3	0.5774	-1.7321	1764.9	164.3
4	0.5000	-2.0000	320.5	176.7
5	0.4472	-2.2361	289.1	173.1
9	0.3333	-3.0000	227	166.6
10	0.3162	-3.1623	181.6	181.0
12	0.2887	-3.4641	235.7	171.6
16	0.2500	-4.0000	189.9	160.5
20	0.2236	-4.4721	168.5	179.5
Normal	1.0	0.0	170	170

Note: for the Shew-XB chart, 2.75 is the "k" value in the control limits $\overline{X_0} \pm kS_0/\sqrt{n}$.

example, a process involving complex analytical measurements. A single extreme outlying observation may trigger an out-of-control signal while in fact the process is in-control, thus increasing the false alarm rate and decreasing the in-control ARL of the control chart. A good model for generating normally distributed processes with occasional outliers is the contaminated normal distribution, the CDF of which is

$$N_P(\theta, \sigma^2) = (1 - p)N(\theta, 1) + pN(\theta, \sigma^2), \quad (26)$$

where $0 \le p \le 1$. We will refer to p and σ^2 as the *percentage of contamination* and the *extremity of contamination*, respectively. When $\theta = 0$, the process is in-control though producing occasional outliers. When $p = 0.0$ and $\theta = 0$, (26) becomes the standard normal CDF. In each simulation run, we generated 500 reference samples, of size $m = 39$ each, from the standard normal distribution. For each reference sample thus generated, we generated 500 test samples, of size $n = 10$ each, from the contaminated normal distribution all with $\theta = 0$ and all the possible combinations of (σ^2, p) where $\sigma^2 = 4, 9, 16$ and $p = 0.01, 0.01, 0.20$. Table 3 shows the simulated values of the two-sided in-control ARLs (in groups of size $n = 10$) of the Shew-XB and the Shew-KS charts for various levels of contamination.

Table 3 shows that the effect of outliers depends on the contamination severity (p, σ^2), and the effect is more pronounced on the Shew-XB than on the Shew-KS chart. Keeping in mind that the in-control ARL of the traditional Shew-XB chart for a process operating with no outliers is 163 (in groups of size $n = 10$), we make the following observations on the results in Table 3.

(i) Under very light percentage $p = 1\%$ and light extremity $\sigma^2 = 4$ of contamination, outliers have no

TABLE 3: Simulated Values of the ARL_0 (in groups of size $n = 10$) of the Two-sided Shew-XB and the Shew-KS Charts for an in-control Processes ($\theta = 0$) with outliers.

%	Extremity $\sigma^2 = 4$		$\sigma^2 = 9$		$\sigma^2 = 16$	
	Shew-XB $k = 2.80$	KS $L = 0.541$	Shew-XB $k = 2.80$	KS $L = 0.541$	Shew-XB $k = 2.80$	KS $L = 0.541$
$P = 0.0\%$	163	171	163	171	163	171
$P = 1\%$	159	171	115	171	70	171
$P = 5\%$	90	147	39	148	21	143
$P = 10\%$	61	144	22	127	12	118
$P = 15\%$	41	120	15	120	8	107
$P = 20\%$	33	112	11	94	6	88

Note: the case when $P = 0.0\%$ represents a process operating under a standard normal distribution with no outliers.

effect on both charts as the in-control ARLs of the two charts do not change.

(ii) Under very light percentage $p = 1\%$ but moderate extremity $\sigma^2 = 9$ of contamination, outliers have a noticeable effect on the traditional Shew-XB chart as its ARL_0 drops to 115, which entails about $163/115 = 1.4$ times as many false alarms as the expected ARL_0 of 163. When the extremity of contamination grows to $\sigma^2 = 16$, outliers have substantial effect on the Shew-XB chart as its ARL_0 drops to 70, which entails about 2.3 times as many false alarms. In contrast, the light percentage of contamination $p = 1\%$ has no effect on the Shew-KS neither when $\sigma^2 = 9$ nor when $\sigma^2 = 16$.

(iii) Under a moderate percentage $p = 10\%$ and light extremity $\sigma^2 = 4$ of contamination, outliers have noticeable effect on the Shew-XB as its ARL_0 drops to 61, which entails about 2.6 times as many false alarms as the expected ARL_0 of 163. In contrast, the Shew-KS triggers $171/144 = 1.2$ times as many false alarms. When the extremity of contamination grows to $\sigma^2 = 9$, outliers have a greater effect on the Shew-XB chart as its ARL_0 drops to 22, entailing about 7.4 times as many false alarms. In contrast, the Shew-KS triggers 1.4 times as many false alarms. With severe extremity of contamination, $\sigma^2 = 16$, the ARL_0 of the Shew-XB chart drops to 12, entailing 13.8 times as many false alarms. In contrast, the Shew-KS triggers 1.5 times as many false alarms.

(iv) Outliers can have a more dramatic effect on the traditional Shew-XB chart when the percentage of contamination is as large as $p = 20\%$. With light extremity of contamination, $\sigma^2 = 4$, the ARL_0 of the Shew-XB chart drops to 33, entailing about 4.9 times as many false alarms as the expected ARL_0 of 163. In contrast, the Shew-KS triggers 1.5 times as many false alarms. With moderate extremity of contamination, $\sigma^2 = 9$, the ARL_0 of the Shew-XB chart drops to 11, entailing 14.8 times as many false alarms. In contrast, the Shew-KS triggers 1.8 times as many false alarms. With severe extremity of contamination, $\sigma^2 = 16$, the ARL_0 of the Shew-XB chart drops to just 6, entailing

about 27 times as many false alarms. In contrast, the Shew-KS triggers 1.9 times as many false alarms.

To sum up the results of Table 3, we conclude that for monitoring processes contaminated by outliers, one should not use the traditional Shew-XB, unless the percentage and the severity of contamination are both very light, around $(\sigma^2, p) = (4, 0.01)$. Otherwise, the traditional Shew-XB would trigger many folds of false alarm signals as those for an uncontaminated process. Outliers have some effect on the Shew-KS chart when the percentage of contamination is as high as $(\sigma^2, p) = (16, 0.20)$.

6.3. A Simulation Study for Efficiency. To compare two control charts, we adjust their control limits so that their in-control ARLs become approximately equal and then compare their out-of-control ARLs at various levels of change in the monitored quality characteristic. The chart with the smaller out-of-control ARL is considered to be more efficient.

In this section, we perform a simulation study to compare the efficiencies of the Shew-XB and the Shew-KS charts. The competing charts are compared for processes operating under a normal distribution with a standard deviation of 1.0, a Cauchy distribution and a Laplace distribution. Equation (23) gives the PDF of the Cauchy distribution with center θ (=median, mean does not exist) and scale β. Equation (24) gives the PDF of the Laplace distribution with center θ (=mean = median) and scale β. In (23), the scale β is set to equal 0.2605 so that the Cauchy distribution with center 0 has a probability of 0.05 to the right of 1.645, the same as that of the standard normal distribution. Since the Laplace distribution has variance $=2\beta^2$, the scale β in (24) is set to be $\beta = 1/\sqrt{2}$ so that the Laplace distribution has a standard deviation of 1.0. Efficiency comparisons are made when the median θ of the process is shifted from the in-control value of 0.0 to 1.0 in increments of 0.2. We used a training sample size $m = 39$ and a test sample size $n = 10$ in all comparisons. As mentioned in Sections 4 and 5, the ARLs of both charts are calculated by first getting the exact conditional ARLs and then using one million simulation runs to get the unconditional ARL by averaging the conditional ARLs. Tables 4, 5, and 6 show the simulated values of the two-sided ARLs (in groups of size $n = 10$). The in-control ARLs (when $\theta = 0.0$) of the competing control

TABLE 4: ARLs (in groups of size 10) of the Two-sided Shew-XB and the Shew-KS Charts under a Normal Distribution. Training Sample Size $m = 39$ and Test Sample Size $n = 10$.

	Shew XB $k = 2.75$	Shew-KS $CL = 0.541$
Median = 0.0	168.0	171.0
0.2	109.1	125.9
0.4	35.6	57.3
0.6	9.	15.4
0.8	3.4	6.2
1.0	1.8	3.0

Note: for the Shew-XB chart, 2.75 is the "k" value in the control limits $\overline{X}_0 \pm kS_0/\sqrt{n}$.

TABLE 5: ARLs (in groups of size 10) of the Two-sided Shew-XB and the Shew-KS Charts under a Laplace distribution. Training Sample Size $m = 39$ and Test Sample Size $n = 10$.

	Shew XB $k = 2.80$	Shew-KS $CL = 0.541$
Median = 0.0	170.0	171.0
0.2	126.0	100.1
0.4	58.0	25.0
0.6	20.0	5.3
0.8	9.3	2.3
1.0	5.0	1.6

Note: for the Shew-XB chart, 2.80 is the "k" value in the control limits $\overline{X}_0 \pm kS_0/\sqrt{n}$.

TABLE 6: ARLs (in groups of size 10) of the Two-sided Shew-XB and the Shew-KS Charts under a Cauchy distribution. Training Sample Size $m = 39$ and Test Sample is of Size $n = 10$.

	Shew-XB $k = 6.70$	Shew-KS $CL = 0.541$
Median = 0.0	171.0	171.0
0.2	165.1	38.7
0.4	165.1	3.2
0.6	161.2	1.4
0.8	160.5	1.1
1.0	159.4	1.0

Note: for the Shew-XB chart, 6.70 is the "k" value in the control limits $\overline{X}_0 \pm kS_0/\sqrt{n}$.

charts are made approximately equal by adjusting the control limits under each distribution.

Examinations of Tables 4, 5, and 6 lead to the following findings.

(i) Table 4: For monitoring processes operating under a normal distribution, the Shew-KS is less efficient (has larger out-of-control ARLs) than the traditional Shew-XB chart.

(ii) Table 5: For monitoring processes operating under a Laplace distribution, the proposed Shew-KS is more efficient (has smaller out-of-control ARLs) than the

traditional Shew-XB chart at all shifts in the process center.

(iii) Table 6: For monitoring processes operating under a Cauchy distribution, the Shew-KS becomes dramatically more efficient than the traditional Shew-XB chart at all shifts in the process center. For example, the Shew-KS chart is quicker than the tradition Shew-XB chart by about 4-times, 52-times, 115-times, 146-times, and 159-times to signal at respective shifts of $\theta = 0.2, 0.4, 0.6, 0.8$, and 1.0 in the process center.

To sum up, the results in Tables 4, 5, and 6 lead to the following recommendations.

To monitor processes operating under moderate or heavy-tailed underlying distributions (heavier than those of the normal), the proposed Shew-KS is more efficient than the traditional Shew-XB chart. This is in addition to the advantage that the Shew-KS chart maintains same control limits over the class of (symmetric or asymmetric) continuous distributions. If one is sure that the process underlying distribution is normal, then the traditional Shew-XB chart is recommended over the Shew-KS.

7. Summary and Suggestions for Further Research

In this paper, a distribution-free (or nonparametric) Shewhart-type statistical quality control chart is developed for detecting broad changes in the underlying probability distribution of a process. We assume the availability of a random sample, called training sample, taken when the process was operating in-control. At each sampling instance, we take a random sample from the process output and calculate a modified version of the two-sample Kolmogorov-Smirnov test statistic, which will serve as the charting statistic. A signal is given if the charting statistic falls outside the control limits. Unlike the traditional distribution-based control charts (such as the Shew-XB), the proposed chart maintains the same in-control ARL_0 value over the class of all (symmetric or asymmetric) continuous distributions. Consequently, the control limits of the proposed chart need not be adjusted according to an assumed underlying process distribution. Given the training sample, the conditional ARL of the proposed chart is computed exactly using the binomial probability distribution. The unconditional ARL can then be estimated by simulations. A preliminary simulation study shows that the proposed Shew-KS chart is more efficient than the Shew-XB chart if the process underlying distribution has tails heavier than those of the normal. If the underlying process distribution can be assumed normal, then the Shew-XB chart is more efficient, as expected. The simulation study also indicates that the proposed chart is more robust against increased skewness and/or outliers in the process output.

Further simulation studies are needed to expand the efficiency comparisons of the proposed Shew-KS chart with charts other than Shew-XB. Tabulated values of the control limits are needed for the implementation of the proposed chart. It is worthwhile to investigate how the Kolmogorov-Smirnov statistic can be used with other charting schemes, such as the exponentially weighted moving average (EWMA) and the cumulative sum (CUSUM.)

References

[1] S. Chakraborti, P. van der Laan, and S. T. Bakir, "Nonparametric control charts: an overview and some results," *Journal of Quality Technology*, vol. 33, no. 3, pp. 304–315, 2001.

[2] S. Chakraborti and M. A. Graham, "Nonparametric control charts," in *Encyclopedia of Statistics in Quality and Reliability*, Wiley, New York, NY, USA, 2007.

[3] S. T. Bakir, "Quality control charts for detecting a general change in a process," in *Proceedings of the Section on Quality and Productivity (ASA '97)*, pp. 53–56, American Statistical Association, 1997.

[4] C. Zou and F. Tsung, "Likelihood ratio-based distribution-free EWMA control charts," *Journal of Quality Technology*, vol. 42, no. 2, pp. 174–196, 2010.

[5] G. J. Ross and N. M. Adams, "Two nonparametric control charts for detecting arbitrary distribution changes," *Journal of Quality Technology*, vol. 44, no. 2, pp. 102–116, 2012.

[6] C. Park and M. R. Reynolds Jr., "Nonparametric procedures for monitoring a location parametric based on linear placement statistics," *Sequential Analysis*, vol. 6, no. 4, pp. 303–323, 1987.

[7] J. Orban and D. A. Wolfe, "A class of distribution-free two-sample tests based on placements," *Journal of the American Statistical Association*, vol. 77, pp. 666–670, 1982.

[8] P. Hackl and J. Ledolter, "A control chart based on ranks," *Journal of Quality Technology*, vol. 23, pp. 117–124, 1991.

[9] T. R. Willemain and G. C. Runger, "Designing control charts using an empirical reference distribution," *Journal of Quality Technology*, vol. 28, no. 1, pp. 31–38, 1996.

[10] S. Chakraborti, P. van der Laan, and M. A. van de Wiel, "A class of distribution-free control charts," *Journal of the Royal Statistical Society C*, vol. 53, no. 3, pp. 443–462, 2004.

[11] W. J. Conover, *Practical Nonparametric Statistics*, Wiley, New York, NY, USA, 3rd edition, 1999.

[12] S. Chakraborti, "Run length, average run length and false alarm rate of shewhart X-bar chart: exact derivations by conditioning," *Communications in Statistics Part B*, vol. 29, no. 1, pp. 61–81, 2000.

[13] S. Chakraborti, "Parameter estimation and design considerations in prospective applications of the $\overline{X}$ chart," *Journal of Applied Statistics*, vol. 33, no. 4, pp. 439–459, 2006.

[14] M. B. Vermaat, R. A. Ion, R. J. M. M. Does, and C. A. J. Klaassen, "A comparison of Shewhart individuals control charts based on normal, non-parametric, and extreme-value theory," *Quality and Reliability Engineering International*, vol. 19, no. 4, pp. 337–353, 2003.

[15] W. A. Jensen, L. A. Jones-Farmer, C. W. Champ, and W. H. Woodall, "Effects of parameter estimation on control chart properties: a literature review," *Journal of Quality Technology*, vol. 38, no. 4, pp. 349–364, 2006.

[16] N. L. Johnson, S. Kotz, and N. Balakrishnan, *Continuous Univariate Distributions*, vol. 2, Wiley, New York, NY, USA, 1995.

2

Risk-Based Allowed Outage Time and Surveillance Test Interval Extensions for Angra 1

Sonia M. Orlando Gibelli,[1] P. F. Frutuoso e Melo,[2] and Sérgio Q. Bogado Leite[1]

[1] *Comissão Nacional de Energia Nuclear, DRS/CGRC, 22294-900 Rio de Janeiro, RJ, Brazil*
[2] *Programa de Engenharia Nuclear, COPPE/UFRJ, 21941-972 Rio de Janeiro, RJ, Brazil*

Correspondence should be addressed to Sonia M. Orlando Gibelli, sonia@cnen.gov.br

Academic Editor: Mohammad Modarres

In this work, Probabilistic Safety Assessment (PSA) is used to evaluate Allowed Outage Times (AOT) and Surveillance Test Intervals (STI) extensions for three Angra 1 nuclear power plant safety systems. The interest in such an analysis lies on the fact that PSA comprises a risk-based tool for safety evaluation and has been increasingly applied to support both the regulatory and the operational decision-making processes. Regarding Angra 1, among other applications, PSA is meant to be an additional method that can be used by the utility to justify Technical Specification relaxation to the Brazilian regulatory body. The risk measure used in this work is the Core Damage Frequency, obtained from the Angra 1 Level 1 PSA study. AOT and STI extensions are evaluated for the Safety Injection, Service Water and Auxiliary Feedwater Systems using the SAPHIRE code. In order to compensate for the risk increase caused by the extensions, compensatory measures as (1) test of redundant train prior to entering maintenance and (2) staggered test strategy are proposed. Results have shown that the proposed AOT extensions are acceptable for two of the systems with the implementation of compensatory measures whereas STI extensions are acceptable for all three systems.

1. Introduction

Traditionally, Technical Specifications (TS) such as limiting conditions of operation, which include system/component AOT and STI, have been established based only on deterministic analysis [1, 2] and engineering judgment [2]. However, the experience with plant operation indicates that some elements of the requirements may be unnecessarily restrictive, and a few may not be conducive to safety [2], stressing the need to review them based on probabilistic models capable of assessing the incremental risks associated with their modifications.

In the last decades, PSAs have been elaborated and used not only to support risk-informed regulation but also to evaluate new plant designs, among other applications. Due to its broad modeling capability, which includes system functions and common-cause failure events (CCF), PSA is especially suitable for the analysis of TS modifications. Risk-based methods to improve TS requirements are meant to (1) evaluate the risk impact of TS modifications in such a

way as to objectively justify them and (2) provide risk-based information for the regulatory decision-making process [1].

This work presents an evaluation of AOT and STI extensions for three Angra 1 safety systems [3] through the use of its PSA Level 1 study, namely (1) Safety Injection System (SIS), (2) Service Water System (SWS), and (3) Auxiliary Feedwater System (AFWS). The SIS is a two-train standby system; the SWS is a two-train system but with three pumps (one of which is a swing) where one pump is in service during normal operation and the other two are in standby mode, and the AFWS is a standby system with two motor-operated pumps plus a turbine-driven pump as diversity. They were chosen to cover the types of typical safety systems of a Westinghouse two-loop PWR design. The calculations are carried out by means of the SAPHIRE code [4] with Angra 1 PSA data as the baseline input. The development of Angra 1 PSA resulted in an average estimation for the Core Damage Frequency (CDF) value of $4.015E - 05$ per reactor-year, originated from internal events and including the external event flood, although typically Level 1 PSAs evaluate the core

TABLE 1: AOT and STI proposed extensions.

System	AOT	STI
Safety injection system (SIS)	24 hours → 168 hours	1 month → 3 months
Service water system (SWS)	48 hours → 168 hours	1 month → 3 months
Auxiliary feedwater system (AFWS)	48 hours → 168 hours	1 month → 3 months

damage frequency by considering only internal accident scenarios [5]. The proposed extensions are shown in Table 1.

At first, the analyses of the AOT and STI extensions are carried out separately. However, at the end of the study, simultaneous analyses of TS modifications for two systems are also evaluated. Nevertheless, contributions to risk originated by interactions between AOT and STI are out of the scope of this work.

The risk measure adopted in this work is the CDF that can be obtained from a PSA Level 1, as part of its results. TS modifications resulting in small risk increments, that is, increments smaller than $1.0E - 06$/reactor-year, are considered acceptable whenever the related CDF is less than $1.0E - 04$/reactor-year. However, for CDF increments greater than $1.0E - 06$/reactor-year, the acceptability of TS modifications depends on an evaluation process that should be performed in accordance with the applicable safety criteria. In this work, the U.S. Nuclear Regulatory Commission (NRC) safety criteria for TS risk-based evaluation are adopted [6]. As part of this study, two types of compensatory measures are proposed to compensate for risk increments associated with TS modifications: (1) test of the system redundant component, right before entering the AOT and (2) modification of the current test strategy from sequential to staggered, when applicable.

We present in the following a discussion of the state of the art of the subject, concerning the use of probabilistic approaches for the discussion of allowable outage time and surveillance test interval extensions.

Reference [7] presents results of studies of interactions between AOT and STI. The quantification of the interactions is developed in terms of risk, through the use of PSA methods. For such, an approach for modifications of AOT and STI and their effects in risk is used, taking into account the interactions between the two parameters. The work is divided into several steps and aims to present approaches that can encompass risk measures from the component level to the CDF risk level. However, the study presented in this paper concentrates the analysis only in the component level. According to its conclusions, it would be necessary to include a system-level approach or above CDF in order to make it possible to include test strategies and common-cause failures. For such, the authors developed an algorithm to deal with interactions between AOT and STI.

The methodological approach presented in [8] includes the calculation of the risk impact of a TS modification proposal, through the use of PSA. The calculations had been developed for the Seabrook and south Texas plants. The risk measures used for carrying out the study are the system's unavailabilities and the CDF. The acceptance criteria adopted in the study approve changes whose modifications in the risk do not exceed 10%. The difference between this approach

and our work lies mainly in the adoption, in the latter, of compensatory measures to neutralize the risk impact increase associated to the TS modification.

Reference [9] deals with the comparison between the risk increase associated with AOT extension and the risk associated with plant shutdown. Examples are shown for the Residual Heat Removal and Service Water systems of a BWR. The study suggests the use of the compensatory measure and test of the redundant train, for the decision-making process between continued operation with AOT extension and plant shutdown.

Reference [10] discusses the interactions between AOT and ST interval requirements by using probabilistic methods. The proposed methodology encompasses (a) the definition of AOT and STI interactions; (b) their quantification in terms of risk using PSA methods; (c) an approach for evaluating simultaneous AOT and STI modifications; (d) an assessment of strategies for giving flexibility to plant operation through simultaneous changes on AOT and STI using tradeoff-based risk criteria.

Reference [11] deals with STI optimization based on PSA methods. The approach is divided into three levels: component, system, and plant. The study concentrates on the system level application that, according to the authors, has presented results that differ from the existing technical specification STI requirements. Sequential and staggered testing strategies are used. Test strategies are introduced through the development of fault trees that include several time-dependent variables related not only to the test interval, but also to the repair time and duration of the test. The cited work uses PSA methods and Markov processes [12] to model dependences in the component and system levels.

Reference [13] presents an analysis of time-dependent unavailabilities of periodically tested components under various test and repair policies in which component renewals may eventually take place. Cost functions are developed under three different preventive maintenance policies, including test, maintenance, repair, and accident costs. The roles of different costs and aging parameters are explicitly obtained for several models, mainly in the case of an extended Weibull failure rate.

Reference [14] presents a section dedicated to technical specifications in respect of limiting conditions of operation, requirements of tests, and the use of PSA to present the concepts for the evaluation of what would be "optimum," in terms of AOT and STI associated risk. The work cites the use of PSA related to the treatment of common-cause failures. It also emphasizes the relevance to distinguish the single-event AOT from the cumulative AOT (for example, yearly AOT). The paper also evaluates the risk associated with the STI variations and the test-limit risk. The work presents a calculation proposal of AOT extensions and their comparison with the acceptance risk criteria. The adoption of compensatory measures to compensate possible risk increases is not included in the work.

Reference [15] uses a method for evaluating the risk associated with AOT for several plant configurations, based on risk measures. The risks associated with various plant configurations considered in the study are compared with an adopted risk criterion, and the results obtained for the various proposed configurations are compared among each other. However, a methodology of compensatory measures is not introduced for configurations that include AOT extensions, when risk exceeds the acceptable ones, according to the criterion.

Reference [16] presents a proposal of simultaneous optimization of parameters related to risk-based test and maintenance and functions of cost, modeled through genetic algorithms in the system level. The work presents an example of application of the methodology for the high-pressure injection system. The results present values of costs and unavailabilities of valves and pumps, establishing a correspondence with test intervals and periods of preventive maintenance for the same valves and pumps.

Reference [17] proposes a new method for explicit modeling of single-component failure event within multiple common-cause failure groups simultaneously. This method is based on a modification of the frequently utilized beta factor parametric model. The motivation for developing this method lays in the fact that one of the most widespread softwares for fault tree and event tree modeling as part of the probabilistic safety assessment do not comprise the option for simultaneous assignment of single-failure event to multiple common-cause failure groups.

Reference [18] deals with common-cause failure probabilities in fault-tree analyses including testing and time dependencies of standby safety systems. Modeling and quantification of common-cause failures of redundant standby safety systems can be implemented by implicit or explicit fault-tree techniques. The paper derives common-cause event probabilities for both methods for systems with time-related CCFs modeled through generic multiple failure rates. The impact of test interval periods and test staggering strategy are included. An economic model provides insights into the impacts of various parameters: the optimal test interval increases with the increase in redundancy and testing cost and decreases with the increase of accident cost and initiating event rates. Staggered testing with additional tests allows the estimation of the longest optimal test intervals.

As part of a risk-informed reviewing of technical specifications, [19] considers a method for determining risk-balanced allowed outage times for a VVER440 plant. The method was tentatively applied to the emergency core cooling system including accumulators, low-pressure injection, and recirculation. Two different risk measures are interesting in studying AOTs: the AOT single event risk and the average yearly risk [2]. Both are required to stay within predetermined criteria. The longest outage time that satisfies both constraints has been established as the risk-based AOT.

Reference [20] presents the development and application of a multiple objective genetic algorithm to perform the simultaneous optimization of periodic test intervals (TI) and test strategies, both included in test planning (TP). Lessons learned from the high pressure injection system results show that the double-loop multiple-objective evolutionary algorithm is able to find the Pareto set of solutions.

Reference [21] presents a proposal of maintenance risk management through the development of a pilot study which evaluates the risk of the plant during maintenance activities, using PSA methods. The article presents a modeling for common-cause failures, without, however, presenting an application for extensions of AOT and STI. The scope of the mentioned work includes a discussion of risk monitor, PSA modeling, risk measures, and acceptance criteria as well as the role of regulatory bodies.

Reference [1] discusses a method for risk-informed optimization of allowed outage times to be used in the reviewing process of technical specifications of a Finnish VVER440 nuclear plant. The method takes into account realistic component repair times and their changes with AOTs, the possibility of common-cause failures and the risk increase in extended power operation versus forced shutdown. The method has been used to review the AOTs of the plant emergency core-cooling pumps. The results suggest that the AOTs of single failures could be shortened, while the AOTs of CCFs should be changed from immediate shutdown to three days to repair. Shutdown risks and the possibility of CCFs were found to have a major effect on optimal AOTs.

Reference [22] presents the analysis of surveillance test interval by Markov processes for shutdown systems in CANDU nuclear power plants. In order to comply with regulatory requirements, the system availability is evaluated taking into account component failure rate data and the benefits of the tests. There are many factors that should be considered in determining the surveillance test intervals for shutdown systems, and these include the desired target availability, the actual availability, the probability of spurious trips, the test duration, and the adverse effects of testing, such as wearout, introduction of human errors, and additional costs. The paper uses a Markov model to quantify the effect of surveillance test duration and interval on the system unavailability and spurious trip probability. The model can also be used to analyze the variation of CDF in respect of changes in the test interval once combined with the conditional core damage model derived from event trees and fault trees of the plant PSA.

In order to calculate the risk impact caused by testing and maintenance (AOT and STI) by means of PSA, several efforts have been carried out internationally. Component and system level evaluations were found in the literature, among which only a few have chosen CDF as a risk measure. Some of the reviewed works emphasize the evaluation of interactions between the contributions of testing and maintenance. Other studies have focused on comparing the risk of plant shutdown with the risk associated with continued operation after the expiry of the AOT and STI limits. To compensate the risk increase, these works suggest the use of compensatory measures as for example, the test of the redundant train before starting maintenance activities. Works that use genetic algorithms for optimization of TS considering cost-related parameters have also been found in the literature.

The originality of our work lies in the proposed PSA modeling to reflect the use of compensatory measures, namely test of the redundant train and/or modification of the testing strategy from sequential to staggered to compensate the increase in risk caused by AOT and STI extensions.

For that purpose, a specific methodology was developed to fit the fault simulation (or Corrective Maintenance (CM)) of system trains, whose redundancies are affected in what concerns the calculation of common-cause failures [23]. This methodology includes not only the simulation of the test of the redundant train, but also the treatment of the related common-cause failures, which must be changed to depict the newly tested train condition. Furthermore, the STI extension is also modeled, as well as the compensation for the possible introduced increase in risk, by means of modeling effects in risk when the test strategy is switched from sequential to staggered, in case it is feasible. The calculations were performed using the SAPHIRE computer code [4], taking the Angra 1 PSA as the input data. The SAPHIRE code, used by the NRC, was adopted by both Angra 1 utility and the regulatory body as a tool for calculating Angra 1 PSA, which justifies its use in our work.

This paper is organized as follows: Section 2 addresses the risk impact by considering both the AOT and STI contributions to the total CDF. Common-cause failures are treated in this context in Section 3. Initially, a 1-out-of-3 : G system is analyzed and then the same analysis is detailed for a two-component system. Next, the AOT and STI modeling are discussed in Section 4 and, finally, Section 5 details the compensatory measures that are used, which are related to the test of the redundant train, as well as to the staggered and sequential test strategies. Section 6 deals with the results obtained, by first discussing the current technical specifications for Angra 1 and then presenting the system calculation results. Overall conclusions and recommendations are presented in Section 7.

2. Risk Impact

2.1. AOT Risk Impact. It is well known that component unavailability is associated with risk increase and can occur either due to Corrective Maintenance (CM) or Preventive Maintenance (PM). This work deals only with the CM type of component unavailability. The AOT of a component under maintenance is established in such a way as to provide enough time to repair it without incurring in undue risk. In order to evaluate the risk associated with the AOT, the following aspects should be considered:

(i) risk increase;

(ii) duration;

(iii) frequency of occurrence.

Based on these aspects, three types of risk impacts associated with the AOT should be controlled: (1) CDF increment,

(2) the single AOT risk impact, and (3) the yearly AOT risk impact.

The single-event risk (r) is a function of both CDF increment and duration (d) of the component unavailability. The single event risk can be expressed by [2]

$$r = (\text{CDF}_1 - \text{CDF}_B) \cdot d, \tag{1}$$

where CDF_1 is the risk level when the component is known to be down or unavailable, and CDF_B is the baseline risk, obtained from the PSA level 1 analysis.

The yearly AOT risk (R) is defined as the single event risk multiplied by the frequency of occurrence (f)

$$R = f \cdot (\text{CDF}_1 - \text{CDF}_B) \cdot d. \tag{2}$$

The literature on risk analysis [5] presents the treatment of the calculation of single event and yearly average AOT contributions, when compared with the acceptance criteria

$$
\begin{aligned}
r \leq r_c & \quad d \leq \frac{r_c}{\Delta R} \\
R \leq R_c & \quad d \leq \frac{R_c}{\Delta R \cdot f}
\end{aligned}
\quad \text{Criterion } d \leq \min\left[\frac{r_c}{\Delta R}; \frac{R_c}{\Delta R \cdot f}\right],
\tag{3}
$$

where r_c = single event risk criterion, R_c = yearly risk criterion.

When it comes to AOT extension, both risk-type contributions must be evaluated. Whether the annual frequency of a component entering an AOT is greater than one, the yearly risk contribution will be also greater than the single-event contribution. However, this is more likely to happen when dealing with PM, which is associated with a programmed maintenance schedule. On the contrary, regarding CM assessment, the frequency of occurrence of unscheduled maintenances is expected to be close to the component failure rate, which is much lesser than one. According to the NRC risk criteria, the single-component event risk (r_c) should not be greater than $5.0E - 07$/reactor-yr and there is no established criteria for the yearly averaged risk (R_c) [24]. As this work aims to analyze component failure, which leads to a corrective maintenance, the single-event criteria is applied.

2.2. STI Risk Impact. The risk contribution associated with the component test interval is mostly related to the possibility that the component fails during the period between two consecutive tests. Since the components under consideration belong to standby safety systems, component failures are understood as standby time-related failures. An exemption lies on pump A of the SWS, which is in service during normal operation. For calculation purposes, we consider the three pumps of this system belonging to the same common-cause group.

If the test is efficient, the component failure probability (Q) drops to zero immediately after the test and starts to increase as a function of time. The average unavailability of a periodically tested component is a function of both the failure rate (λ) and the test interval time (T) and can be expressed by [2]

$$Q \approx \frac{1}{2}\lambda T. \qquad (4)$$

The increase in CDF associated with the test interval extension is

$$\Delta CDF = CDF_{ESTI} - CDF_B, \qquad (5)$$

where CDF_{ESTI} is the CDF taking into account the extended test interval.

3. Treatment of Common-Cause Failures

In Angra-1 PSA, the treatment of common-cause failures is carried out by means of the Multiple Greek Letters (MGL) Model [5], which is considered the most general extension of the Beta Factor Model. In order to simulate the failure of one component in an m-component system, it is useful to utilize the Basic Parameter Model [25]. The concepts underlying this model and its relation with the MGL model are summarized below. Consider a common-cause group consisting of three identical components A, B, and C. Defining event X_I as the single independent failure of component X, C_{XY} as the common-cause failure of components X and Y (and not Z), and C_{XYZ} as the common cause failure of components X, Y, and Z, then the total failure of component X can be expressed by

$$X_T = X_I \cup C_{XY} \cup C_{XZ} \cup C_{XYZ}. \qquad (6)$$

Also, if

$Q_1 = P[X_I]$ = failure probability of component X from independent causes,

$Q_2 = P[C_{XY}]$ = common-cause failure probability of components X and Y (and not Z),

$Q_3 = P[C_{XYZ}]$ = common-cause failure probability of components X, Y, and Z, and since the events are mutually exclusive, then

$$Q_T = P[X_T] = Q_1 + 2Q_2 + Q_3. \qquad (7)$$

Similarly, for a two-component system the total failure of X is expressed by

$$Q_T = Q_1 + Q_2. \qquad (8)$$

The MGL general equation that expresses the common-cause failure probability among k particular components belonging to a common-cause group with m components, Q_k, is [5]

$$Q_k^{(m)} = \frac{1}{\binom{m-1}{k-1}}\left(\prod_{i=1}^{k}\rho_i\right)(1-\rho_{k+1})Q_T, \qquad (9)$$

where $1 \leq k \leq m$, $\rho_1 = 1$, $\rho_2 = \beta$, $\rho_3 = \gamma$, $\rho_4 = \delta, \ldots, \rho_{m+1} = 0$, and Q_T is the component total failure probability. It can be seen that (7) and (8) are readily obtained from equation (9). Moreover, for the case of a two-component system, the MGL Model is reduced to the Beta Factor Model, a widely used model for common-cause analysis based on a single parameter (β), in addition to the component total failure probability. The Beta Factor Model for a two-component common-cause group is expressed by

$$Q_k^{(2)} = \begin{cases} (1-\beta) \cdot Q_T & k = 1 \\ \beta \cdot Q_T & k = 2. \end{cases} \qquad (10)$$

3.1. Failure Probability of a Three-Component System. A 1-out-of-3 : G system is considered failed when all three components have failed. For that case, neglecting cut sets of type $\{C_{AB}, C_{AC}\}$ as explained in [25], the expanded fault tree can be represented by

$$\{A_I, B_I, C_I\}; \{A_I, C_{BC}\}; \{B_I, C_{AC}\}; \{C_I, C_{AB}\}; \{C_{ABC}\}. \qquad (11)$$

The system (S) failure probability $S = A_T \cap B_T \cap C_T$ will be then

$$Q_S = P[S] = Q_1^3 + 3Q_1Q_2 + Q_3. \qquad (12)$$

For a three-component system, the conditional failure probability, given that component A has failed, can be expressed by

$$P\left[\frac{S}{A_T}\right] = \frac{P[A_T \cap B_T \cap C_T]}{P[A_T]} = \frac{Q_S}{Q_T}. \qquad (13)$$

Developing the conditional probabilities for the addition of the minimal cut sets, one can obtain the expression for the Basic Parameter Model:

$$\frac{Q_S}{Q_T} = Q_1^2\frac{Q_1}{Q_T} + 2Q_1\frac{Q_2}{Q_T} + Q_2\frac{Q_1}{Q_T} + \frac{Q_3}{Q_T}. \qquad (14)$$

For practical considerations, taking into account that we are considering a 1-out-of-3 : G logic, and using the approximation $Q_1 \approx Q_T$, then (14) reduces to

$$\frac{Q_S}{Q_T} \approx Q_1^2 + 3Q_2 + \frac{Q_3}{Q_T}. \qquad (15)$$

Next, by using $Q_2 = (1/2)\beta(1 - \gamma)Q_T$ and $Q_3 = \beta\gamma Q_T$, then (15) becomes

$$\frac{Q_S}{Q_T} \approx Q_1^2 + \frac{3}{2}\beta(1 - \gamma)Q_T + \beta\gamma. \qquad (16)$$

Applying (16) for Angra 1 SWS, considering the adopted values for $\beta = 0.02$, $\gamma = 0.63$, and assuming Q_1 and $Q_T < 10^{-3}$, it is reasonable to consider the approximation

$$\frac{Q_S}{Q_T} \approx \beta\gamma, \qquad (17)$$

where Q_S/Q_T represents the system failure given one component has failed and $\beta\gamma$ represents the common-cause failures. A detailed discussion on this subject can be found in Appendix E of [25].

Table 2: Modifications on pump probabilities to simulate CM.

System/pump	Modification of independent failures	Modification of common-cause failures
SIS—train A pump	$Q_{FS} \rightarrow 1$ (true) $(Q_1^{(2)})$ $Q_{FR} \rightarrow 1$ (true) $Q_{MA} \rightarrow 1$ (true)	$Q_{CC} \rightarrow \beta \, (Q_2^{(2)})$
SWS—train A pump	$Q_{FR} \rightarrow 1$ (true) $(Q_1^{(3)})$	$Q_{CCABC} \rightarrow \beta\gamma \, (Q_3^{(3)})$ $Q_{CCAB} \rightarrow 0 \, (Q_2^{(3)})$ $Q_{CCAC} \rightarrow 0 \, (Q_2^{(3)})$ $Q_{CCBC} \rightarrow 0 \, (Q_2^{(3)})$
AFWS—train A motor-operated pump	$Q_{FS} \rightarrow 1$ (true) $(Q_1^{(2)})$ $Q_{FR} \rightarrow 1$ (true) $Q_{MA} \rightarrow 1$ (true)	$Q_{CC} \rightarrow \beta \, (Q_2^{(2)})$

Q_{FS} stands for the pump A failure to start probabilities, Q_{FR} stands for the pump A failure to run probabilities, and Q_{MA} stands for the pump A unavailability due to maintenance.

3.2. Failure Probability of a Two-Component System. Similarly, considering a 1-out-of-2 : G system comprised by components A and B, the system conditional failure probability, given component A has failed, can be expressed by

$$P\left[\frac{S}{A_T}\right] = \frac{P[A_T \cap B_T]}{P[A_T]} = \frac{Q_S}{Q_T}. \qquad (18)$$

Equation (18) can also be expressed as the sum of the minimum cut sets, which results in

$$\frac{Q_S}{Q_T} = \frac{Q_1^2}{Q_T} + \frac{Q_2}{Q_T}. \qquad (19)$$

For practical considerations $Q_1 \approx Q_T$ and (19) is reduced to

$$\frac{Q_S}{Q_T} \approx Q_1 + \beta. \qquad (20)$$

4. AOT and STI Extension Modeling

4.1. Current Technical Specifications. The technical specifications taken into account in this work are part of the Final Safety Analysis Report (FSAR) [3]. It is worth mentioning that Angra 1 is a two-loop plant, where most of the systems are typically a two-train type, that is, with two pumps, one in each train. An example of that is the SIS whose pumps are submitted to 24-hour allowed outage time in case of failure of one of them. Despite that fact, exemptions of system designs are also treated in this work as the SWS and the AFWS.

The SWS is a two-train system with a third swing pump, all belonging to the same common-cause group. However, one pump must be operating during normal plant operation. The SWS allowed outage time, given that one pump is failed, is 48 hours. In this system one pump is sufficient for the post-accident core-cooling operation.

The AFWS comprises two motor-operated pumps and one turbine-driven pump. However, only the two motor-operated pumps belong to the same common-cause group, being the turbine-driven diversity of the system. Therefore, for common-cause evaluation only the motor-operated pumps are taken into account. The allowed outage time for loss of one motor-operated pump is 48 hours.

The SIS, SWS, and AFWS surveillance test intervals are one month. Concerning test strategy, the three SWS pumps are tested sequentially while the SIS and AFWS surveillance tests are staggered. Therefore, in order to compensate risk, the only candidate system to a test strategy modification to the staggered type is the SWS.

It should be mentioned that the SIS, SWS, and AFWS pump tests are performed on-line. This means that at anytime they might be demanded, they will be ready to operate.

4.2. AOT Extension Modeling. AOT extensions, for each one of the mentioned systems, are analyzed assuming a CM or a train failure in the corresponding system. By doing this, a component failure is simulated by setting its independent failure probability to one (or true) in the SAPHIRE code, and common-cause failures are treated according to (15) or (20). Since the pumps are the most important components concerning TS in the safety systems here analyzed, the extension proposals are only applied to them. Table 2 shows the modifications to be implemented on the pump probabilities, including the common-cause failures of the SIS, SWS, and AFWS to simulate CM, according to the methodology previously described. It can be noticed that, since during normal plant operation the SWS has one train in service, the analysis of that in-service train failure implies setting the pump failure-to-run probability equal to 1 (true). For the other systems, Q_1 from (20) is a combination of the pump independent unavailability modes (i.e., maintenance and failure-to-start and failure-to-run modes). Moreover, for common-cause analysis purposes, the AFWS is considered a two-train system, as explained before.

4.3. STI Extension Modeling. In order to reflect STI extensions, the calculation of CDF_{ESTI} should include modifications on the pump failure to start and common-cause failure to start unavailabilities. Therefore, for a two-component

TABLE 3: Modifications on pump probabilities to simulate STI extensions as proposed in Table 1.

System/pumps	Failure to start	Common-cause failure
SIS	$Q'_{FS} \rightarrow 3 \cdot Q_{FS} \; (Q_1^{(2)})$	$Q'_{CCFS} \rightarrow 3 \cdot Q_{CCFS} \; (Q_2^{(2)})$
SWS	$Q'_{FS} \rightarrow 3 \cdot Q_{FS} \; (Q_1^{(3)})$	$Q'_{CCFS} \rightarrow 3 \cdot Q_{CCFS} \; (Q_3^{(3)})$
AFWS	$Q'_{FS} \rightarrow 3 \cdot Q_{FS} \; (Q_1^{(2)})$	$Q'_{CCFS} \rightarrow 3 \cdot Q_{CCFS} \; (Q_2^{(2)})$

Q'_x stands for pump probabilities for the extended time period.

system, the pump unavailabilities can be expressed by (using (4) and (9) and the approximation $Q_1 \approx Q_T$)

$$Q_1^{(2)} \approx \frac{1}{2}\lambda T_E,$$

$$Q_2^{(2)} \approx \frac{1}{2}\beta\lambda T_E, \qquad (21)$$

where T_E is the extended test interval and β is the common-cause factor.

Similarly, in case of a three-component system, the equations expressing the test extension are:

$$Q_1^{(3)} \approx \frac{1}{2} \cdot \lambda \cdot T_E,$$

$$Q_2^{(3)} \approx \frac{1}{4}\lambda \cdot \beta \cdot (1 - \gamma) \cdot T_E, \qquad (22)$$

$$Q_3^{(3)} \approx \frac{1}{2}\lambda \cdot \beta \cdot \gamma \cdot T_E.$$

Thus, test interval extensions can be simulated by multiplying both the pump failure to start and common-cause probabilities by an "x" factor that represents the ratio between the extended test interval and the current one. Table 3 shows that, in this work, this factor is 3, since we want to extend the current pumps STI from one to 3 months.

5. Compensatory Measures

When TS modification results in small increments in CDF, compensatory measures can be applied to compensate or balance the undue risk, in such a way that the value of the total risk is kept within acceptable levels. In this work, the compensatory measures applied are (1) test of the redundant train right before entering the AOT and (2) implementation of a staggered testing strategy, if applicable, for compensating both AOT and STI extensions.

5.1. Test of the Redundant Train. Both the single-event risk and the yearly risk are increased due to the unavailability of one train during a certain period of time, d. However, given a component failure, the overall risk can be reduced or compensated if the redundant component is submitted to a new additional test. The effect of this test is to lower the unavailability of the tested component, which is considered to be zero right after the test is performed. Then, the risk associated with the tested component starts again to increase until the next test is performed or the component is demanded.

Equation (4) shows how the unavailability of a tested component behaves in terms of its test interval, T. It means

that, in case of a new additional test right before entering the AOT period, T can be replaced by d_{AOT}, where d_{AOT} stands for the extended duration of the failed component unavailability. Based on that, the redundant component failure to start unavailability can be expressed by

$$Q_{FS} \approx \frac{1}{2}\lambda \cdot d_{AOT}. \qquad (23)$$

In addition, common-cause failures of the tested train should be replaced by

$$Q_{CCFS}^{(2)} \approx \frac{1}{2}\beta \cdot \lambda \cdot d_{AOT}, \qquad (24)$$

where $Q_{CCFS}^{(2)}$ is the probability of common-cause failure to start of a two-component system, β is the beta factor for starting failures related to standby components, and we have made the approximation $Q_T \approx Q_{FS}$ in (9). It should be stressed here that $0 < d \leq d_{AOT}$, in general, where d has been used in (1) and (2) [2].

Actually, the only 3-pump system treated in this work is the SWS. This system has a particularity of being a 2-train system, but with an extra swing pump. During normal operation, one of the pumps must be running, which means that the new additional testing on standby pumps can only be applied to 2 pumps.

In addition to the Technical Specifications surveillance requirements that include test intervals, a new additional test can be performed to the redundant component, right after a component is considered failed, as a compensatory measure to the total risk. This extra test should be carried out right before entering AOT and not before AOT expires. The idea is to consider the redundant component "as good as new" right after this additional new test.

Table 4 shows the conditions used in simulating the test of the redundant pump. The unavailability of the tested component in this case is divided by four, to reflect the reduction in the pump test interval, from the original 4 weeks to the duration of the AOT (168 hours).

5.2. Staggered versus Sequential Test Strategies. Normally, TS is not prescriptive with respect to the test strategy to be adopted by the utility for the plant safety systems. However, when two redundant pumps are sequentially tested, the probability of introducing the same type of human error in both pumps increases when compared to the staggered testing strategy. The advantage in adopting staggered testing is to reduce the number of failures caused by human errors during the test performance. Consequently, the common-cause failure probabilities are reduced when the test strategy

TABLE 4: Test of the redundant train pump.

Test of the redundant train pump	Pump A: commence of AOT	Pump B: test is performed right before entering AOT	Pump C: test is performed right before entering AOT	Modifications on pump common-cause failure to start probabilities
Two- or three-component system	$Q_{FS} \to 1$ (true) $Q_{FR} \to 1$ (true) $Q_{MA} \to 1$ (true)	$Q'_{FS} = Q_{FS}/4$ $Q_{MA} \to 0$ (false)	$Q'_{FS} = Q_{FS}/4$ $Q_{MC} \to 0$ (false)	$Q'_{CCFS} = Q_{CCFS}/4$

TABLE 5: Modification of pump probabilities to simulate staggered testing strategy.

System	Modification of common-cause failure to start probability to reflect staggered testing
SIS—Pumps A and B	$Q'_{CCFS} = 1/2 \cdot Q_{CCFS}$
SWS—Pumps A and B, or B and C, or A and C	$Q'_{CCFS} = 1/2 \cdot Q_{CCFS}$
SWS—Pumps A and B and C	$Q'_{CCFS} = 1/3 \cdot Q_{CCFS}$
AFWS—A and B motor-operated pumps	$Q'_{CCFS} = 1/2 \cdot Q_{CCFS}$

applied to redundant components is switched from sequential to staggered testing.

In terms of the Alpha Factor Model, for systems submitted to sequential test strategy, the common-cause failure probability among k particular components belonging to a common-cause group with m components, $Q_k^{(m)}$, is given by [25]

$$Q_k^{(m)} = \frac{k}{\binom{m-1}{k-1}} \frac{\alpha_k^{Seq}}{\alpha_t} Q_T, \tag{25}$$

where $k = 1, 2, \ldots, m$, and $\alpha_t = \sum_{k=1}^{m} k\alpha_k^{Seq}$.

For systems submitted to staggered test strategy, on the other side, $Q_k^{(m)}$ is given by

$$Q_k^{(m)} = \frac{1}{\binom{m-1}{k-1}} \alpha_k^{Stag} Q_T. \tag{26}$$

Q_k in the Basic Parameter model is affected by the testing strategy adopted, since for staggered testing, the number of times a group of k components is tested depends on the response to the failure observed, whereas for sequential testing all components in the group are tested at each test episode. This yields the following relation for the staggered and sequential estimators of Q_k [23]:

$$\frac{Q_k^{Stag}}{Q_k^{Seq}} = \frac{1}{k}. \tag{27}$$

Therefore, as an example, for a two-component system, when the test strategy is modified from sequential to staggered, the common-cause failure related to failure to start is reduced by a factor of two. This can be explained by the fact that staggered tests increase the number of tests "against" the common-cause failures.

According to (25) and (26), independent failures or Q_1 expressions do have different calculations depending on the

test strategy. However, in this work, these differences in test strategies concerning independent failures were taken into account in the Angra 1 PSA database.

Table 5 shows the necessary modifications of pump common-cause failure probabilities to switch the test strategy from sequential to staggered testing, when applicable.

6. Results

The calculations were carried out by the SAPHIRE code to simulate pump AOT and STI extensions for the Angra 1 SIS, SWS, and AFWS. The results indicate, most of the times, the need to introduce compensatory measures to bring the risk within the appropriate acceptance criterion.

Single-event and yearly risk results are presented for the three systems and their respective pump AOT extensions, with and without compensatory measures. It should be noticed that difficulties in obtaining Angra 1 specific data for pump unavailability due to PM and CM, led to the adoption of their failure rates as the frequencies for the calculation of the average yearly risk associated with the AOT extensions. Considering that in this work only CM contributions are taken into account and with the pump failure rates being much less than 1, one can conclude that the single event AOT is more important than the yearly risk for the AOT risk acceptance decision-making process.

Regarding the STI extensions, the risk level of the system considering the extension is compared with the CDF.

According to current TS and operational practices in Angra 1, the possibility of implementing individual AOT and STI extensions for the SIS, SWS, and AFWS taking into account the introduction of compensatory measures is presented in Table 6. We observe that, since the staggered testing strategy is already adopted for the SIS and AFWS, this compensatory measure is not applicable to these systems.

Table 7 presents the SIS results for the analysis of AOT extension, additional test of the redundant pump, and STI extension. In this table, single-event and yearly AOT risks

TABLE 6: Possibility of AOT and STI extensions for the SIS, SWS, and AFWS.

System	Test strategy is staggered	Multiple Greek letters (MGL)	Compensatory measure: test of the redundant train	Compensatory measure: staggered test strategy
SIS	Yes	β	Yes	No
SWS	No	β and γ	Yes	Yes
AFWS	Yes	β	Yes	No

TABLE 7: SIS results.

	r	R (yr^{-1})	ΔCDF (yr^{-1})	CDF$_1$ (yr^{-1})	f (yr^{-1})	d (yr)
(1)	$5.0E-07$	$1.29E-07$	$2.6E-05$	$6.6E-05$	0.26	$1.92E-02$
(2)	$2.0E-07$	$5.2E-08$	$1.0E-05$	$5.05E-05$	0.26	$1.92E-02$
(3)			$2.0E-08 < 1.0E-06$	$4.017E-05$		

(1) Analysis of the AOT extension.
(2) Analysis of the test of the redundant train.
(3) Analysis of the STI extension.

TABLE 8: SWS results.

	r	R (yr^{-1})	ΔCDF (yr^{-1})	CDF$_1$ (yr^{-1})	f (yr^{-1})	d (yr)
(1)	$3.0E-05$	$7.4E-06$	$1.54E-03$	$1.58E-03$	0.25	$1.92E-02$
(2)			$1.0E-06 < 2.6E-06 < 1.0E-05$	$4.27E-05$		
(3)			$2.0E-07 < 1.0E-06$	$4.03E-05$		

(1) Analysis of the AOT extension.
(2) Analysis of the STI extension.
(3) Analysis of staggered testing.

are obtained using a baseline CDF$_B$ of $4.015E-05$ per reactor-year, as calculated by the SAPHIRE code using Angra 1 PSA Level 1 results. Also, the frequency of occurrence of AOTs appearing in this and the following tables are derived from the respective pump failure rates [26]. We observe that the incremental core damage frequency obtained for the AOT extension, $2.6E-05$/yr, is greater than the acceptance criterion, despite the single-event contribution obtained for the extension, $5.0E-7$/yr is equal to the criterion r_c. We conclude that this extension is not acceptable for the SIS without compensatory measures. Upon the simulation of the redundant train test, both the increment of the CDF and the single-event contribution diminish, as can be seen in Table 7. However, ΔCDF is now equal to $1.0E-05$/yr, which is exactly the boundary of the acceptance criterion. This means that despite the implementation of the compensatory measure "test of train B," just prior to the period of the AOT, other measures could be considered in risk-based regulatory decision making, such as the availability of redundant trains of other safety systems to compensate for this increased risk. Finally, the result displayed in Table 7 for the STI extension from one month to three months shows that the value obtained for the CDF, $2.0E-08$/yr, is acceptable in terms of risk analysis, according to the criterion without the need of introduction of compensatory measures.

Table 8 shows the SWS results for the analysis of AOT extension, STI extension, and introduction of the staggered testing strategy. One can easily see that the value obtained for the CDF$_1$ upon the AOT extension, $1.58E-03$/yr, is greater than the baseline, which indicates the need of a compensatory measure. Using the method presented in Table 4 for the test of the redundant train yields a result of $3.82E-05$/yr for CDF$_1$ (not shown in Table 8) that characterizes a decrease in CDF which, according to the criterion, can always be allowed. In other words, the test of the redundant pump is enough to compensate the increase in CDF caused by the AOT extension.

The value obtained for the ΔCDF upon the STI extension, $2.6E-06$/yr, is acceptable according to the criterion, since Angra 1 CDF$_B$ is less than $1.0E-04$/yr. The introduction of staggered testing can even reduce this risk increment, as can be seen in Table 8. The value obtained for ΔCDF with the staggered testing strategy was $2.0E-07$/yr.

Table 9 presents the AFWS results for the analysis of AOT extension, STI extension, additional test of the redundant motor-operated pump, and additional test of the redundant motor-operated pump including the turbine-driven pump. The result of the AOT extension yields a ΔCDF of $3.0E-04$/yr, which is unacceptable without compensatory measures. Likewise, the simulation of the test of the redundant motor-operated pump is not enough to compensate the AOT extension, since both the increase in ΔCDF and the single-event risk do not meet the established criterion. However, the introduction of the additional test of the turbine-driven pump can also be taken as a compensatory measure. The increase in ΔCDF in this case, which is $2.42E-04$/yr, still remains unacceptable. Therefore, within the scope of this work an AOT extension for the AFWS should not be allowed

Table 9: AFWS results.

	r	R (yr^{-1})	ΔCDF (yr^{-1})	CDF$_1$ (yr^{-1})	f (yr^{-1})	d (yr)
(1)	$5.8E-06$	$5.1E-06$	$3.0E-04$	$3.4E-04$	0.88	$1.92E-02$
(2)	$5.0E-06$	$4.4E-06$	$2.63E-04$	$3.03E-04$	0.88	$1.92E-02$
(3)	$4.6E-06$	$4.1E-06$	$2.42E-04$	$2.82E-04$	0.88	$1.92E-02$
(4)			$1.0E-06 < 1.4E-06 < 1.0E-05$	$4.15E-05$		

(1) Analysis of the AOT extension.
(2) Analysis of the test of the redundant motor-operated pump.
(3) Analysis of the test of the redundant motor-operated pump and the turbine-driven pump.
(4) Analysis of the STI extension.

Table 10: Overview of the AOT and STI extensions.

System	AOT extension to 168 h without compensatory measures	AOT extension to 168 h with compensatory measures	STI extension to 3 months without compensatory measures	STI extension to 3 months with compensatory measures
SIS	No	Yes (restrictions applied)	Yes (no restrictions)	Not applicable
SWS	No	Yes (no restrictions)	Yes (restrictions applied)	Yes (no restrictions)
AFWS	No	No	Yes (restrictions applied)	Not applicable

Table 11: Results of simultaneous extensions.

	r	R (yr^{-1})	ΔCFD (yr^{-1})	CDF$_1$ (yr^{-1})	f (yr^{-1})	d (yr)
(1)	$1.6E-07$	$4.2E-08$	$8.5E-06$	$4.86E-05$	0.26	$1.92E-02$
(2)			$2.0E-07$	$4.03E-05$		
(3)	$2.0E-07$	$5.1E-08$	$1.0E-05$	$5.05E-05$	0.26	$1.92E-02$

(1) Simultaneous AOT extensions for SIS and SWS.
(2) Simultaneous STI extensions for SIS and SWS.
(3) Simultaneous AOT and STI extensions (SIS).

under any conditions. At last, the STI extension for the AFWS can be allowed due to the fact that the ΔCDF increase is acceptable, as can be seen in Table 9.

An overview of the results of the AOT and STI extensions are presented in Table 10. The term "no restrictions" in this table means that the corresponding ΔCDF, as calculated for the extension, is smaller than $1.0E-06$ per reactor year, while "restrictions applied" means that $1.0E-06 < \Delta$CDF $< 1.0E-05$, when CDF$_B$ is less than $1.0E-04$, which is the case of Angra 1.

Although plant configuration control is not in the scope of this work, we have analyzed a few combinations of simultaneous AOT and STI extensions, based on the overview of results presented in Table 10. Thus, as a very first step in developing a program for plant configuration control that allows the establishment of a risk-based planning for maintenance activities, three combinations of simultaneous AOT or STI extensions have been calculated. Table 11 presents the results obtained with the SAPHIRE code for (1) simultaneous AOT extensions for the SIS and SWS; (2) simultaneous STI extensions for the SIS and SWS (3) simultaneous AOT and STI extensions for the SIS. Interactions between AOT and STI extensions have not been taken into account in this analysis. In what concerns item (1) of Table 11, the single-event risk r value of $1.6E-07$/yr is less than the criterion r_c value of $5.0E-07$/yr and CDF lies between $1.0E-05$/yr and $1.0E-06$/yr, which makes

this configuration acceptable for the SIS and SWS AOT extensions to 168 h. For item (2), the increment value of CDF, $2.0E-07$/yr, indicates that, based on risk analysis, the simultaneous STI extensions for the SIS and SWS are acceptable. Finally, regarding item (3), the risk of single event (r) value of $2.0E-07$/yr is less than the criterion, which would make this configuration acceptable by this point of view. However, the calculated ΔCDF of $1.0E-05$/yr is in the limit between acceptance and rejection indicating the need, in the regulatory decision-making process, to consider other aspects such as the availability of redundant trains of other safety systems, to compensate this limiting value of ΔCDF.

7. Conclusions

The results obtained in this work show that AOT and STI extensions for the SIS, SWS, and AFWS of Angra 1 power plant are feasible without incurring in unacceptable increase in the plant total risk, mostly after the implementation of compensatory measures.

AOT and STI extensions for these systems result in different impacts on the total CDF. While AOT extensions can only be accepted for the SIS and SWS upon the implementation of compensatory measures, STI extensions are acceptable for all three systems without the need of compensatory measures. Clearly, in the decision-making process of a TS modification, other aspects such as operational experience, lessons learned from previous TS modifications, and traditional engineering judgment are also to be considered, in addition to the risk analysis performed.

AOT extensions are meant to allow time flexibility to perform adequate component maintenance and repair, which in turn reduces both the AOT frequency and unplanned plant shutdowns. STI extensions, on the other side, can be implemented with virtually no significant contribution to CDF, thus substantially reducing an unnecessary burden of the plant team in carrying out a large number of unnecessary tests, so that their attention can be concentrated on activities more relevant to safety. Reducing the number of tests also reduces the number of occurrences of unplanned plant shutdowns caused by test-induced transients.

In what concerns TS modifications, sensitivity analyses may be necessary to address the role of key assumptions adopted during the preparation of the study, which act as a support to uncertainty analysis. Experience on sensitivity analyses developed for modifications of risk-based TS shows that the risk associated with them is relatively insensitive to uncertainties when compared, for instance, to the effect on risk from uncertainties in assumptions regarding plant design changes, or regarding significant changes to plant operating procedures [24]. Nevertheless, a sensitivity analysis of the risks associated with the components in question is recommended. Such an analysis can be done through the use of risk importance measures that may be relative or absolute and have the purpose of classifying the significance of components or systems in terms of their contributions to the overall risk. Importance measures have direct application to plant configuration control in measuring the significance of the unavailability effect of a single component that has been isolated for maintenance.

The most utilized importance measures for assessing nuclear plant components and their main applications are [5].

(1) Birnbaum is defined as follows. the rate of change in total risk of the system with respect to changes in a risk element's basic probability (or frequency). It indicates the sensitivity of the minimal cut set upper bound with respect to a change in the basic event probability. It is sensitive to the component position in the fault-tree structure.

(2) Fussell-Vesely is an indication of the fraction of the minimal cut set upper bound probability (or sequence frequency) that involves the cut sets containing the basic event of interest. In an aging regime, it can be interpreted as the amount of a component allowed degradation of performance as a function of risk increase. Also shows the importance of the long term averaged performance of a component (thus, it is not appropriate for measuring the importance of a set of similar components instantaneously taken out of service).

(3) RRW (risk reduction worth) is an indication of how much the minimal cut set upper bound would decrease if the basic event never occurred. In other words it expresses the risk change when the component is clearly available.

(4) RAW (risk achievement worth) is an indication of how much the minimal cut set upper bound would go up if the basic event always occurred. In other words, it gives the risk increase when the component is unavailable for maintenance or due to failure.

At first, regarding the changes in TS measured in this work, the most appropriate measures of importance for the sensitivity analysis appears to be the RAW and Birnbaum, for extensions of the AOT and STI, respectively. Nevertheless, it is recommended here to also consider the Fussell-Vesely importance measure.

However, new implications on importance measuring are to be taken into account. New developments and roles of different importance measures have been pointed out [27], concerning the decision-making process on permanent and temporary configurations, technical specifications, online risk monitoring, and also ranking safety significance of systems, structures, components, and human actions. A proposal on the use of path sets, instead of cut sets has been made [28], which shows that in this manner, importance of preventing top events is addressed instead.

As earlier mentioned, this work represents an important step towards plant configuration control, which is designed to operate in an efficient and effective use of plant resources, or safety systems. Therefore, and recommended the development of a configuration control program in which the following objectives must be achieved is desirable [29]:

(i) management of the configuration of components that are simultaneously unavailable;

(ii) management of the standby components that are operable;

(iii) management of the duration of the configuration (CFI);

(iv) management of the frequency with which the configuration occurs;

(v) management of AOT and STI interactions [7].

The calculation method presented in this work, which includes the use of compensatory measures and comparison with risk criteria, is the basic calculating tool for the management of the goals presented above. Furthermore, configuration control strategies involve the control of risk levels and risk contributions similar to those defined here and addressed during the development of this study.

Another recommendation is the inclusion of Preventive Maintenance (PM) in the AOT analysis. In this case, it is essential to develop a specific data base of plant operational experience that clearly makes the distinction between the PM and CM unavailabilities. It is worth mentioning that the collection of operational data must be targeted for PSA use [30].

The preservation of the defense in-depth principle and the observation of engineering limitations should be also emphasized. The quantitative criteria described in the regulatory guidelines [24, 31] are used to ensure that any risk increase is within acceptable limits. However, this does not exclude traditional considerations for the decision-making process, to ensure that changes comply with the rules and regulations. Practical considerations are an integrated part of the judgment concerning the acceptability of the implementation of modifications.

It is important to address here the issue of PSA itself. PSA modeling limitations have been long discussed. One issue recently raised [32] concerns on the wide use of the so-called fault tree linking method for performing the evaluation of accident scenario frequencies [33]. This approach relies on the fact that for each initiating event, all pertinent fault trees related to the accident sequences are linked through a fault tree with an AND gate whose cut sets are generated and analyzed. Reference [33] points out that this may not be accurate. The example discussed shows that the CDF may be in error around a factor of 5. The paper proposes the use of binary decision diagrams (BDD) [34]. The BDD of a formula is a compact encoding of the truth table of this formula. It would be interesting to investigate, in this context, the role of the opposite approach to perform a PSA: the one of large event trees and small fault trees, known as event trees with boundary conditions (or explicit method) [32, 35].

Another feature to be considered is the set of PSA truncation limits [36]. According to [37], it should be adequate to retain the minimal cut sets that contribute 90–99% to the point estimate CDF. However, a tighter control could be necessary to take into account smaller probability/frequency cut sets that might have substantially larger uncertainty factors compared with those that dominate the point estimate CDF [35].

The issues raised in [32, 33] have a deep influence on the discussion of the importance measures to be used, as it is clearly mentioned in both references.

Acknowledgment

The invaluable support of Eletrobras Termonuclear S. A. on plant information is deeply acknowledged. The third author is gratefully indebted to Conselho Nacional de Desenvolvimento Científico e Tecnológico (CNPq) of Brazil, for financial support to this work.

References

[1] S. P. Sirén and K. E. Jänkällä, "Risk-informed optimization of allowed outage times at Loviisa NPP," in *Proceedings of the 8th International Conference on Probabilistic Safety Assessment and Management (IAPSAM '06)*, New Orleans, Fla, USA, 2006.

[2] P. Samanta, I. Kim, T. Mankamo et al., "Handbook of methods for risk-based analyses of technical specifications," US Nuclear Regulatory Commission NUREG/CR-6141, Washington, DC, USA, 1994.

[3] Eletrobrás Termonuclear, "Final Safety Analysis Report CNAAA," Unity 1. Rev. 33, Rio de Janeiro, Brazil, 2005.

[4] NRC, "Systems analysis programs for hands-on integrated reliability evaluations (SAPHIRE)," US Nuclear Regulatory Commission NUREG/CR-6116, Washington, DC, USA, 1998.

[5] M. Modarres, *Risk Analysis in Engineering: Techniques, Tools, and Trends*, Taylor & Francis, Boca Raton, Fla, USA, 2006.

[6] NRC, "Regulatory guide 1.174: an approach for using PSA in risk-informed decisions on plant-specific changes to the licensing basis," US Nuclear Regulatory Commission, Washington, DC, USA, 2002.

[7] S. Martorell, G. Serradell, G. Verdú, and P. Samanta, "Probabilistic analysis of the interaction between allowed outage time and surveillance test interval requirements," in *IAEA Advances in Reliability Analysis and Probabilistic Safety Assessment for Nuclear Reactors. IAEA-TECDOC-737*, pp. 204–212, Vienna, Austria, 1994.

[8] K. N. Fleming and R. P. Murphy, "Lessons learned in applying PSA methods to technical specifications optimization," in *IAEA Advances in Reliability Analysis and Probabilistic Safety Assessment for Nuclear Reactors. IAEA-TECDOC-737*, pp. 185–191, Vienna, Austria, 1994.

[9] T. Mankamo, I. S. Kim, and P. K. Samanta, "Risk-based evaluation of allowed outage times (AOTs): considering risk of shutdown," in *IAEA. Advances in Reliability Analysis and Probabilistic Safety Assessment for Nuclear Reactors. IAEA-TECDOC-737*, pp. 216–222, Vienna, Austria, 1994.

[10] S. A. Martorell, V. G. Serradell, and P. K. Samanta, "Improving allowed outage time and surveillance test interval requirements: a study of their interactions using probabilistic methods," *Reliability Engineering and System Safety*, vol. 47, no. 2, pp. 119–129, 1995.

[11] M. Čepin and B. Mavko, "Probabilistic safety assessment improves surveillance requirements in technical specifications," *Reliability Engineering and System Safety*, vol. 56, no. 1, pp. 69–77, 1997.

[12] M. L. Shooman, *Probabilistic Reliability: An Engineering Approach*, Robert E Krieger, Malabar, Fla, USA, 1990.

[13] J. K. Vaurio, "On time-dependent availability and maintenance optimization of standby units under various maintenance policies," *Reliability Engineering and System Safety*, vol. 56, no. 1, pp. 79–89, 1997.

[14] I. B. Wall, J. J. Haugh, and D. H. Worlege, "Recent applications of PSA for managing nuclear power plant safety," *Progress in Nuclear Energy*, vol. 39, no. 3-4, pp. 367–425, 2001.

[15] M. Čepin and S. Martorell, "Evaluation of allowed outage time considering a set of plant configurations," *Reliability Engineering and System Safety*, vol. 78, no. 3, pp. 259–266, 2002.

[16] S. Martorell, A. Sánchez, S. Carlos, and V. Serradell, "Simultaneous and multi-criteria optimization of TS requirements and maintenance at NPPs," *Annals of Nuclear Energy*, vol. 29, no. 2, pp. 147–168, 2002.

[17] D. Kancev and M. Cepin, "A new method for explicit modelling of single failure event within different common cause failure groups," *Reliability Engineering and System Safety*, vol. 103, pp. 84–93, 2012.

[18] J. K. Vaurio, "Common cause failure probabilities in standby safety system fault tree analysis with testing—scheme and timing dependencies," *Reliability Engineering and System Safety*, vol. 79, no. 1, pp. 43–57, 2003.

[19] T. M. J. Kivirinta and K. E. Jänkällä, "Determining risk-balanced allowed outage times for Loviisa power plant," in *Proceedings of the 8th International Conference on Probabilistic Safety Assessment and Management (IAPSAM '06)*, New Orleans, Fla, USA, 2006.

[20] S. Martorell, S. Carlos, J. F. Villanueva et al., "Use of multiple objective evolutionary algorithms in optimizing surveillance requirements," *Reliability Engineering and System Safety*, vol. 91, no. 9, pp. 1027–1038, 2006.

[21] X. He, J. Tong, and J. Chen, "Maintenance risk management in Daya Bay nuclear power plant: PSA model, tools and

applications," *Progress in Nuclear Energy*, vol. 49, no. 1, pp. 103–112, 2007.

[22] S. Cho and J. Jiang, "Analysis of surveillance test interval by Markov process for SDS1 in CANDU nuclear power plants," *Reliability Engineering and System Safety*, vol. 93, no. 1, pp. 1–13, 2008.

[23] NRC, "Procedures for treating common cause failures in safety and reliability studies," US Nuclear Regulatory Commission NUREG/CR-4780, Washington, DC, USA, 1987.

[24] NRC, "Regulatory guide 1.177: an approach for plant-specific risk-informed decision-making: technical specifications," US Nuclear Regulatory Commission, Washington, DC, USA, 1998.

[25] NRC, "Guidelines on modeling common-cause failures in probabilistic risk assessment," US Nuclear Regulatory Commission NUREG/CR-5485, Washington, DC, USA, 1998.

[26] S. M. Ross, *Introduction to Probability Models*, Academic Press, San Diego, Calif, USA, 1993.

[27] J. K. Vaurio, "Developments in importance measures for risk-informed ranking and other applications," in *Proceedings of the 8th International Conference on Probabilistic Safety Assessment and Management (IAPSAM '06)*, New Orleans, Fla, USA, 2006.

[28] R. W. Youngblood, "Risk significance and safety significance," *Reliability Engineering and System Safety*, vol. 73, no. 2, pp. 121–136, 2001.

[29] NRC, "Study of operational risk-based configuration control," US Nuclear Regulatory Commission NUREG/CR-5641, BNL-NUREG-52261, Washington, DC, USA, 1991.

[30] NRC, "Handbook of parameter estimation for probabilistic risk assessment," US Nuclear Regulatory Commission NUREG/CR-6823, SAND2003-3348P, Washington, DC, USA, 2003.

[31] NRC, "Risk-informed decision-making: technical specifications. Standard review plant," US Nuclear Regulatory Commission NUREG-0800, Washington, DC, USA, 2007.

[32] S. Epstein and A. Rauzy, "Can we trust PRA?" *Reliability Engineering and System Safety*, vol. 88, no. 3, pp. 195–205, 2005.

[33] NRC, "Probabilistic risk assessment procedures guide," US Nuclear Regulatory Commission NUREG/CR-2300, Washington, DC, USA, 1982.

[34] R. E. Bryant, "Graph based algorithms for Boolean function manipulation," *IEEE Transactions on Computers*, vol. 35, no. 8, pp. 677–691, 1986.

[35] L. F. S. Oliveira, P. F. Frutuoso e Melo, J. E. P. Lima, and I. L. Stal, "An application of the explicit method for analysing intersystem dependencies in the evaluation of event trees," *Nuclear Engineering and Design*, vol. 90, no. 1, pp. 25–41, 1985.

[36] M. Čepin, "Analysis of truncation limit in probabilistic safety assessment," *Reliability Engineering and System Safety*, vol. 87, no. 3, pp. 395–403, 2005.

[37] NRC, "Probabilistic safety analysis procedures guide," US Nuclear Regulatory Commission NUREG/CR-2815, Washington, DC, USA, 1985.

On the Mean Residual Life Function and Stress and Strength Analysis under Different Loss Function for Lindley Distribution

Sajid Ali

Department of Decision Sciences, Bocconi University, via Roenthen 1, 20136 Milan, Italy

Correspondence should be addressed to Sajid Ali; sajidali.qau@hotmail.com

Academic Editor: Shey-Huei Sheu

Purpose. Mathematical properties of Lindley distribution are derived under different loss functions. These properties include mean residual life function, Lorenz curve, stress and strength characteristic, and their respective posterior risk via simulation scheme. *Methodology*. Bayesian approach is used for the reliability characteristics. Results are compared on the basis of posterior risk. *Findings*. Using prior information on the parameter of Lindley distribution, Bayes estimates for reliability characteristics are compared under different loss functions. *Practical Implications*. Since Lindley distribution is a mixture of gamma and exponential distribution, so Bayesian estimation of reliability characteristics will have a great implication in reliability theory. *Originality*. A real life application to waiting time data at the bank is also described for the developed procedures. This study is useful for researcher and practitioner in reliability theory.

1. Introduction

Exponential distribution is frequently used as a lifetime distribution in statistics and applied areas; the Lindley distribution has been ignored in the literature since 1958. Lindley distribution originally developed by Lindley [1] and some classical statistic properties are investigated by Ghitany et al. [2]. Sankaran [3] introduced a discrete version of Lindley distribution known as discrete Poisson-Lindley distribution, and Ghitany and Al-Mutairi [4] described some estimation methods. The distribution of zero-truncated Poisson-Lindley was introduced by Ghitany et al. [5] who used the distribution for modeling count data in the case where the distribution has to be adjusted for the count of missing zeros. Zamani and Ismail [6] introduced negative binomial distribution as an alternative to zero-truncated Poisson-Lindley distribution. Recently, Ghitany et al. [7] introduced a two-parameter weighted Lindley distribution and pointed that Lindley distribution is particularly useful in modelling biological data from mortality studies.

The rest of the study is organized as follows. Section 2 deals with the derivation of posterior distribution using different noninformative and informative priors. Using different loss functions, the Bayes estimators and their respective posterior risks are discussed in Section 3. Elicitation of hyperparameter is also discussed in Section 3. Simulation study of Bayes estimates of mean residual life and their posterior risks is performed in Section 4. Lorenz curve discussion for Lindley distribution is given in Section 5 while stress and strength reliability characteristics and simulation study under different loss functions is discussed/performed in Section 6. Real life application is illustrated in Section 7. Finally, Section 8 deals with a conclusion and some future remarks.

2. Likelihood Function and Posterior Distributions

The posterior distribution summarizes available probabilistic information on the parameters in the form of prior distribution and the sample information contained in the likelihood function. The likelihood principle suggests that the information on the parameter should depend only on its posterior distribution. Bayesian scientist's job is to assist the investigator to extract features of interest from the posterior distribution. In this section, we will use the Lindley model as sampling distribution mingles with noninformative priors for the derivation of posterior distribution. A random variable X

TABLE 1: Bayes estimator and posterior risk under different loss functions.

Loss function	Bayes estimator (BE)	Posterior risk (PR)
$L_1 = \text{SELF} = (\theta - d)^2$	$E(\theta \mid \mathbf{x})$	$\text{Var}(\theta \mid \mathbf{x})$
$L_2 = \text{WSELF} = \dfrac{(\theta - d)^2}{\theta}$	$\left(E(\theta^{-1} \mid \mathbf{x})\right)^{-1}$	$E(\theta \mid \mathbf{x}) - \left(E(\theta^{-1} \mid \mathbf{x})\right)^{-1}$
$L_3 = M/Q\ \text{SELF} = \left(1 - \dfrac{d}{\theta}\right)^2$	$\dfrac{E(\theta^{-1} \mid \mathbf{x})}{E(\theta^{-2} \mid \mathbf{x})}$	$1 - \dfrac{E(\theta^{-1} \mid \mathbf{x})^2}{E(\theta^{-2} \mid \mathbf{x})}$
$L_4 = \text{PLF} = \dfrac{(\theta - d)^2}{d}$	$\sqrt{E(\theta^2 \mid \mathbf{x})}$	$2\left[\sqrt{E(\theta^2 \mid \mathbf{x})} - E(\theta \mid \mathbf{x})\right]$
$L_5 = \text{SLLF} = (\log\theta - \log d)^2$	$\exp(E(\log\theta \mid \mathbf{x}))$	$\text{Var}(\log\theta \mid \mathbf{x})$
$L_6 = \text{ELF} = \left[\dfrac{d}{\theta} - \log\dfrac{d}{\theta} - 1\right]$	$\left(E(\theta^{-1} \mid \mathbf{x})\right)^{-1}$	$E(\log\theta \mid \mathbf{x}) - \log\left(E(\theta^{-1} \mid \mathbf{x})\right)$
$L_7 = \text{KLF} = \left(\sqrt{\dfrac{d}{\theta}} - \sqrt{\dfrac{\theta}{d}}\right)^2$	$\sqrt{\dfrac{E(\theta \mid \mathbf{x})}{E(\theta^{-1} \mid \mathbf{x})}}$	$2[E(\theta \mid \mathbf{x})E(\theta^{-1} \mid \mathbf{x}) - 1]$

TABLE 2: BEs and their respective PRs of MRLF under SELF.

n	UP			JP		
θ	7.142857	1.2	0.114114	7.142857	1.2	0.114114
20	16.6609 (8.74099)	1.16041 (0.049399)	0.108763 (0.000527)	17.1312 (8.650769)	1.2008 (0.047705)	0.113795 (0.000504)
40	16.9256 (4.50602)	1.17773 (0.025448)	0.111713 (0.000277)	17.1644 (4.44596)	1.1982 (0.025012)	0.11429 (0.000271)
60	16.9761 (3.02169)	1.18603 (0.017205)	0.112294 (0.000186)	17.1357 (2.99484)	1.19976 (0.017109)	0.11402 (0.000184)
80	17.0298 (2.28011)	1.19015 (0.012994)	0.112893 (0.000141)	17.1499 (2.26492)	1.20048 (0.012883)	0.114194 (0.000139)
100	17.0348 (1.82524)	1.19247 (0.010436)	0.113075 (0.000113)	17.1309 (1.81551)	1.20075 (0.010364)	0.114117 (0.000112)
n	LMP			GP		
θ	7.142857	1.2	0.114114	7.142857	1.2	0.114114
20	16.6296 (8.574081)	1.15994 (0.044900)	0.110424 (0.000504)	17.1139 (8.52409)	1.19828 (0.045907)	0.115161 (0.000489)
40	16.9097 (4.40598)	1.17747 (0.025014)	0.112539 (0.000270)	17.1556 (4.40518)	1.19692 (0.025013)	0.114963 (0.000266)
60	16.9654 (2.903167)	1.18586 (0.017105)	0.112844 (0.000185)	17.1298 (2.906730)	1.19891 (0.017032)	0.114467 (0.000182)
80	17.0218 (2.25801)	1.19002 (0.012794)	0.113305 (0.000140)	17.1455 (2.25698)	1.19984 (0.012796)	0.114528 (0.000139)
100	17.0284 (1.81523)	1.19236 (0.010354)	0.113405 (0.000113)	17.1274 (1.81519)	1.20023 (0.010353)	0.114385 (0.000112)

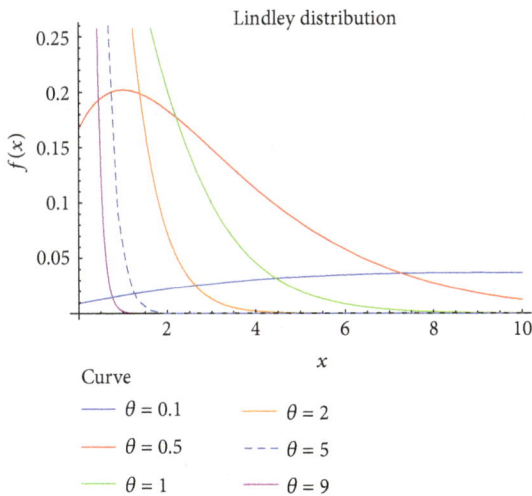

FIGURE 1: The Lindley distribution.

is said to possess a Lindley distribution if it has the following form:

$$f(x) = \frac{\theta^2}{\theta + 1}(1 + x)e^{-\theta x}, \quad x, \theta > 0. \quad (1)$$

It is obvious from Figure 1 that behavior of Lindley distribution is close to exponential or gamma distribution.

The likelihood function for a random sample $x_1, x_2, \ldots, x_n$ which is taken from Lindley distribution is

$$L(\mathbf{x}, \theta) = \frac{\theta^{2n}}{(\theta + 1)^n}\prod_{i=1}^{n}(1 + x_i)e^{-\theta\sum_{i=1}^{n}x_i}, \quad \theta > 0. \quad (2)$$

2.1. Posterior Distribution Using Noninformative Priors

2.1.1. Uniform Prior. An argument in the favor of uniform prior is that when the data are sufficiently informative, so that likelihood function is sharply peaked, then it really does not matter what prior is used, since all reasonably smooth prior densities will lead to approximately the same posterior density. The uniform density in most cases is convenient to simplify calculations of the posterior. This argument supports the uniform prior only in those cases where it produces approximately the same conclusions as the highly imprecise prior constructed from a sufficiently large class of prior densities. If the data are highly informative, the uniform prior

TABLE 3: BEs and their respective PRs of MRLF under WSELF.

n	UP			JP		
θ	7.142857	1.2	0.114114	7.142857	1.2	0.114114
20	16.1511 (0.509768)	1.11927 (0.041144)	0.10415 (0.004613)	16.6062 (0.504952)	1.1582 (0.040598)	0.10896 (0.004538)
40	16.6632 (0.262396)	1.15649 (0.021239)	0.109296 (0.002417)	16.8981 (0.262306)	1.17658 (0.021156)	0.111815 (0.002415)
60	16.7998 (0.176282)	1.17169 (0.014341)	0.110663 (0.001631)	16.9577 (0.176031)	1.18525 (0.014309)	0.112362 (0.001627)
80	16.8969 (0.132920)	1.17933 (0.010824)	0.111659 (0.001234)	17.0159 (0.132909)	1.18956 (0.010819)	0.112945 (0.001234)
100	16.9283 (0.106526)	1.18378 (0.008691)	0.112084 (0.000991)	17.0238 (0.105716)	1.1920 (0.008573)	0.113117 (0.000990)
n	LMP			GP		
θ	7.142857	1.2	0.114114	7.142857	1.2	0.114114
20	16.1217 (0.504878)	1.11883 (0.040112)	0.105691 (0.004437)	16.5891 (0.504848)	1.15569 (0.042586)	0.110343 (0.004418)
40	16.6478 (0.261905)	1.15624 (0.021123)	0.110091 (0.002414)	16.8893 (0.261276)	1.17531 (0.021111)	0.112492 (0.002417)
60	16.7894 (0.176031)	1.17152 (0.014307)	0.111199 (0.001624)	16.9518 (0.176019)	1.18440 (0.014308)	0.112811 (0.001625)
80	16.889 (0.132795)	1.1792 (0.010822)	0.112063 (0.001232)	17.0115 (0.132902)	1.18892 (0.010819)	0.11328 (0.001228)
100	16.9220 (0.105446)	1.18367 (0.008489)	0.112409 (0.000991)	17.0202 (0.105156)	1.19148 (0.008452)	0.113385 (0.000990)

TABLE 4: BEs and their respective PRs of MRLF under MSELF.

n	UP			JP		
θ	7.142857	1.2	0.114114	7.142857	1.2	0.114114
20	15.6991 (3979.6)	1.0835 (0.313983)	0.100281 (0.998953)	16.1011 (3504.08)	1.11726 (0.314576)	0.104331 (0.998814)
40	16.4163 (4489.67)	1.13671 (0.494297)	0.107086 (0.998747)	16.6369 (4476.16)	1.1554 (0.506637)	0.109393 (0.998662)
60	16.6305 (4645.4)	1.15802 (0.571236)	0.109126 (0.998682)	16.7819 (4477.83)	1.17094 (0.562509)	0.110729 (0.998622)
80	16.7679 (4749.79)	1.16888 (0.611298)	0.110479 (0.998637)	16.8833 (4749.34)	1.17875 (0.625842)	0.111709 (0.998591)
100	16.8244 (4790.7)	1.17533 (0.635275)	0.111128 (0.998616)	16.9174 (4718.19)	1.18331 (0.609666)	0.112125 (0.998578)
n	LMP			GP		
θ	7.142857	1.2	0.114114	7.142857	1.2	0.114114
20	15.6738 (3509.61)	1.08313 (0.312572)	0.101543 (0.99881)	16.0855 (3491.33)	1.11502 (0.306844)	0.105556 (0.998771)
40	16.4021 (4477.74)	1.13648 (0.493397)	0.107804 (0.998572)	16.6286 (4469.05)	1.15419 (0.50569)	0.11003 (0.998638)
60	16.6207 (4437.01)	1.15786 (0.560576)	0.109627 (0.998624)	16.7763 (4369.97)	1.17011 (0.52163)	0.11116 (0.998606)
80	16.7604 (4743.29)	1.16876 (0.610779)	0.110863 (0.998523)	16.8790 (4745.63)	1.17813 (0.610592)	0.112035 (0.998528)
100	16.8182 (4785.43)	1.17523 (0.604848)	0.11144 (0.998573)	16.9140 (4768.19)	1.18281 (0.596919)	0.112386 (0.998568)

may produce reasonable inferences. The uniform prior for θ is defined as

$$p(\theta) \propto 1, \quad \theta > 0. \tag{3}$$

The posterior distribution of parameter θ for the given data using (2) and (3) is

$$p(\theta \mid \mathbf{x}) = \frac{\left(\left(\theta^{2n}\right)/(\theta+1)^n\right) e^{-\theta \sum_{i=1}^{n} x_i}}{\left(\left(\theta^{2n}\right)/(\theta+1)^n\right) e^{-\theta \sum_{i=1}^{n} x_i} d\theta}, \quad \theta > 0. \tag{4}$$

2.1.2. Jeffreys Prior. Jeffreys was motivated by invariance requirements and suggested a solution to provide a non-informative prior. He used differential geometry method. The requirements are invariance under 1-1 transformations and invariance under sufficient statistics. One-dimensional version of Jeffreys prior has been justified from many different viewpoints. Jeffreys [8] proposed a formal rule for obtaining a noninformative prior as follows. If θ is a k-vector valued parameter, then, JP of θ is $g(\theta) \propto \sqrt{|\det I(\theta)|}$ where $I(\theta)$ is a kxk Fisher's (information) matrix whose (i, j)th

element is $-E\left[\partial^2 \ln L(\theta \mid \mathbf{x})/\partial\theta_i\partial\theta_j\right]$ $i, j = 1, 2, \ldots, k$. Fisher's information matrix is not directly related to the notation of lack of information. The connection comes from the role of Fisher's matrix in asymptotic theory. Jeffreys noninformative priors based on Fisher's information matrix often lead to a family of improper priors. The Jeffreys prior of the parameter θ is

$$p(\theta) \propto \frac{\sqrt{\theta^2 + 4\theta + 2}}{\theta(1+\theta)}, \quad \theta > 0,$$

$$p(\theta \mid x)$$

$$= \frac{\left(\left(\theta^{2n-1}\right)/(\theta+1)^{n+1}\right) \sqrt{\theta^2 + 4\theta + 2} e^{-\theta \sum_{i=1}^{n} x_i}}{\int_0^\infty \left(\left(\theta^{2n-1}\right)/(\theta+1)^{n+1}\right) \sqrt{\theta^2 + 4\theta + 2} e^{-\theta \sum_{i=1}^{n} x_i} d\theta}, \quad \theta > 0. \tag{5}$$

2.2. Posterior Distribution Using Informative Prior. In case of an informative prior, the use of prior information is equivalent to adding a number of observations to a given sample size and, therefore, leads to a reduction of the variance/posterior

TABLE 5: BEs and their respective PRs of MRLF under PLF.

n	UP			JP		
θ	7.142857	1.2	0.114114	7.142857	1.2	0.114114
20	16.9211 (0.520576)	1.1815 (0.042188)	0.111158 (0.004791)	17.3777 (0.493072)	1.2205 (0.039405)	0.115987 (0.004383)
40	17.0582 (0.265186)	1.18848 (0.021509)	0.112945 (0.002464)	17.2934 (0.258053)	1.20859 (0.020784)	0.115468 (0.002356)
60	17.0648 (0.177533)	1.19326 (0.014463)	0.11312 (0.001653)	17.2229 (0.174329)	1.20683 (0.012135)	0.114822 (0.001604)
80	17.0966 (0.133627)	1.1956 (0.010893)	0.113516 (0.001246)	17.2158 (0.131813)	1.20584 (0.010708)	0.114803 (0.001218)
100	17.0883 (0.10698)	1.19684 (0.008735)	0.113575 (0.000999)	17.1838 (0.105815)	1.20506 (0.008616)	0.114608 (0.000981)
n	LMP			GP		
θ	7.142857	1.2	0.114114	7.142857	1.2	0.114114
20	16.8904 (0.485218)	1.18104 (0.034205)	0.112775 (0.004302)	17.3612 (0.492507)	1.21811 (0.033652)	0.117262 (0.004203)
40	17.0425 (0.256431)	1.18823 (0.020514)	0.11376 (0.002342)	17.2848 (0.256428)	1.20734 (0.020449)	0.116118 (0.002310)
60	17.0543 (0.173642)	1.19309 (0.011465)	0.113666 (0.001602)	17.2171 (0.173498)	1.20599 (0.011464)	0.115259 (0.001583)
80	17.0886 (0.131689)	1.19547 (0.010704)	0.113926 (0.001214)	17.2114 (0.131609)	1.20520 (0.010704)	0.115132 (0.001207)
100	17.0819 (0.105019)	1.19673 (0.008536)	0.113903 (0.000979)	17.1803 (0.105017)	1.20454 (0.008527)	0.114871 (0.0009735)

TABLE 6: BEs and their respective PRs of MRLF under SLLF.

n	UP			JP		
θ	7.142857	1.2	0.114114	7.142857	1.2	0.114114
20	16.4002 (0.031314)	1.13929 (0.036423)	0.106377 (0.043796)	16.8701 (0.031313)	1.17967 (0.036422)	0.111405 (0.043784)
40	16.7929 (0.015685)	1.16697 (0.018282)	0.110483 (0.021993)	17.0316 (0.015687)	1.18743 (0.018286)	0.113059 (0.021026)
60	16.8873 (0.010466)	1.1788 (0.012203)	0.111469 (0.014688)	17.0468 (0.010471)	1.19253 (0.012205)	0.113194 (0.014687)
80	16.9629 (0.007851)	1.1847 (0.009157)	0.112271 (0.011023)	17.083 (0.007854)	1.19503 (0.009158)	0.113571 (0.011024)
100	16.9813 (0.006283)	1.1881 (0.007329)	0.112576 (0.008823)	17.0774 (0.006282)	1.19638 (0.007329)	0.113618 (0.008826)
n	LMP			GP		
θ	7.142857	1.2	0.114114	7.142857	1.2	0.114114
20	16.3694 (0.031254)	1.13883 (0.036408)	0.108011 (0.043296)	16.8526 (0.031289)	1.17711 (0.036412)	0.112809 (0.043172)
40	16.7772 (0.015670)	1.16672 (0.018277)	0.111303 (0.021016)	17.0228 (0.015674)	1.18614 (0.018204)	0.113742 (0.021026)
60	16.8767 (0.010459)	1.17862 (0.012201)	0.112016 (0.014542)	17.0410 (0.010454)	1.19166 (0.012202)	0.113645 (0.014523)
80	16.955 (0.007847)	1.18457 (0.009156)	0.112681 (0.011023)	17.0786 (0.007845)	1.19438 (0.009153)	0.113908 (0.010987)
100	16.9749 (0.006280)	1.18799 (0.007328)	0.112905 (0.008822)	17.0739 (0.006286)	1.19586 (0.007329)	0.113887 (0.008800)

risk of the Bayes estimates. Bansal [9] discussed a method to evaluate the relevance of a prior information in terms of the number of additional observation supposed to be added to a given sample size. We used the gamma and conjugate informative prior for analysis.

2.2.1. Gamma Prior. Posterior distribution using informative gamma prior (GP) is

$$p(\theta \mid x)$$
$$= \frac{\left(\left(\theta^{2n+a-1}\right)/(\theta+1)^n\right)e^{-\theta\left(\sum_{i=1}^n x_i+b\right)}}{\int_0^\infty \left(\left(\theta^{2n+a-1}\right)/(\theta+1)^n\right)e^{-\theta\left(\sum_{i=1}^n x_i+b\right)}d\theta}, \quad a, b, \theta > 0.$$
$$(6)$$

2.2.2. Likelihood Matching Prior (Conjugate Prior). The likelihood matching prior (LMP) for Lindley distribution is

$$p(\theta) \propto \frac{\theta^{2a}}{(\theta+1)^a}e^{-\theta b}, \quad a, b, \theta > 0,$$
$$(7)$$

and the posterior distribution using LMP is

$$p(\theta \mid x)$$
$$= \frac{\left(\left(\theta^{2n+2a}\right)/(\theta+1)^{n+a}\right)e^{-\theta\left(\sum_{i=1}^n x_i+b\right)}}{\int_0^\infty \left(\left(\theta^{2n+a-1}\right)/(\theta+1)^n\right)e^{-\theta\left(\sum_{i=1}^n x_i+b\right)}d\theta}, \quad a, b, \theta > 0.$$
$$(8)$$

3. Bayes Estimators and Posterior Risk under Different Loss Functions

This section spotlight is on the derivation of the Bayes estimator (BE) under different loss functions and their respective posterior risk (PR). The results are compared for noninformative as well as informative priors. If the decision is a choice of an estimator, then, the Bayes decision is a Bayes estimator. The Bayes estimators are evaluated under squared error loss function (SELF), weighted squared error loss function (WSELF), precautionary loss function (PLF), modified (quadratic) squared error loss function (M/Q SELF), logarithmic loss function (SLLF), entropy loss function (ELF),

TABLE 7: BEs and their respective PRs of MRLF under ELF.

n	UP			JP		
θ	7.142857	1.2	0.114114	7.142857	1.2	0.114114
20	16.1511 (5.57928)	1.11927 (0.243079)	0.10415 (0.150269)	16.6062 (5.53632)	1.1582 (0.231109)	0.10896 (0.150136)
40	16.6632 (5.63416)	1.15649 (0.299794)	0.109296 (0.141658)	16.8981 (5.62276)	1.17658 (0.254403)	0.111815 (0.141275)
60	16.7998 (5.64793)	1.17169 (0.322941)	0.110663 (0.139528)	16.9577 (5.64668)	1.18525 (0.326034)	0.112362 (0.136368)
80	16.8969 (5.65816)	1.17933 (0.334435)	0.111659 (0.137915)	17.0159 (5.62723)	1.18956 (0.315758)	0.112945 (0.136815)
100	16.9283 (5.6611)	1.18378 (0.341065)	0.112084 (0.127363)	17.0238 (5.67237)	1.1920 (0.324957)	0.113117 (0.12645)
n	LMP			GP		
θ	7.142857	1.2	0.114114	7.142857	1.2	0.114114
20	16.1217 (5.53559)	1.11883 (0.22428)	0.105691 (0.147276)	16.5891 (5.53325)	1.15569 (0.207762)	0.110343 (0.148622)
40	16.6478 (5.6233)	1.15624 (0.249369)	0.110091 (0.149015)	16.8893 (5.62316)	1.17531 (0.232234)	0.112492 (0.137587)
60	16.7894 (5.64668)	1.17152 (0.322650)	0.111199 (0.148555)	16.9518 (5.61599)	1.1844 (0.314585)	0.112811 (0.145671)
80	16.889 (5.62725)	1.1792 (0.314213)	0.112063 (0.138788)	17.0115 (5.62175)	1.18892 (0.310674)	0.11328 (0.139026)
100	16.9220 (5.66035)	1.18367 (0.320887)	0.112409 (0.128663)	17.0202 (5.65197)	1.19148 (0.35406)	0.113385 (0.124952)

TABLE 8: BEs and their respective PRs of MRLF under KLF.

n	UP			JP		
θ	7.142857	1.2	0.114114	7.142857	1.2	0.114114
20	16.404 (0.063125)	1.13965 (0.073520)	0.106432 (0.088576)	16.8667 (0.063124)	1.17931 (0.073519)	0.111351 (0.088753)
40	16.7939 (0.031494)	1.16706 (0.036732)	0.110498 (0.044231)	17.0307 (0.031159)	1.18734 (0.036731)	0.113046 (0.044226)
60	16.8877 (0.020986)	1.17884 (0.024480)	0.111475 (0.029485)	17.0464 (0.020987)	1.19249 (0.024481)	0.113188 (0.029405)
80	16.9632 (0.015733)	1.18473 (0.018357)	0.112274 (0.022107)	17.0828 (0.015732)	1.19501 (0.018359)	0.113568 (0.022108)
100	16.9815 (0.012586)	1.18811 (0.014684)	0.112579 (0.017685)	17.0773 (0.012585)	1.19636 (0.014685)	0.113616 (0.017682)
n	LMP			GP		
θ	7.142857	1.2	0.114114	7.142857	1.2	0.114114
20	16.3737 (0.063005)	1.1392 (0.073491)	0.108032 (0.088577)	16.8494 (0.063076)	1.17679 (0.073498)	0.112726 (0.087329)
40	16.7783 (0.031144)	1.16681 (0.036724)	0.111309 (0.044177)	17.0220 (0.031132)	1.18606 (0.036726)	0.113721 (0.043938)
60	16.8772 (0.020973)	1.17867 (0.024476)	0.112019 (0.029403)	17.0406 (0.020103)	1.19163 (0.024469)	0.113636 (0.029356)
80	16.9553 (0.015725)	1.1846 (0.018355)	0.112683 (0.022108)	17.0784 (0.015723)	1.19436 (0.018358)	0.113902 (0.022035)
100	16.9751 (0.012581)	1.18801 (0.014683)	0.112906 (0.017624)	17.0737 (0.012582)	1.19585 (0.014681)	0.113883 (0.017639)

and K-Loss function. K-loss function proposed by Wasan [10] is well fitted for a measure of inaccuracy for an estimator of a scale parameter of a distribution defined on $R^+ = (0, \infty)$; this loss function is called K-loss function (KLF). Kanefuji and Iwase [11] used KLF for the estimation of a scale parameter with a known coefficient of variations. Table 1 (by [12]) will show the Bayes estimators and their posterior risks for the above-mentioned loss function.

3.1. Elicitation of Hyperparameter(s). Even though many authors have pointed a need for a formal and comprehensive process for elicitation of hyperparameters, there is no standard method. For elicitation, mainly two points are considered; the functional form of the prior distribution and hyperparameter(s), that is why a natural conjugate prior distribution has been generally recommended because its functional form is identical to likelihood function and posterior distribution can be determined by the way of

conjugacy. To determine hyperparameter, we adopted the method discussed by Ali et al. [12].

4. Mean Residual Life Function

For a continuous distribution with the density $f(x)$ and cumulative distribution function $F(x)$, the mean residual life function is defined as

$$m(x) = E(X - x \mid X > x) = \frac{1}{1 - F(x)} \int_x^\infty [1 - F(t)] \, dt. \tag{9}$$

Bayramoglu and Gurler [13] study the mean residual life function of k out of n system with nonidentical components while Govil and Aggarwal [14] and Abdous and Berred [15] compared for different distributions like gamma, exponential, Pareto, uniform, truncated normal, Maxwell,

TABLE 9: Bayes estimator and posterior risk of stress and strength parameter under SELF.

n	UP			JP		
R	0.026364	0.0604	0.998771	0.026364	0.0604	0.998771
20, 40	0.028511 (0.000077)	0.064361 (0.000288)	0.998763 $(1.44921 * 10^{-7})$	0.027985 (0.000078)	0.063217 (0.000282)	0.998728 $(1.40568 * 10^{-7})$
60, 80	0.027029 (0.000030)	0.061592 (0.000109)	0.998756 $(6.22876 * 10^{-8})$	0.026962 (0.000030)	0.061386 (0.000109)	0.998749 $(6.19239 * 10^{-8})$
80, 100	0.026832 (0.000023)	0.061133 (0.000084)	0.998754 $(4.89937 * 10^{-8})$	0.026797 (0.000023)	0.061068 (0.000084)	0.998749 $(4.88093 * 10^{-8})$
80, 60	0.026663 (0.000029)	0.060702 (0.000105)	0.998735 $(6.74031 * 10^{-8})$	0.026796 (0.000029)	0.061019 (0.000105)	0.998738 $(6.75851 * 10^{-8})$
100, 60	0.026575 (0.000026)	0.060422 (0.000095)	0.998733 $(6.31669 * 10^{-8})$	0.026766 (0.000026)	0.060865 (0.000095)	0.99874 $(6.31061 * 10^{-8})$
n	LMP			GP		
R	0.026364	0.0604	0.998771	0.026364	0.0604	0.998771
20, 40	0.028581 (0.000076)	0.064860 (0.000281)	0.99874 $(1.40336 * 10^{-7})$	0.027987 (0.000078)	0.063762 (0.000280)	0.99871 $(1.37326 * 10^{-7})$
60, 80	0.027051 (0.000029)	0.061760 (0.000109)	0.998748 $(6.18727 * 10^{-8})$	0.026956 (0.000029)	0.061625 (0.000108)	0.998743 $(6.1525 * 10^{-8})$
80, 100	0.026848 (0.000022)	0.061323 (0.000083)	0.998748 $(4.87444 * 10^{-8})$	0.026792 (0.000022)	0.061255 (0.000083)	0.998745 $(4.85798 * 10^{-8})$
80, 60	0.026676 (0.000028)	0.061014 (0.000105)	0.998729 $(6.69801 * 10^{-8})$	0.026781 (0.000028)	0.061306 (0.000104)	0.998733 $(6.73044 * 10^{-8})$
100, 60	0.026585 (0.000026)	0.060732 (0.000094)	0.998728 $(6.28193 * 10^{-8})$	0.026748 (0.000026)	0.061143 (0.000094)	0.998736 $(6.23054 * 10^{-8})$

TABLE 10: Bayes estimator and posterior risk of stress and strength parameter under WSELF.

n	UP			JP		
R	0.026364	0.0604	0.998771	0.026364	0.0604	0.998771
20, 40	0.025616 (0.002896)	0.059646 (0.004714)	0.998762 $(1.4508 * 10^{-7})$	0.025146 (0.002839)	0.058595 (0.004622)	0.998728 $(1.40713 * 10^{-7})$
60, 80	0.025913 (0.001116)	0.059791 (0.001801)	0.998756 $(6.2361 * 10^{-8})$	0.025849 (0.001113)	0.059591 (0.001795)	0.998749 $(6.1996 * 10^{-8})$
80, 100	0.025971 (0.000860)	0.059748 (0.001384)	0.998754 $(4.90522 * 10^{-8})$	0.025938 (0.000859)	0.059685 (0.001383)	0.998749 $(4.88671 * 10^{-8})$
80, 60	0.025571 (0.001092)	0.058960 (0.001742)	0.998735 $(6.74825 * 10^{-8})$	0.025699 (0.001091)	0.059267 (0.001742)	0.998738 $(6.74652 * 10^{-8})$
100, 60	0.025581 (0.000993)	0.058846 (0.001576)	0.998733 $(6.32415 * 10^{-8})$	0.025765 (0.009991)	0.059276 (0.001568)	0.99874 $(6.31859 * 10^{-8})$
n	CP			GP		
R	0.026364	0.0604	0.998771	0.026364	0.0604	0.998771
20, 40	0.025679 (0.002801)	0.060117 (0.004543)	0.99874 $(1.40486 * 10^{-7})$	0.025148 (0.002839)	0.059091 (0.004571)	0.99871 $(1.37462 * 10^{-7})$
60, 80	0.025934 (0.001111)	0.059954 (0.001786)	0.998748 $(6.19446 * 10^{-8})$	0.025843 (0.001110)	0.059822 (0.001702)	0.998743 $(6.15958 * 10^{-8})$
80, 100	0.025987 (0.000860)	0.059934 (0.001382)	0.998748 $(4.8802 * 10^{-8})$	0.025933 (0.000859)	0.059869 (0.001382)	0.998745 $(4.86369 * 10^{-8})$
80, 60	0.025584 (0.001092)	0.059262 (0.001741)	0.998729 $(6.70581 * 10^{-8})$	0.025684 (0.001091)	0.059548 (0.001737)	0.998733 $(6.70385 * 10^{-8})$
100, 60	0.025591 (0.000993)	0.059146 (0.001558)	0.998728 $(6.28928 * 10^{-8})$	0.025747 (0.00990)	0.059549 (0.001553)	0.998736 $(6.23805 * 10^{-8})$

TABLE 11: Bayes estimator and posterior risk of stress and strength parameter under MSELF.

n	UP			JP		
R	0.026364	0.0604	0.998771	0.026364	0.0604	0.998771
20, 40	0.023332 (0.999986)	0.055666 (0.999815)	0.998762 (0.003708)	0.023021 (0.999987)	0.054887 (0.999813)	0.998728 (0.003812)
60, 80	0.024906 (0.999984)	0.058118 (0.999798)	0.998756 (0.003728)	0.024851 (0.999984)	0.057935 (0.999800)	0.998749 (0.003729)
80, 100	0.025178 (0.999984)	0.058442 (0.999796)	0.998754 (0.003734)	0.025149 (0.999984)	0.058385 (0.999797)	0.998749 (0.003734)
80, 60	0.024596 (0.999985)	0.057357 (0.999806)	0.998735 (0.003790)	0.024705 (0.999984)	0.057630 (0.999803)	0.998738 (0.003780)
100, 60	0.024689 (0.999984)	0.057390 (0.999806)	0.998733 (0.003796)	0.024846 (0.999984)	0.057777 (0.999802)	0.99874 (0.003776)
n	LMP			GP		
R	0.026364	0.0604	0.998771	0.026364	0.0604	0.998771
20, 40	0.023375 (0.999986)	0.056022 (0.999811)	0.99874 (0.003706)	0.023023 (0.999987)	0.055252 (0.999802)	0.998709 (0.003666)
60, 80	0.024924 (0.999984)	0.058258 (0.999797)	0.998748 (0.003721)	0.024847 (0.999984)	0.058141 (0.999796)	0.998743 (0.003727)
80, 100	0.025191 (0.999984)	0.058613 (0.999794)	0.998748 (0.003732)	0.025144 (0.999984)	0.058552 (0.999795)	0.998745 (0.003731)
80, 60	0.024607 (0.999985)	0.057626 (0.999803)	0.998728 (0.003780)	0.024692 (0.999984)	0.057881 (0.99980)	0.998733 (0.003759)
100, 60	0.024697 (0.999984)	0.057659 (0.999803)	0.998728 (0.003712)	0.024831 (0.999984)	0.058025 (0.99980)	0.998736 (0.003708)

TABLE 12: Bayes estimator and posterior risk of stress and strength parameter under PLF.

n	UP			JP		
R	0.026364	0.0604	0.998771	0.026364	0.0604	0.998771
20, 40	0.029827 (0.002631)	0.066557 (0.004393)	0.998763 ($1.451 * 10^{-7}$)	0.029348 (0.002628)	0.065490 (0.004345)	0.998728 ($1.4747 * 10^{-7}$)
60, 80	0.027570 (0.001082)	0.062474 (0.001764)	0.998756 ($6.23652 * 10^{-8}$)	0.027506 (0.001087)	0.062271 (0.001769)	0.998749 ($6.24015 * 10^{-8}$)
80, 100	0.027252 (0.000841)	0.061814 (0.001363)	0.998754 ($4.90548 * 10^{-8}$)	0.027219 (0.000843)	0.061751 (0.001361)	0.998749 ($4.88704 * 10^{-8}$)
80, 60	0.027198 (0.001070)	0.061565 (0.001725)	0.998735 ($6.74885 * 10^{-8}$)	0.027326 (0.001059)	0.061873 (0.001707)	0.998738 ($6.75706 * 10^{-8}$)
100, 60	0.027064 (0.000979)	0.061206 (0.001568)	0.998733 ($6.3247 * 10^{-8}$)	0.027248 (0.000966)	0.061638 (0.001546)	0.99874 ($6.32386 * 10^{-8}$)
n	LMP			GP		
R	0.026364	0.0604	0.998771	0.026364	0.0604	0.998771
20, 40	0.0298893 (0.002617)	0.067019 (0.004317)	0.99874 ($1.42516 * 10^{-7}$)	0.029351 (0.002627)	0.066001 (0.004417)	0.99871 ($1.37503 * 10^{-7}$)
60, 80	0.027591 (0.001080)	0.062633 (0.001748)	0.998748 ($6.19503 * 10^{-8}$)	0.027500 (0.001080)	0.062503 (0.001746)	0.998743 ($6.16024 * 10^{-8}$)
80, 100	0.027268 (0.000839)	0.061999 (0.001354)	0.998748 ($4.88056 * 10^{-8}$)	0.027214 (0.000834)	0.061934 (0.001347)	0.998745 ($4.86409 * 10^{-8}$)
80,60	0.027211 (0.001058)	0.061868 (0.001707)	0.998729 ($6.70654 * 10^{-8}$)	0.027311 (0.001060)	0.062150 (0.001689)	0.998733 ($6.69898 * 10^{-8}$)
100, 60	0.027074 (0.000958)	0.061508 (0.001543)	0.998728 ($6.28994 * 10^{-8}$)	0.027231 (0.000957)	0.061908 (0.001529)	0.998736 ($6.23856 * 10^{-8}$)

TABLE 13: Bayes estimator and posterior risk of stress and strength parameter under SLLF.

n	UP			JP		
R	0.026364	0.0604	0.998771	0.026364	0.0604	0.998771
20, 40	0.027033 (0.11001)	0.061970 (0.077534)	0.998762 $(1.4526 * 10^{-7})$	0.026509 (0.109867)	0.060832 (0.077376)	0.998728 $(1.40892 * 10^{-7})$
60, 80	0.026465 (0.042628)	0.060684 (0.029901)	0.998756 $(6.24387 * 10^{-8})$	0.026398 (0.042621)	0.060480 (0.029899)	0.998749 $(6.20736 * 10^{-8})$
80, 100	0.026398 (0.032862)	0.060436 (0.023035)	0.998754 $(4.91134 * 10^{-8})$	0.026363 (0.032859)	0.060372 (0.023033)	0.998749 $(4.89283 * 10^{-8})$
80, 60	0.026109 (0.042241)	0.059822 (0.029329)	0.998735 $(6.7568 * 10^{-8})$	0.026243 (0.042241)	0.060138 (0.029323)	0.998738 $(6.75607 * 10^{-8})$
100, 60	0.026071 (0.038466)	0.059625 (0.026609)	0.998733 $(6.33217 * 10^{-8})$	0.026262 (0.038458)	0.060068 (0.026605)	0.99874 $(6.32621 * 10^{-8})$
n	LMP			GP		
R	0.026364	0.0604	0.998771	0.026364	0.0604	0.998771
20, 40	0.027104 (0.109771)	0.062474 (0.077367)	0.99874 $(1.40666 * 10^{-7})$	0.026511 (0.109817)	0.061372 (0.077337)	0.99871 $(1.37639 * 10^{-7})$
60, 80	0.026487 (0.042620)	0.060853 (0.029893)	0.998748 $(6.20223 * 10^{-8})$	0.026392 (0.042619)	0.060718 (0.029804)	0.998743 $(6.16733 * 10^{-8})$
80, 100	0.026414 (0.032860)	0.060626 (0.023029)	0.998748 $(4.88632 * 10^{-8})$	0.026358 (0.032858)	0.060559 (0.023030)	0.998745 $(4.8698 * 10^{-8})$
80, 60	0.026123 (0.042240)	0.060133 (0.029322)	0.998729 $(6.71435 * 10^{-8})$	0.026227 (0.042235)	0.060425 (0.029307)	0.998733 $(6.7069 * 10^{-8})$
100, 60	0.026081 (0.038459)	0.059934 (0.026602)	0.998728 $(6.29729 * 10^{-8})$	0.026244 (0.038455)	0.060347 (0.026580)	0.998736 $(6.24607 * 10^{-8})$

and lognormal distributions. The mean residual life function for Lindley distribution is

$$m(x) = \frac{\theta + 2 + \theta x}{\theta(\theta + 1 + \theta x)}. \tag{10}$$

Ghitany et al. [2, 5] point out the following remarks.

$m(0) = \mu$.

$m(x)$ is a decreasing function in x and θ and $1/\theta < m(x) < (\theta + 2)/(\theta(\theta + 1)) = \mu$.

From $m(x)$, one can easily observe that mean residuals life is a diminishing function of time because the distribution belongs to exponential family and for larger parameter value it is close to zero. To evaluate the Bayes estimates and their risk, since the integral appears in both numerator and denumerator, we required a suitable approximate method to obtain Bayes estimates and respective posterior risks. The simplest method is Lindley's [16] approximation method, which approaches the ratio of the integrals as a whole and produces a single numerical results. Thus, Lindley approximation (LA) given by Lindley [16] for obtaining the Bayes estimator and posterior risk of θ (Mathematica can be used for the solution of integral but takes large time as compared to LA). Many researchers have used this approximation for solving the ratio of integrals for different numbers of parameters for lifetime distributions; see among others Howlader and Hossain [17], Singh et al. [18], and Preda et al. [19].

If n is sufficiently large, the ratio of the integral of the form according to Lindley [16] can be computed as

$$I(x) = E[u(\theta)] = \frac{\int_\theta u(\theta) \exp[l(\theta, \mathbf{x}) + g(\theta)] d\theta}{\int_\theta \exp[l(\theta, \mathbf{x}) + g(\theta)] d\theta}, \quad \theta > 0, \tag{11}$$

where $u(\theta)$ = function of θ only; $l(\theta, \mathbf{x})$ = log of likelihood; $g(\theta)$ = log of prior of θ, can be evaluated as

$$I(x) = u(\hat{\theta}) + 0.5\left[(\hat{u}_{\theta\theta} + 2\hat{u}_\theta \hat{p}_\theta)\hat{\sigma}_{\theta\theta}\right] + 0.5\left[(\hat{u}_\theta \hat{\sigma}_{\theta\theta})(\hat{L}_{\theta\theta\theta}\hat{\sigma}_{\theta\theta})\right], \tag{12}$$

where $\hat{\theta}$ = MLE of θ = $(-(\overline{x} - 1) + \sqrt{(\overline{x} - 1)^2 + 8\overline{x}})/2\overline{x}, \overline{x} > 0$; $\hat{u}_\theta = \partial u(\hat{\theta})/\partial\theta$; $\hat{u}_{\theta\theta} = \partial^2 u(\hat{\theta})/\partial\theta^2$; $\hat{L}_{\theta\theta\theta} = \partial^3 l(\hat{\theta})/\partial\theta^3$; $\hat{p}_\theta = \partial g(\hat{\theta})/\partial\theta$; $\hat{\sigma}_{\theta\theta} = -1/\hat{L}_{\theta\theta}$; $\hat{L}_{\theta\theta} = \partial^2 l(\hat{\theta})/\partial\theta^2$; the simulation study of mean residual life function (MRLF) for different loss functions under different prior is using $t = 3$ (see Tables 2, 3, 4, 5, 6, 7, and 8).

Using SELF for θ = 7.142857, Bayes estimates are overestimated while for θ = 1.2 and 0.114114 results are underestimated and by increasing sample size these approaches to true parameter values. This behaviour can be observed in all loss functions except PLF. Making comparison between symmetric and asymmetric loss functions, one can

TABLE 14: Bayes estimator and posterior risk of stress and strength parameter under ELF.

n	UP			JP		
R	0.026364	0.0604	0.998771	0.026364	0.0604	0.998771
20, 40	0.025616 (7.27524)	0.059646 (5.60043)	0.998762 (0.002477)	0.025146 (7.21335)	0.058595 (5.53674)	0.998728 (0.002446)
60, 80	0.025913 (7.28493)	0.059791 (5.61898)	0.998756 (0.002490)	0.025849 (7.27996)	0.059591 (5.56257)	0.998749 (0.002450)
80, 100	0.025971 (7.28523)	0.059748 (5.62379)	0.998754 (0.002494)	0.025938 (7.28483)	0.059685 (5.60591)	0.998749 (0.002443)
80, 60	0.025571 (7.31176)	0.058960 (5.64728)	0.998735 (0.002532)	0.025699 (7.30167)	0.059267 (5.62381)	0.998738 (0.002535)
100, 60	0.025581 (7.31282)	0.058846 (5.65251)	0.998733 (0.002535)	0.025765 (7.30984)	0.059276 (5.63837)	0.99874 (0.002522)
n	LMP			GP		
R	0.026364	0.0604	0.998771	0.026364	0.0604	0.998771
20, 40	0.025679 (7.21076)	0.060117 (5.48446)	0.99874 (0.002452)	0.025148 (7.21315)	0.059091 (5.47148)	0.99871 (0.002452)
60, 80	0.025934 (7.23392)	0.059954 (5.53147)	0.998748 (0.002405)	0.025843 (7.23390)	0.059822 (5.53139)	0.998743 (0.002406)
80, 100	0.025987 (7.28404)	0.059934 (5.60153)	0.998748 (0.002366)	0.025933 (7.28403)	0.059868 (5.60174)	0.998745 (0.002312)
80, 60	0.025584 (7.30072)	0.059262 (5.61983)	0.998729 (0.002345)	0.025684 (7.30076)	0.059548 (5.61233)	0.998733 (0.002353)
100, 60	0.025591 (7.30237)	0.059146 (5.63426)	0.998728 (0.002246)	0.025747 (7.30235)	0.059549 (5.63259)	0.998736 (0.002240)

TABLE 15: Bayes estimator and posterior risk of stress and strength parameter under KLF.

n	UP			JP		
R	0.026364	0.0604	0.998771	0.026364	0.0604	0.998771
20, 40	0.027025 (0.226074)	0.061958 (0.158076)	0.998762 ($2.9052 * 10^{-7}$)	0.026527 (0.225789)	0.060863 (0.15776)	0.998728 ($2.81784 * 10^{-7}$)
60, 80	0.026465 (0.086164)	0.060685 (0.060249)	0.998756 ($1.24877 * 10^{-7}$)	0.026399 (0.086151)	0.060482 (0.060245)	0.998749 ($1.24147 * 10^{-7}$)
80, 100	0.026398 (0.066263)	0.060436 (0.046336)	0.998754 ($9.82267 * 10^{-8}$)	0.026364 (0.066258)	0.060373 (0.046331)	0.998749 ($9.78566 * 10^{-8}$)
80, 60	0.026111 (0.085374)	0.059825 (0.059089)	0.998735 ($1.35136 * 10^{-7}$)	0.026242 (0.085368)	0.060137 (0.059076)	0.998738 ($1.35101 * 10^{-7}$)
100, 60	0.026073 (0.077673)	0.059629 (0.053573)	0.998733 ($1.26643 * 10^{-7}$)	0.026260 (0.077674)	0.060066 (0.053566)	0.99874 ($1.26324 * 10^{-7}$)
n	LMP			GP		
R	0.026364	0.0604	0.998771	0.026364	0.0604	0.998771
20, 40	0.027091 (0.225697)	0.062444 (0.157728)	0.99874 ($2.81332 * 10^{-7}$)	0.026529 (0.225695)	0.061382 (0.157082)	0.99871 ($2.75278 * 10^{-7}$)
60, 80	0.026486 (0.086152)	0.06085 (0.060234)	0.998748 ($1.24045 * 10^{-7}$)	0.026393 (0.086148)	0.060717 (0.060235)	0.998743 ($1.23347 * 10^{-7}$)
80, 100	0.026414 (0.066253)	0.060625 (0.046324)	0.998748 ($9.77264 * 10^{-8}$)	0.026359 (0.066247)	0.060558 (0.046322)	0.998745 ($9.73961 * 10^{-8}$)
80, 60	0.026125 (0.085360)	0.060132 (0.059016)	0.998729 ($1.34287 * 10^{-7}$)	0.026227 (0.085300)	0.060420 (0.059014)	0.998733 ($1.34238 * 10^{-7}$)
100, 60	0.026083 (0.077670)	0.059934 (0.053519)	0.998728 ($1.25946 * 10^{-7}$)	0.026243 (0.077659)	0.060341 (0.053515)	0.998736 ($1.25921 * 10^{-7}$)

TABLE 16: Waiting time (in minutes) before customer service in Bank A.

0.8	2.9	4.3	5.0	6.7	8.2	9.7	11.9	14.1	19.9
0.8	3.1	4.3	5.3	6.9	8.6	9.8	12.4	15.4	20.6
1.3	3.2	4.4	5.5	7.1	8.6	10.7	12.5	15.4	21.3
1.5	3.3	4.4	5.7	7.1	8.6	10.9	12.9	17.3	21.4
1.8	3.5	4.6	5.7	7.1	8.8	11.0	13.0	17.3	21.9
1.9	3.6	4.7	6.1	7.1	8.8	11.0	13.1	18.1	23.0
1.9	4.0	4.7	6.2	7.4	8.9	11.1	13.3	18.2	27.0
2.1	4.1	4.8	6.2	7.6	8.9	11.2	13.6	18.4	31.6
2.6	4.2	4.9	6.2	7.7	9.5	11.2	13.7	18.9	33.1
2.7	4.2	4.9	6.3	8.0	9.6	11.5	13.9	19.0	38.5

TABLE 17: Waiting time (in minutes) before customer service in Bank B.

0.1	1.2	2.3	2.9	3.5	5.3	6.8	8.0	8.5	13.2
0.2	1.8	2.3	3.1	3.9	5.6	7.3	8.5	11.0	13.7
0.3	1.9	2.5	3.1	4.0	5.6	7.5	8.7	12.1	14.5
0.7	2.0	2.6	3.2	4.2	6.2	7.7	9.5	12.3	16.0
0.9	2.2	2.7	3.4	4.5	6.3	7.7	10.7	12.8	16.5
1.1	2.3	2.7	3.4	4.7	6.6	8.0	10.9	12.9	28.0

TABLE 18: K-S distances and associated P value and MLE.

Data set	K-S test	P value	MLE
Bank A	0.065	0.721	0.187
Bank B	0.083	0.826	0.280

easily observes that SELF has smaller posterior risk. Since there is defect in symmetric loss function, that is, SELF assign equal weight to over and under estimation, so we have to look for an alternative choice. In case of asymmetric loss functions; WSELF, SLLF, and PLF can be alternative choices.

5. Lorenz Curve (See Figure 2)

For a positive random variable X, the Lorenz curve is defined by the graph of the ratio

$$L(F(x)) = \frac{E(X \mid X \le x) F(x)}{E(X)}, \tag{13}$$

against $F(x)$ with the properties $L(p) \le p$, $L(0) = 0$, and $L(1) = 1$ for $0 \le p \le 1$. If X represents annual income, then, $L(p)$ is the proportion of total income that accrues to individuals having the $100p\%$ lowest incomes; see Gail and Gastwirth [20] for details of Lorenz curves. For the exponential distribution, it is well known that Lorenz curve is given by

$$L(p) = p[p + (1 - p)\log(1 - p)]. \tag{14}$$

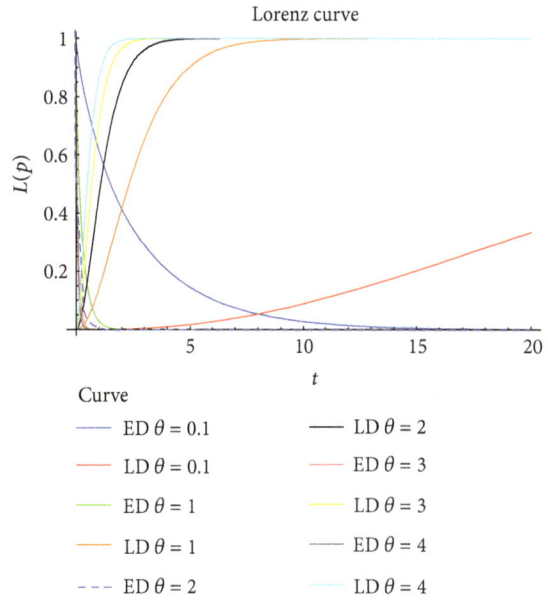

FIGURE 2: Lorenz curve graph.

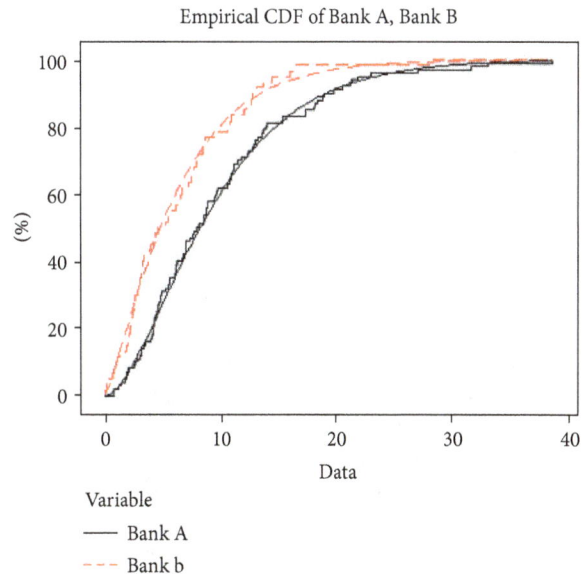

FIGURE 3: Empirical CDF of bank data.

For the Lindley distribution, the Lorenz curve is

$$L(p) = 1 - \frac{\theta(\theta+1)(1-p)}{(\theta+2)(\theta+1+\theta t)}\left(\frac{2}{\theta}+1+2t+\theta t+\theta t^2\right), \tag{15}$$

where $p = F(t)$.

The comparison of Lorenz curve for exponential and Lindley distribution is given in Figures 2 and 3. Lindley distribution performance for Lorenz curve is better as compared to exponential distribution for different values of θ.

TABLE 19: BEs and respective PRs of stress and strength under different prior of real data where $R = 0.354264$.

↓ Loss function|prior →	UP	JP	LMP	GP
SELF	0.354572 (0.001590)	0.35578 (0.001588)	0.354518 (0.001589)	0.355719 (0.001588)
WSELF	0.350132 (0.004439)	0.351317 (0.004436)	0.35008 (0.004438)	0.351257 (0.004432)
MSELF	0.345899 (0.958108)	0.347017 (0.957694)	0.34585 (0.957126)	0.346961 (0.957115)
PLF	0.356806 (0.004470)	0.358004 (0.004448)	0.356754 (0.004470)	0.357944 (0.004449)
SLLF	0.352333 (0.012639)	0.353537 (0.012664)	0.35228 (0.012637)	0.353476 (0.012663)
ELF	0.350132 (2.09262)	0.351317 (2.08584)	0.35008 (2.09292)	0.351257 (2.08618)
KLF	0.352345 (0.025358)	0.353541 (0.023508)	0.352292 (0.025354)	0.353481 (0.025307)

TABLE 20: BEs and respective PRs of mean residual life function under different prior of real data of Bank A.

↓ Loss function|prior →	UP	JP	LMP	GP
SELF	8.40917 (0.477926)	8.45888 (0.475225)	8.40618 (0.477931)	8.45692 (0.475245)
WSELF	8.35267 (0.056499)	8.40204 (0.056847)	8.34973 (0.056459)	8.40007 (0.056485)
MSELF	8.29763 (574.087)	8.34567 (584.203)	8.29478 (573.489)	8.34375 (583.797)
PLF	8.43754 (0.056738)	8.48693 (0.056088)	8.43456 (0.056735)	8.48498 (0.056124)
SLLF	8.38079 (0.006753)	8.43049 (0.006754)	8.37781 (0.006750)	8.42852 (0.006754)
ELF	8.35267 (4.24852)	8.40204 (4.26033)	8.34973 (4.24782)	8.40007 (4.23986)
KLF	8.38088 (0.013528)	8.43041 (0.013531)	8.37791 (0.013523)	8.42845 (0.013524)

6. Stress and Strength Analysis for Lindley Distribution

Let X and Y be two random variables such that X represents "strength" and Y "stress"; then, reliability of the stress and strength model is presented as

$$R = P(Y < x) = \int_{-\infty}^{\infty} \int_{-\infty}^{x} f(x,y)\, dy\, dx, \qquad (16)$$

where $P(Y < X)$ is a relationship which represents the probability that the strength exceeds the stress and $f(x,y)$ is joint p.d.f. of X and Y for example, receptor in human eye operates only if it is simulated by a source where magnitude X is greater than a random lower threshold Y for the eye, so here R is the probability that the receptor operates. In mechanical reliability of a system R gives the probability of a system failure, if the system fails whenever the strength is less than the applied stress.

This quandary has a long narration starting with the revolutionary work of Birnbaum [21] and Birnbaum and McCarty [22]. The stress strength was first introduced by Church and Harris [23]. A comprehensive handling of the different stress strength models can be found in outstanding monograph by Kotz et al. [24]. Some most fresh work on the stress strength model can be found in Kundu and Gupta [25, 26], Raqab and Kundu [27], Kundu and Raqab [28], Krishnamoorthy et al. [29], Eryilmaz [30], and the references cited therein. Recently, Al-Mutairi et al. [31] considered the stress and strength analysis of Lindley distribution using SELF but we are generalizing their work by considering different types of loss functions.

Suppose that X and Y are two independent Lindley rv's with respective parameters θ_1 and θ_2, then,

$$R = P(Y < X) = \int_0^{\infty} P(Y < X \mid Y = y) f(y)\, dy$$

$$= \int_0^{\infty} \int_0^{x} f(x,y)\, dy\, dx$$

$$R = \frac{\theta_1^2 \left(\theta_1(1+\theta_1) + (1+\theta_1)(3+\theta_1)\theta_2 + (3+2\theta_1)\theta_2^2 + \theta_2^3\right)}{(1+\theta_1)(1+\theta_2)(\theta_1+\theta_2)^3}.$$
$$(17)$$

Remark. If $\theta_1 = \theta_2 = \theta$, then, $R = 0.5$.

$$I(x) = E[R = u(\theta_1, \theta_2)]$$

$$= \frac{\int_\theta u(\theta_1,\theta_2) \exp\left[l(\theta_1,\theta_2,\mathbf{x}) + g(\theta_1,\theta_2)\right] d\theta_1 d\theta_2}{\int_\theta \exp\left[l(\theta_1,\theta_2,\mathbf{x}) + g(\theta_1,\theta_2)\right] d\theta_1 d\theta_2}$$

$$\theta_1, \theta_2 > 0,$$
$$(18)$$

where $u(\theta_1,\theta_2)$ = function of θ_1 and θ_2 only; $l(\theta_1,\theta_2,\mathbf{x})$ = log of likelihood; $g(\theta_1,\theta_2)$ = log of prior of θ_1 and θ_2, can be evaluated as

$$I(x) = u\left(\hat{\theta}_1, \hat{\theta}_2\right)$$

$$+ 0.5 \left(\begin{array}{c} \left(\hat{u}_{\theta_1\theta_1} + 2\hat{u}_{\theta_1}\hat{p}_{\theta_1}\right)\hat{\sigma}_{\theta_1\theta_1} \left(\hat{u}_{\theta_1\theta_2} + 2\hat{u}_{\theta_1}\hat{p}_{\theta_2}\right)\hat{\sigma}_{\theta_2\theta_1} \\ + \left(\hat{u}_{\theta_2\theta_1} + 2\hat{u}_{\theta_2}\hat{p}_{\theta_1}\right)\hat{\sigma}_{\theta_1\theta_2} + \left(\hat{u}_{\theta_2\theta_2} + 2\hat{u}_{\theta_2}\hat{p}_{\theta_2}\right)\hat{\sigma}_{\theta_2\theta_2} \end{array} \right)$$

$$+ 0.5 \left(\begin{array}{c} \left(\hat{u}_{\theta_1}\hat{\sigma}_{\theta_1\theta_1} + \hat{u}_{\theta_2}\hat{\sigma}_{\theta_1\theta_2}\right) \\ \left(\hat{L}_{\theta_1\theta_1\theta_1}\hat{\sigma}_{\theta_1\theta_1} + \hat{L}_{\theta_1\theta_2\theta_1}\hat{\sigma}_{\theta_1\theta_1} + \hat{L}_{\theta_2\theta_1\theta_1}\hat{\sigma}_{\theta_2\theta_1} + \hat{L}_{\theta_2\theta_2\theta_1}\hat{\sigma}_{\theta_2\theta_2}\right) \\ + \left(\hat{u}_{\theta_1}\hat{\sigma}_{\theta_2\theta_1} + \hat{u}_{\theta_2}\hat{\sigma}_{\theta_2\theta_2}\right) \\ \left(\hat{L}_{\theta_1\theta_1\theta_1}\hat{\sigma}_{\theta_1\theta_1} + \hat{L}_{\theta_2\theta_2\theta_1}\hat{\sigma}_{\theta_1\theta_1} + \hat{L}_{\theta_1\theta_2\theta_1}\hat{\sigma}_{\theta_2\theta_1} + \hat{L}_{\theta_2\theta_2\theta_2}\hat{\sigma}_{\theta_2\theta_2}\right) \end{array} \right),$$
$$(19)$$

where $\hat{\theta}_j = MLE$ of $\theta_j = (-(\bar{j} - 1) + \sqrt{(\bar{j} - 1)^2 + 8\bar{j}})/$ $2\bar{j}, j = x, y; \bar{x} > 0; \hat{u}_{\theta_i} = \partial u(\hat{\theta}_i)/\partial \hat{\theta}_i; \hat{u}_{\theta_i \theta_j} = \partial^2 u(\hat{\theta}_{i,j})/$ $\partial \hat{\theta}_{i,j}^2; \hat{L}_{\theta_i \theta_j \theta_i} = \partial^3 l(\hat{\theta}_{i,j})/\partial \hat{\theta}_{i,j}^3; \hat{p}_{\theta_i} = \partial g(\hat{\theta}_{i,j})/\partial \hat{\theta}_{i,j}; \hat{\sigma}_{\theta_i \theta_j} =$ $-1/\hat{L}_{\theta_i \theta_j}; \hat{L}_{\theta_i \theta_j} = \partial^2 l(\hat{\theta}_{i,j})/\partial \hat{\theta}_{i,j}^2; \hat{L}_{\theta_i \theta_j \theta_i} = \hat{\sigma}_{\theta_i \theta_j} = 0$ as both x and y are independent $R = 0.026364$ (when $\theta_1 = 0.1$ and $\theta_2 = 1$); $R = 0.0604$ (for $\theta_1 = 1$ and $\theta_2 = 9$) and $R = 0.998771$ (when $\theta_1 = 9$ and $\theta_2 = 0.1$) (see Tables 9, 10, 11, 12, 13, 14, and 15).

The Bayes estimates of stress and strength under SELF, MSELF, and ELF are underestimated. By comparing symmetric and asymmetric loss functions, it is noted that posterior risk of SELF is smaller than asymmetric loss functions. In case of asymmetric loss functions, WSELF and PLF have smaller posterior risk than other available loss functions.

Evaluating the performance of informative and noninformative priors, one can easily observe that informative priors have smaller posterior risk due to the availability of compact information. LM and gamma priors both have approximately the same behaviour depending upon the choice of hyperparameters value. More compact information will lead to correct hyperparameters which will lead to definitely better results and smaller posterior risk than noninformative priors. Although there are some depicts where informative priors have posterior risks greater than noninformative priors which is just due to random generation. Increasing sample size in case of SLLF has an inverse effect.

7. Real Life Application

Ghitany et al. [2] provide waiting times (in minutes) before service of 100 bank customers data set for Lindley distribution. They fitted both Lindley and exponential distributions (both have the same number of parameters) by method of maximized likelihood method and found Lindley distribution provides better fit. The data is given in Tables 16 and 17.

We fit on both data sets Kolmogorov-Smirnov test and found that Lindley distribution is good fitted. The values of K-S test along P value are given in Table 18.

The Bayes estimates of stress and strength reliability under different priors and their posterior risk are evaluated in Tables 19 and 20.

Since the Lindley distribution belongs to the exponential family so the natural conjugate prior is Gamma distribution.

The posterior risks of LMP and GP are approximately the same as compared to noninformative priors. There are some posterior risk values which are greater than noninformative priors. These are just due to hyperparameters value effect that is, more accurate values will lead to the smaller posterior risk. PLF and WSELF loss functions have smaller posterior risk as compared to other loss functions.

8. Conclusion and Suggestions

We consider the Bayesian analysis of the Lindley model via informative and informative priors under different loss functions. Based on posterior distribution, different properties,

we conclude that informative priors (LMP, GP) performance approximately equal and have smaller posterior risk's as compared to the noninformative priors; also Jeffreys prior results are more precised than uniform prior. In other words, we can summarize result as

$$GP(PR) \leq LMP(PR) < JP(PR) < UP(PR). \qquad (20)$$

The choice of loss function as concerned, one can easily observe based on evidence (different properties as discussed above) that PLF, SLLF, and WSELF are suitable than other asymmetrical loss functions. One thing is common as we increase sample size posterior risk comes down. In future, this work can be extended using censored data.

References

[1] D. V. Lindley, "Fiducial distributions and Bayes' theorem," *Journal of the Royal Statistical Society. Series B*, vol. 20, pp. 102–107, 1958.

[2] M. E. Ghitany, B. Atieh, and S. Nadarajah, "Lindley distribution and its application," *Mathematics and Computers in Simulation*, vol. 78, no. 4, pp. 493–506, 2008.

[3] M. Sankaran, "The discrete Poisson-Lindley distribution," *Biometrics*, vol. 26, pp. 145–149, 1970.

[4] M. E. Ghitany and D. K. Al-Mutairi, "Estimation methods for the discrete Poisson-Lindley distribution," *Journal of Statistical Computation and Simulation*, vol. 79, no. 1, pp. 1–9, 2009.

[5] M. E. Ghitany, D. K. Al-Mutairi, and S. Nadarajah, "Zero-truncated Poisson-Lindley distribution and its application," *Mathematics and Computers in Simulation*, vol. 79, no. 3, pp. 279–287, 2008.

[6] H. Zamani and N. Ismail, "Negative binomial-Lindley distribution and its application," *Journal of Mathematics and Statistics*, vol. 6, no. 1, pp. 4–9, 2010.

[7] M. E. Ghitany, F. Alqallaf, D. K. Al-Mutairi, and H. A. Husain, "A two-parameter weighted Lindley distribution and its applications to survival data," *Mathematics and Computers in Simulation*, vol. 81, no. 6, pp. 1190–1201, 2011.

[8] H. Jeffreys, *Theory of Probability*, Oxford University Press, 3rd edition, 1964.

[9] A. K. Bansal, *Bayesian Parametric Inference*, Narosa Publishing House, New Delhi, India, 2007.

[10] M. Wasan, *Parametric Estimation*, McGraw-Hill, New York, NY, USA, 1970.

[11] K. Kanefuji and K. Iwase, "Estimation for a scale parameter with known coefficient of variation," *Statistical Papers*, vol. 39, no. 4, pp. 377–388, 1998.

[12] S. Ali, M. Aslam, and S. M. A. Kazmi, "A study of the effect of the loss function on Bayes Estimate, posterior risk and hazard function for Lindley distribution," *Applied Mathematical Modelling*, vol. 37, no. 8, pp. 6078–6078, 2013.

[13] I. Bayramoglu and S. Gurler, "On the mean residual life function of the k-out-of-n system with nonidentical components," in *International Conference on Mathematical and Statistical Modeling in Honor of Enrique Castillo*, pp. 28–30, Ciudad Real, Spain, June 2006.

[14] K. K. Govil and K. K. Aggarwal, "Mean residual life function for normal, gamma and lognormal densities," *Reliability Engineering*, vol. 5, no. 1, pp. 47–51, 1983.

[15] B. Abdous and A. Berred, "Mean residual life estimation," *Journal of Statistical Planning and Inference*, vol. 132, no. 1-2, pp. 3–19, 2005.

[16] D. V. Lindley, "Approximate Bayesian methods," *Trabajos de Estadistica Y de Investigacion Operativa*, vol. 31, no. 1, pp. 223–245, 1980.

[17] H. A. Howlader and A. M. Hossain, "Bayesian survival estimation of Pareto distribution of the second kind based on failure-censored data," *Computational Statistics and Data Analysis*, vol. 38, no. 3, pp. 301–314, 2002.

[18] R. Singh, S. K. Singh, U. Singh, and G. P. Singh, "Bayes estimator of generalized-exponential parameters under Linex loss function using Lindley's approximation," *Data Science Journal*, vol. 7, pp. 65–75, 2008.

[19] V. Preda, E. Panaitescu, and A. Constantinescu, "Bayes estimators of modified-Weibull distribution parameters using Lindley's approximation," *WSEAS Transactions on Mathematics*, vol. 9, no. 7, pp. 539–549, 2010.

[20] M. H. Gail and J. L. Gastwirth, "A scale-free goodness-of-fit test for the exponential distribution based on the Lorenz curve," *Journal of the American Statistical Association*, vol. 73, no. 364, pp. 787–793, 1978.

[21] Z. W. Birnbaum, "On a use of the Mann-Whitney statistic," in *Proceedings of the 3rd Berkeley Symposium on Mathematical Statistics and Probability*, vol. 1, pp. 13–17, University of California Press, Berkeley, Calif, USA, 1956.

[22] Z. W. Birnbaum and R. C. McCarty, "A distribution-free upper confidence bound for $\Pr(Y < X)$, based on independent samples of X and Y," *Annals of Mathematical Statistics*, vol. 29, pp. 558–562, 1958.

[23] J. D. Church and B. Harris, "The estimation of reliability from stress strength relationships," *Technometrics*, vol. 12, pp. 49–54, 1970.

[24] S. Kotz, Y. Lumelskii, and M. Pensky, *The Stress-Strength Model and Its Generalizations: Theory and Applications*, World Scientific, River Edge, NJ, USA, 2003.

[25] D. Kundu and R. D. Gupta, "Estimation of $P(Y < X)$ for Weibull distributions," *IEEE Transactions on Reliability*, vol. 55, no. 2, pp. 270–280, 2006.

[26] D. Kundu and R. D. Gupta, "Estimation of $P(Y < X)$ for generalized exponential distribution," *Metrika*, vol. 61, no. 3, pp. 291–308, 2005.

[27] M. Z. Raqab and D. Kundu, "Comparison of different estimators of $P(Y < X)$ for a scaled Burr type X distribution," *Communications in Statistics: Simulation and Computation*, vol. 34, no. 2, pp. 465–483, 2005.

[28] D. Kundu and M. Z. Raqab, "Estimation of $R = P(Y < X)$ for three-parameter Weibull distribution," *Statistics and Probability Letters*, vol. 79, no. 17, pp. 1839–1846, 2009.

[29] K. Krishnamoorthy, S. Mukherjee, and H. Guo, "Inference on reliability in two-parameter exponential stress-strength model," *Metrika*, vol. 65, no. 3, pp. 261–273, 2007.

[30] S. Eryilmaz, "On system reliability in stress-strength setup," *Statistics and Probability Letters*, vol. 80, no. 9-10, pp. 834–839, 2010.

[31] D. K. Al-Mutairi, M. E. Ghitany, and D. Kundu, "Inferences on stress-strength reliability from Lindley distribution," to appear in *Communications in Statistics—Theory and Methods*, available from http://home.iitk.ac.in/~kundu/lindley-ss.pdf.

4

Fuzzy RAM Analysis of the Screening Unit in a Paper Industry by Utilizing Uncertain Data

Harish Garg, Monica Rani, and S. P. Sharma

Department of Mathematics, Indian Institute of Technology, Roorkee 247667, Uttarakhand, India

Correspondence should be addressed to Harish Garg, harishg58iitr@gmail.com

Academic Editor: Tadashi Dohi

Reliability, availability, and maintainability (RAM) analysis has helped to identify the critical and sensitive subsystems in the production systems that have a major effect on system performance. But the collected or available data, reflecting the system failure and repair patterns, are vague, uncertain, and imprecise due to various practical constraints. Under these circumstances it is difficult, if not possible, to analyze the system performance up to desired degree of accuracy. For this, Artificial Bee Colony based Lambda-Tau (ABCBLT) technique has been used for computing the RAM parameters by utilizing uncertain data up to a desired degree of accuracy. Results obtained are compared with the existing Fuzzy Lambda-Tau results and we conclude that proposed results have a less range of uncertainties. Also ranking the subcomponents for improving the performance of the system has been done using RAM-Index. The approach has been illustrated through analyzing the performance of the screening unit of a paper industry.

1. Introduction

In any production plant, systems are expected to be operational and available for the maximum possible time so as to maximize the overall production and hence profit. That is each component/system of the entire production plant will run failure free for enhancing the production as well as productivity of the plant and furnish their excellent performance. However, failures are inevitable; a product will fail sooner or later. These failures may be the result of human error, poor maintenance, or inadequate testing and inspection. Therefore, the systems and components undergo several failure-repair cycles that include logistic delays while performing repair leads to the degradation of systems' overall performance [1]. System performance depends on reliability and availability of the system/components, operating environment, maintenance efficiency, operation process and technical expertise of operators, and so forth. To improve the system reliability and availability, implementation of appropriate maintenance strategies play an important role. High performance of these units can be achieved with highly reliable subunits and perfect maintenance. To this effect the knowledge of behavior of system, their component(s) is customary in order to plan and adapt suitable maintenance strategies. Thus, maintainability is also to be a key index to enhance the performance of these systems [2, 3]. On the other hand availability of the system can be improved by improvement in its reliability and maintainability. To maintain the availability of sophisticated systems to a higher level, the systems structure design or system components of higher availability should be required, or both of them are performed simultaneously. Implementation of these methods to improve the system availability or reliability will normally consume resources such as cost, weight, volume, and so forth. Thus, it is very important for decision-makers to fully consider both the actual business and the quality requirements. Thus keeping in view the competitive environment, behavior of such systems can be studied in terms of their reliability, availability, and maintainability (RAM).

RAM as an engineering tool evaluates the equipment performance at different stages in design process ad hence play an important role in controlling both the quantity and quality of the products. They aim at estimating and predicting the probability of the failure and optimizing the operation management related with the provision of the failures, that is, maintenance policies. Factors that affect RAM of a repairable industrial system include machinery operating conditions, maintenance conditions, infrastructural facilities, and so forth [2, 4]. The growing complexity of technological systems as well as rapidly increasing operation and maintenance costs incurred due to loss of operation as a consequence of sudden or sporadic failures have brought to the forefront the aspects of RAM associated with the production/manufacturing systems. The expectation today is that complex equipment and systems should not be free from defects and systematic failures but also perform the required function for a stated time interval and should have a fail-safe behavior in case of critical or catastrophic failures. But, failure is nearly an unavoidable phenomenon in mechanical systems/components. For failure analysis variety of methods exists in literature. These include reliability block diagrams (RBDs), Monte Carlo simulation (MCS), Markov Modeling, failure mode and effect analysis, Petri nets, fault tree analysis, and so forth [2, 3, 5–7]. Most of the repairable mechanical systems exhibit constant repair and failure rate after initial burn-in period of bath tub curve. Out of these both FTA and PN recognized as a powerful tool for estimating the reliability of large scaled systems, where system success or failure is described by the state of the top event. The probability of a top event is a function of the failure probability of a primary event, whose data are collected either from the available historical data or raw data provided by the experts. It is often assumed that all probabilities or probability distributions are known or perfectly determinable. This assumption is not consistent with reality because we are faced with different fundamental uncertainties for reliability modeling and analysis of a system in design stage. Thus the data, available from the past record, is incomplete, imprecise, vague, and conflicting; that is, historical records can only represent the past behavior but may be unable to predict future behavior of the equipment, and that leads to inadequate knowledge of basic failure events. Also, the traditional analytical techniques need large amounts of data, which are difficult to obtain because of various practical constraints such as rare events of components, human errors, and economic considerations for the estimation of failure/repair characteristics of the system. In such circumstances, it is usually not easy to analyze the behavior and performance of these systems up to desired degree of accuracy by utilizing available resources, data, and information. These challenges imply that a new and pragmatic approach is needed to access and analyze RAM of these systems because organizational performance and survivability depends a lot on reliability and maintainability of its components/parts and systems.

To this effect, the composite measure of reliability, availability, and maintainability has been introduced, called as RAM-Index, for measuring the system performance by simultaneously considers all the three key indices which influence the system performance directly. Rajpal et al. [8] developed an artificial neural network (ANN) model, by using historical data, for assessing the effect of input

parameters on this Index of a repairable system. Their index was static in nature while Komal et al. [9] introduced RAM-Index which was time dependent and used historical uncertain data for its evolution. However, almost all the previous studies were carried out by considering the failure rate of the component which follows the exponential distribution that is, a constant failure rate model. Rani et al. [10] have extended this idea for a time varying failure rate and a constant repair rate model. In the present paper, system performance in terms of RAM-index of a repairable industrial systems has been analyzed by considering the time varying failure rate instead of constant failure rate model. To compute the RAM parameters and consequently RAM-Index by utilizing uncertain, limited, and vague data, an approach gave by Knezevic and Odoom [11] may be used. Based on these the behavior analysis of complex repairable industrial systems are analyzed by the researchers in the form of fuzzy membership functions of various reliability parameters [12, 13]. In their approach, PN is used to model the system while fuzzy set theory is used to quantify the uncertain, vague, and imprecise data. But it is analyzed from the studies that when this approach is applied for large and complex systems, the computed reliability indices in the form of fuzzy membership function have wide spread (support), that is, high level of uncertainty exists due to various fuzzy arithmetic operations are used in the computations [14–17] and hence it does not provide the actual trend of the system behavior. In order to reduce the uncertainty level in the analysis, spread for each reliability index must be reduced up to a desired degree of accuracy so that plant personnel may use these indices to analyze the system behavior more closely and take more sound decisions to improve the performance of the plant. For overcoming this drawback and to generalized the approach for large and complex systems, Artificial Bee Colony based Lambda-Tau (ABCBLT) technique is used in this study [17]. ABCBLT technique is hybridized technique in which Lambda-Tau methodology has been used for obtaining the expression of RAM parameters and Artificial Bee Colony (ABC) [18–20] is used to construct their membership function in the form of triangular fuzzy number by using ordinary arithmetic operation instead of fuzzy arithmetic. Major advantage of the ABCBLT technique is that it give compressed search space for each computed reliability index by utilizing available and uncertain data.

Thus, it is observed from the study that RAM parameters of the system may be calculated by utilizing uncertain, vague, and imprecise data. The objective of the present work is to quantify the uncertainties with the help of fuzzy numbers and to develop an approach for assessing the effect of failure and repair pattern on the composite measure of RAM of industrial systems. For this, a time varying failure rate instead of constant failure rate has been considered for analysis. The approach has been demonstrate through a RAM analysis of the screening unit of a paper industry using ABC and Lambda-Tau methodology. The computed results are compared with the fuzzy Lambda-Tau results. The sensitivity analysis of the components has been done by using proposed RAM-Index analysis for finding the components as

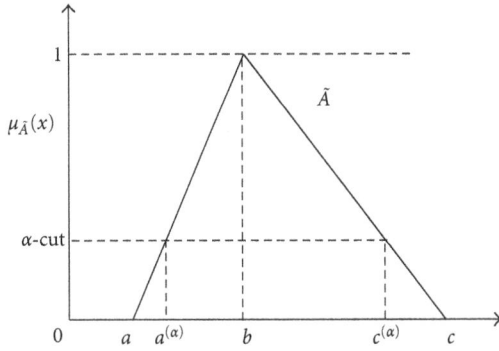

FIGURE 1: Triangular fuzzy number of fuzzy set $\tilde{A}$.

TABLE 1: RAM parameters of the system [10, 17].

Reliability parameters	Expression
Reliability	$R_s(t) = \exp\left[-\left(\dfrac{t}{\theta}\right)^{\beta}\right]$
Availability	$A_s(t) = \exp\left\{-\left(\dfrac{t}{\theta}\right)^{\beta} - \dfrac{t}{\tau}\right\}$ $\times \left[1 + \dfrac{1}{\tau}\int_0^t \exp\left\{\left(\dfrac{t}{\theta}\right)^{\beta} + \dfrac{t}{\tau}\right\} dt\right]$
Maintainability	$M_s(t) = 1 - \exp\left(\dfrac{-t}{\tau}\right)$

per the preferential order for improving the production as well as productivity of the system. The rest of the paper is organized as follow: Section 2 describe the basic notations and terms related to fuzzy set theory which will help to analyze the system performance. The generalized RAM-Index has been defined in Section 3 while the ABCBLT technique has been discussed in Section 4. Section 5 discusses the case study of the screening unit of the paper industry for RAM analysis and along with their results. Finally concrete conclusion has been discussed in Section 6.

2. Basic Concepts of Fuzzy Set Theory

Problems in the real world quite often turn out to be complex owing to an element of uncertainty either in the parameters which define the problem or in the situations in which the problem occurs. Although probability theory has been an effective tool to handle uncertainty, it can be applied only to situations whose characteristics are based on random processes, that is, processes in which the occurrence of events is strictly determined by chance. However, in reality, there turn out to be problems, a large class of them whose uncertainty is characterized by a nonrandom process. Here, the uncertainty may arise due to partial information about the problem or due to information which is not fully reliable or due to partial information about the problem. This problem was overcome by using the notion of the fuzzy set introduced by Zadeh in 1965 [21] in the evaluation of the reliability of a system.

2.1. Fuzzy Set. The concept of fuzzy set was introduced by Zadeh [21] in 1965, which can be defined on the universe of discourse U as $\tilde{A} = \{\langle x, \mu_{\tilde{A}}(x)\rangle \mid x \in U\}$, where $\mu_{\tilde{A}}$ is the membership function of the fuzzy set $\tilde{A}$ defined as $\mu_{\tilde{A}} : U \to [0,1]$ and $\mu_{\tilde{A}}(x)$ indicates the degree of membership of x in $\tilde{A}$ and its value lies between zero and one. When a set is an ordinary set, its membership function can take on only two values 0 and 1, with $\chi_A(x) = 1$ or 0 according as x does or does not belong to A. $\chi_A(x)$ is referred to as the characteristic function of the set A.

2.2. Convex Fuzzy Set. A fuzzy set $\tilde{A}$ in universe U is convex if and only if the membership functions $\mu_{\tilde{A}}$ of $\tilde{A}$ is fuzzy-convex, that is,

$$\mu_{\tilde{A}}(\lambda x_1 + (1-\lambda)x_2)$$
$$\geq \min\left(\mu_{\tilde{A}}(x_1), \mu_{\tilde{A}}(x_2)\right), \quad \forall x_1, x_2 \in U, \ 0 \leq \lambda \leq 1. \tag{1}$$

2.3. Fuzzy Number. A fuzzy number $\tilde{A}$ is a convex normalized fuzzy set $\tilde{A}$ of the real line $\mathbb{R}$ such that

(i) it exists exactly one $x_0 \in \mathbb{R}$ with $\mu_{\tilde{A}}(x_0) = 1$.

(ii) $\mu_{\tilde{A}}$ is piecewise continuous;

and its membership function is defined as

$$\mu_{\tilde{A}}(x) = \begin{cases} f_A(x); & a \leq x \leq b, \\ 1; & x = b, \\ g_A(x); & b \leq x \leq c, \\ 0; & \text{otherwise}, \end{cases} \tag{2}$$

where $0 \leq \mu_{\tilde{A}}(x) \leq 1$ and $a, b, c \in R$ such that $a \leq b \leq c$, and two functions $f_A, g_A : R \to [0,1]$ are called the sides of fuzzy number. The function f_A is nondecreasing continuous functions and the function g_A is nonincreasing continuous functions.

2.4. α-Cut of the Fuzzy Set. α-cut of the fuzzy set $\tilde{A}$, denoted by $A^{(\alpha)}$, is a crisp set which consists of elements of $\tilde{A}$ having at least degree α as is defined mathematically as

$$A^{(\alpha)} = \left\{x \in U : \mu_{\tilde{A}}(x) \geq \alpha\right\}, \tag{3}$$

where α is the parameter in the range $0 \leq \alpha \leq 1$. Every α-cut of a fuzzy number is a closed interval and a family of such intervals describes completely a fuzzy number under study. Hence we have $A^{(\alpha)} = [A_L^{(\alpha)}, A_U^{(\alpha)}]$, where

$$A_L^{(a)}(x) = \inf\left\{x \in R : \mu_{\tilde{A}}(x) \geq \alpha\right\},$$
$$A_U^{(a)}(x) = \sup\left\{x \in R : \mu_{\tilde{A}}(x) \geq \alpha\right\}. \tag{4}$$

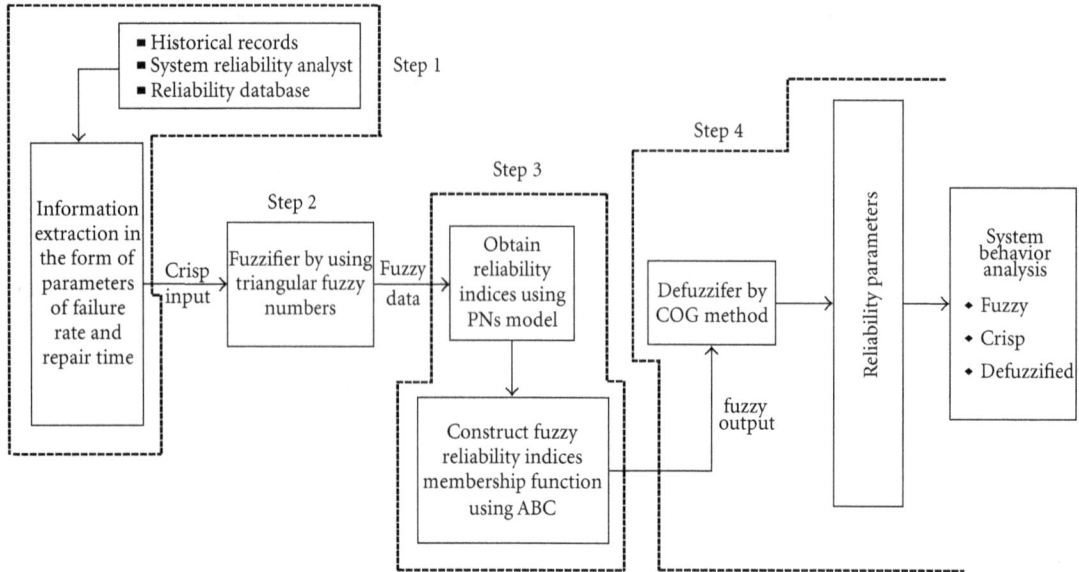

FIGURE 2: Flow chart of the ABCBLT methodology.

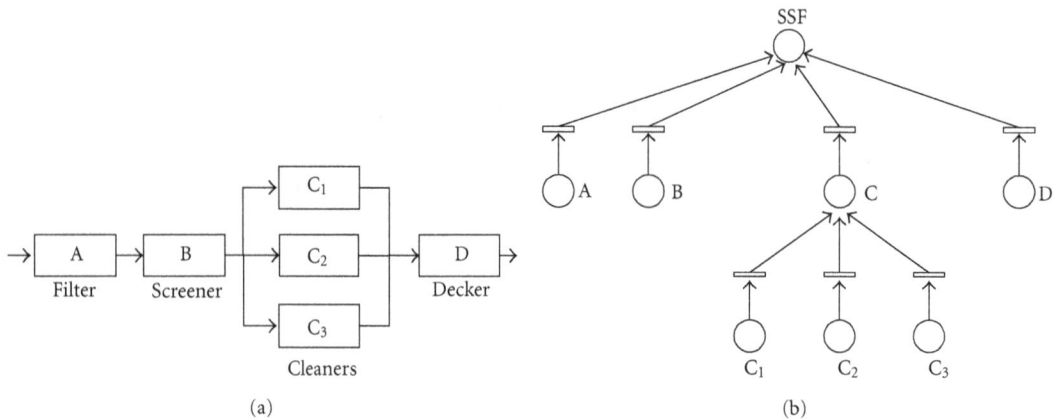

FIGURE 3: Screening system: (a) RBD and (b) PN model.

2.5. Triangular Fuzzy Number and Interval Arithmetic Operations. The concept of membership function is most important aspect in fuzzy set theory. They are used to represent various fuzzy sets. Many membership functions such as normal, triangular, trapezoidal can be used to represent fuzzy numbers. However, triangular membership functions (TMF) are widely used for calculating and interpreting reliability data because of their simplicity and understandability [22, 23]. The decision of selecting triangular fuzzy numbers (TFNs) lies in their ease to represent the membership function effectively and to incorporate the judgement distribution of multiple experts. This is not true for complex membership functions, such as trapezoidal one, and so forth. For instance, imprecise or incomplete information such as low/high failure rate that is about 4 or between 5 and 7, is well represented by TMF. In the present paper triangular membership function is used as it not only conveys the behavior of system parameters but also reflect the dispersion of the data adequately. A triangular fuzzy number $\tilde{A}$ with

parameters $a \leq b \leq c$, denoted as $\tilde{A} = \langle [(a,b,c); \mu] \rangle$ in the set of the real number $\mathbb{R}$ and its membership function is given as

$$\mu_{\tilde{A}}(x) = \begin{cases} \dfrac{x-a}{b-a}; & a \leq x \leq b, \\ 1; & x = b, \\ \dfrac{c-x}{c-b}; & b \leq x \leq c, \\ 0; & \text{otherwise.} \end{cases} \tag{5}$$

The α-cuts of the triangular fuzzy set is defined in the closed interval form as below and shown in graphically in Figure 1:

$$A_\alpha = \left[a^{(\alpha)}, c^{(\alpha)}\right] = [(b-a)\alpha + a, -(c-b)\alpha + c]. \tag{6}$$

The basic arithmetic operations, that is, addition, subtraction, multiplication and division, of fuzzy numbers depends upon the arithmetic of the interval of confidence. The four main arithmetic operation on two triangular fuzzy

TABLE 2: Basic expression of the Lambda-Tau methodology.

Gate	λ_{AND}	τ_{AND}	λ_{OR}	τ_{OR}
Expression	$\prod_{j=1}^{n} \lambda_j \left[\sum_{i=1}^{n} \prod_{\substack{i=1 \\ i \neq j}}^{n} \tau_i \right]$	$\dfrac{\prod_{i=1}^{n} \tau_i}{\sum_{j=1}^{n} \left[\prod_{\substack{i=1 \\ i \neq j}}^{n} \tau_i \right]}$	$\sum_{i=1}^{n} \lambda_i$	$\dfrac{\sum_{i=1}^{n} \lambda_i \tau_i}{\sum_{i=1}^{n} \lambda_i}$

(1) Objective function: $f(\mathbf{x})$, $\mathbf{x} = (x_1, x_2, \ldots, x_d)$
(2) Generate an initial bee population (solution) x_h where $x_h = (x_{h1}, x_{h2}, \ldots, x_{hd})$ and number of employed bees are equal to onlooker bees
(3) Evaluate fitness value
(4) Initialize cycle = 1
(5) For each employed bee
    (a) Produce new food source position v_{hj} in the neighborhood of x_{hj} by $v_{hj} = x_{hj} + u(x_{hj} - x_{kj})$
        where k is a solution in the neighborhood of selected parameter j, u is random number in the range $[-1,1]$
    (b) Evaluate the fitness value at new source v_{hj}
    (c) If new position is better than previous position then memorizes the new position
(6) End For
(7) Calculate the probability values $p_h = f_h / \sum_{h=1}^{N} f_h$ for the solution where N is the total number of food sources
(8) For each onlooker bee
    (a) Chooses a food source depending on p_h for the solutions x_h
    (b) Produce new food source positions v_h from the populations x_h depending upon p_h and evaluate their fitness
    (c) If new position better than previous position, then memorizes the new position
(9) End For
(10) If there is any abandoned solution that is, if employed bee becomes scout then replace its position with a new random source positions
(11) Memorize the best solution achieved so far
(12) cycle = cycle + 1
(13) If termination criterion is satisfied then stop otherwise go to step 5

ALGORITHM 1: Pseudo code of the ABC algorithm.

sets $\tilde{A}$ and $\tilde{B}$ described by the α-cuts are given below for the following intervals:

$$A^{(\alpha)} = \left[A_1^{(\alpha)}, A_3^{(\alpha)} \right], \qquad B^{(\alpha)} = \left[B_1^{(\alpha)}, B_3^{(\alpha)} \right], \quad \alpha \in [0,1] \tag{7}$$

(i) Addition: $\tilde{A} + \tilde{B} = [A_1^{(\alpha)} + B_1^{(\alpha)}, A_3^{(\alpha)} + B_3^{(\alpha)}]$.

(ii) Subtraction: $\tilde{A} - \tilde{B} = [A_1^{(\alpha)} - B_3^{(\alpha)}, A_3^{(\alpha)} - B_1^{(\alpha)}]$.

(iii) Multiplication: $\tilde{A} \cdot \tilde{B} = [P^{(\alpha)}, Q^{(\alpha)}]$,

where $P^{(\alpha)} = \min(A_1^{(\alpha)} \cdot B_1^{(\alpha)}, A_1^{(\alpha)} \cdot B_3^{(\alpha)}, A_3^{(\alpha)} \cdot B_1^{(\alpha)}, A_3^{(\alpha)} \cdot B_3^{(\alpha)})$ and $Q^{(\alpha)} = \max(A_1^{(\alpha)} \cdot B_1^{(\alpha)}, A_1^{(\alpha)} \cdot B_3^{(\alpha)}, A_3^{(\alpha)} \cdot B_1^{(\alpha)}, A_3^{(\alpha)} \cdot B_3^{(\alpha)})$

(iv) Division: $\tilde{A} \div \tilde{B} = \tilde{A} \cdot 1/\tilde{B}$, if $0 \notin \tilde{B}$.

3. Generalized RAM-Index

In order to keep the production and productivity of the system high, it is necessary that the system should operate for long run period without failure. But unfortunately failure is an unavoidable phenomenon in an industrial systems. The failure of subsystem or unit will reduce the efficiency of the system and hence maintainability is essential for it.

So it is necessary for the system analyst that in order to increase the performance of the system, current condition of equipments and subsystems should be changed according to time and the need of effective maintenance program. But the problem to the system analyst is that how to find the component on which more attention should be given to save money, manpower, and time. This problem can be resolved by using the RAM analysis using proposed RAM-Index. The proposed RAM-Index has been valid for a component whose failure rate follows the Weibull distribution while repair time follows the exponential distribution. Major advantage of this index is that by varying the component's failure and repair rate parameters the corresponding effect on the system performance has been observed.

Thus, the generalized RAM-Index for analyzing the performance of the system is given in (8)

$$\text{RAM}(t) = w_1 \times R_s(t) + w_2 \times A_s(t) + w_3 \times M_s(t), \tag{8}$$

where R_s, A_s, and M_s are, respectively, the reliability, availability, and maintainability of the system whose expressions are given in the Table 1 and $w_i \in (0,1)$, $i = 1, 2, 3$ are weights such that $\sum_{i=1}^{3} w_i = 1$. The value of $w = [0.36, 0.30, 0.34]$ has been used for calculating RAM-Index which is same as used by the researchers [8, 9].

(a)

(b)

(c)

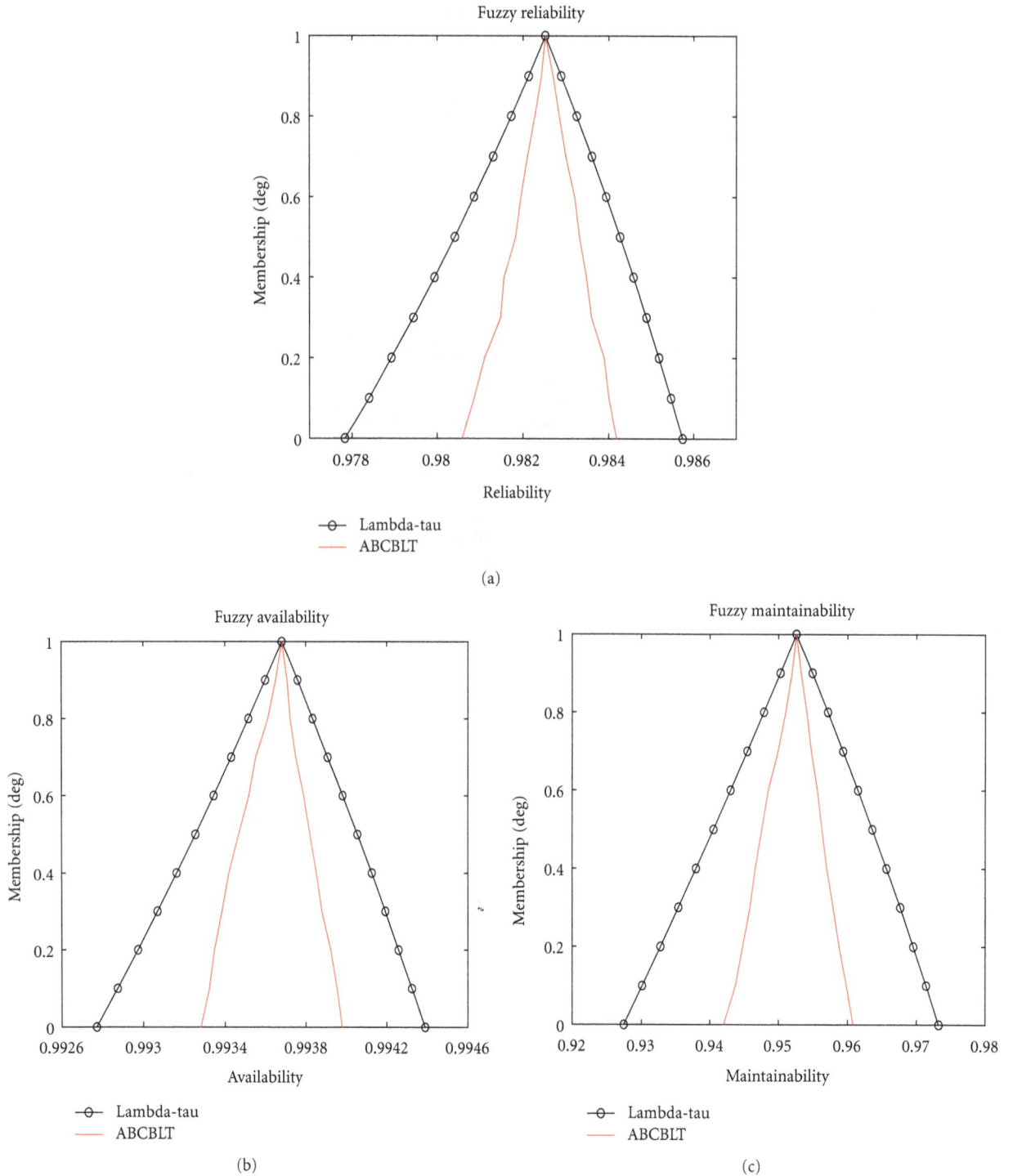

FIGURE 4: Fuzzy RAM parameters plot at ±15% spreads.

Equation (8) can be rewritten in more elaborative form as

$$\text{RAM}(t) = w_1 \times \exp\left[-\left(\frac{t}{\theta}\right)^{\beta}\right] + w_3 \times 1 - \exp\left(\frac{-t}{\tau}\right)$$

$$+ w_2 \times \exp\left\{-\left(\frac{t}{\theta}\right)^{\beta} - \frac{t}{\tau}\right\}$$

$$\times \left[1 + \frac{1}{\tau}\int_0^t \exp\left\{\left(\frac{t}{\theta}\right)^{\beta} + \frac{t}{\tau}\right\} dt\right].$$

(9)

Since historical data is imprecise and vague, so have some sort of uncertainties and consequently RAM parameters and their corresponding RAM-index also have some sorts of

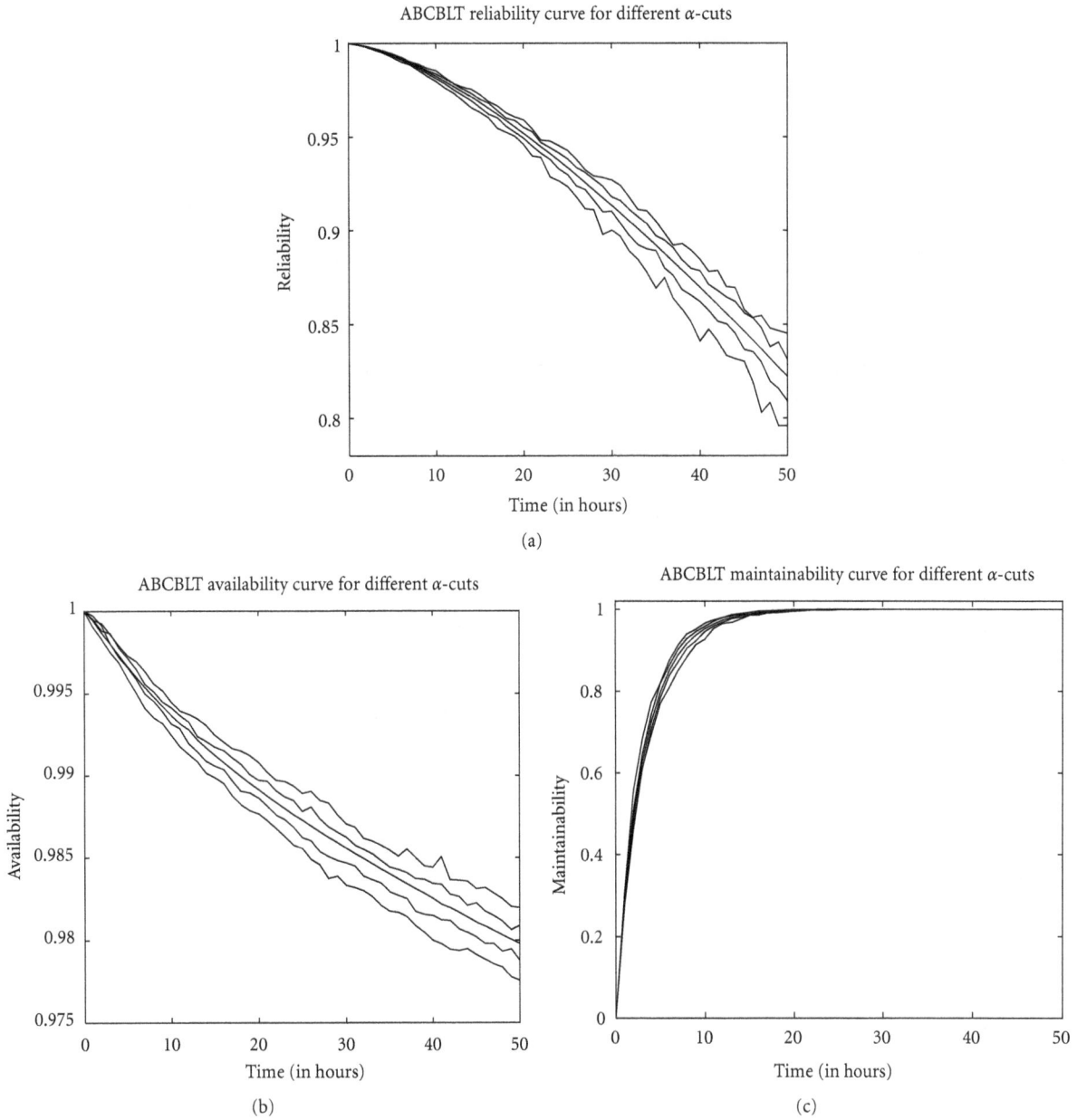

FIGURE 5: Long run period of RAM parameters at different α cuts.

uncertainties. For removing this, fuzzy set theory can be used to represent the uncertain data by taking crisp data into triangular fuzzy numbers. By using quantified data, RAM parameters and consequently RAM-Index becomes a triangular fuzzy membership function which can be expressed as

$$\widetilde{\text{RAM}}(t) = (\text{RAM}_L(t), \text{RAM}_M(t), \text{RAM}_R(t)). \quad (10)$$

It is clear from the (9) that at any time "t", RAM$(t) \in (0, 1)$.

4. ABCBLT Technique

Lambda-Tau is a traditional methodology in which FTA is used to model the system. The basic expression used for analyzing the system based on constant failure rate model, that is, failure rate and repair time, follows the exponential distribution are summarized in Table 2. Knezevic and Odoom [11] extend this idea through PNs and fuzzy set theory for analyzing the various reliability parameters of the complex repairable industrial systems in the form of fuzzy membership functions. In their theory, PN is used for modeling the system while fuzzy takes care of impreciseness present in the data. They calculated the various reliability

TABLE 3: Input data of the system.

Components	Failure data		Repair time
	Weibull parameters		MTTR
	scale (θ)	shape (β)	τ (hrs)
Filter	337	1.33	2.0
Screener	315	1.54	4.0
Cleaner	470	1.88	2.0
Deckers	252	1.76	5.0

TABLE 4: Decrease in spread for the parameters.

Technique	Computed spread for reliability indices		
	Reliability	Availability	Maintainability
FLT	0.00790434	0.00161782	0.04578209
ABCBLT	0.00361200	0.00069999	0.01876803
	Decrease in spread in % from FLT to ABCBLT		
	54.30358511	56.73251659	59.00573783

parameters in the form of fuzzy membership functions which are used for behavior analysis of the repairable industrial system [2]. Their approach is limited as the number of components of the system increases or system structure becomes more complex, the computed reliability indices in the form of fuzzy membership function have wide spread [14, 16, 17] due to various fuzzy arithmetic operations used in the calculations. Thus this approach is not suitable for the behavior analysis of large and complex repairable industrial systems when data is imprecisely known and represented by fuzzy numbers.

To generalize the approach for a complex industrial system, an effective technique is needed which should reduce the uncertainty level so that plant personnel may use these indices to analyze the system's behavior in more promising way for improving the system's performance. ABCBLT [17] is a hybridized technique in which Lambda-Tau methodology is used for obtaining the expression of various RAM parameters and artificial bee colony is used to solve the nonlinear programming problem for constructing their membership function by using ordinary arithmetic operations instead of fuzzy arithmetic operations. Triangular membership functions has been used for fuzzifying the data because it is easy to handle the information and form an accurate results in reliability engineering.

Strategy followed through this approach has been described through flowchart in Figure 2 and their details are given hereafter.

Step 1 (Information extraction phase). In the first phase of the system, the information related to failure rate and repair time of the system components are collected from the various resources such as logbooks, historical records, sheets, and so forth.

Step 2 (Fuzzifying the data). As the extracted data is either out of date or does not represent the actual failure of the system it leads the problem of uncertainty in the

current failure rates and repair times. So, to handle the uncertainties in the analysis, the obtained collected (crisp) data is fuzzified into triangular fuzzy numbers (TFNs) having known spread (support) as suggested by decision maker/design maintenance expert/system reliability analyst in the form of the spread ($\pm 15\%$, $\pm 25\%$, and $\pm 50\%$).

Step 3 (Calculate RAM parameters). In this step, system is modelled with the help of Petri nets and based on that minimal cut sets are obtained by using matrix method. By using these cut sets and the expression of the Lambda-Tau methodology, RAM parameters of the system, listed in Table 1, are obtained. Instead of constructing the membership functions by using fuzzy arithmetic function, an ordinary arithmetic and optimization technique have been used for avoiding the high level of uncertainties existing in the computed reliability indices. For this a nonlinear programming problem (11) has been formulated by utilizing the quantified fuzzy $\theta's$ and $\tau's$. Thus, the lower and upper boundary values of reliability indices are computed at cut level α by solving

$$\text{minimize/maximize} \quad \widetilde{F}(\theta_1, \theta_2, \ldots, \theta_n, \tau_1, \tau_2, \ldots, \tau_m) \text{ or}$$
$$\widetilde{F}(t/\theta_1, \theta_2, \ldots, \theta_n, \tau_1, \tau_2, \ldots, \tau_m)$$
$$\text{subject to} \quad \mu_{\theta_i}(x) \geq \alpha,$$
$$\mu_{\tau_j}(x) \geq \alpha,$$
$$0 \leq \alpha \leq 1,$$
$$i = 1, 2, \ldots, n,$$
$$j = 1, 2, \ldots, m,$$
$$(11)$$

where $\widetilde{F}(\theta_1, \theta_2, \ldots, \theta_n, \tau_1, \tau_2, \ldots, \tau_m)$ and $\widetilde{F}(t/\theta_1, \theta_2, \ldots, \theta_n, \tau_1, \tau_2, \ldots, \tau_m)$ are time independent and dependent fuzzy reliability indices. The obtained minimum and maximum value of $\widetilde{F}$ are denoted by $F_{\min}$ and $F_{\max}$, respectively.

The membership function values of $\widetilde{F}$ at $F_{\max}$ and $F_{\min}$ are both α, that is,

$$\mu_{\widetilde{F}}(F_{\max}) = \mu_{\widetilde{F}}(F_{\min}) = \alpha. \quad (12)$$

Since the problem is nonlinear in nature so it requires an efficient technique to solve this problem. Variety of methods and algorithms exists for optimization of such problems and applied in various technological fields. In this paper ABC [18–20, 24] is used as a tool to find out the optimal solution of the above optimization problems, since ABCs have the advantages of memory, multi-character, local search, and solution improvement mechanism, it is able to discover an excellent optimal solution. The procedure of ABC algorithm is described in Algorithm 1. The objective function for maximization problem and the reciprocal of the objective function for minimization problem is taken as the fitness function. The termination criterion has been used either to a maximum number of generations or order of relative error equal to 10^{-6}, whichever is achieved first.

TABLE 5: Crisp and defuzzified values of RAM parameters at different spreads.

Parameters	Crisp values	Defuzzified values at spread		
		±15%	±25%	±50%
Reliability	0.98251623	Lambda-Tau: 0.98215233	0.98145762	0.97708411
		ABCBLT: 0.98249979	0.98214257	0.97947532
Availability	0.99367854	Lambda-Tau: 0.99362948	0.99353884	0.99304228
		ABCBLT: 0.99363929	0.99354976	0.99302078
Maintainability	0.95261602	Lambda-Tau: 0.95147966	0.94946055	0.94010346
		ABCBLT: 0.95184158	0.95117712	0.94583763

TABLE 6: Change in defuzzified values of parameters (in magnitude).

	Change in the reliability indices values (in %) from		
Parameters	I to II	I to III	II to III
Reliability	0.03703755	0.00167325	0.03537740
Availability	0.00493721	0.00394996	0.00098728
Maintainability	0.11928835	0.08129613	0.03803759

I: crisp, II: Lambda-Tau III: ABCBLT.

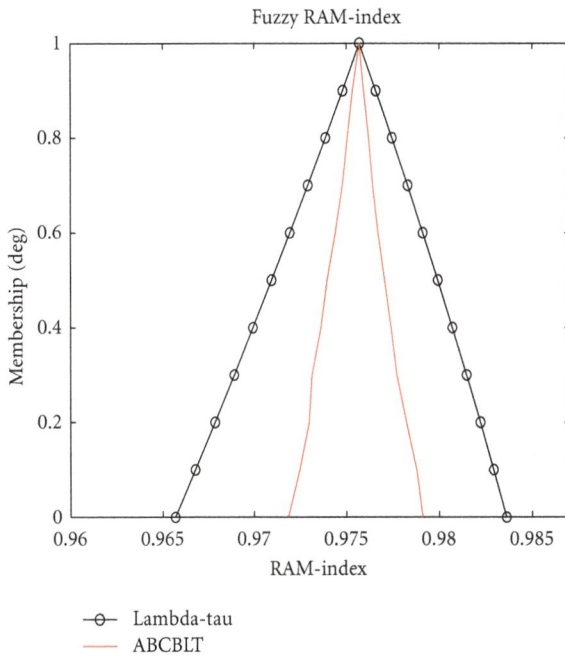

FIGURE 6: Fuzzy RAM-Index plot at ±15% spread.

Step 4 (Defuzzification). As the collected output from the Step 3 of the methodology is of fuzzy output. But in real life, as most of the actions or decisions implemented by human or machines are binary or crisp it is necessary to convert the fuzzy output in crisp output. The process of converting the fuzzy output to crisp is called defuzzification. Out of existence of the several method for defuzzification, center of gravity (COG) method is selected due to its property that it is equivalent to mean of data and so it is very appropriate for reliability calculations [25]. If the membership function

$\mu_{\widetilde{A}}(x)$ of the output fuzzy set $\widetilde{A}$ is described on the interval $[x_1, x_2]$, then COG defuzzification $\bar{x}$ can be defined as

$$\bar{x} = \frac{\int_{x_1}^{x_2} x \cdot \mu_{\widetilde{A}}(x)dx}{\int_{x_1}^{x_2} \mu_{\widetilde{A}}(x)dx}. \tag{13}$$

5. An Illustration with Application

In the present study a paper plant situated in northern part of India producing 200 tons of paper per day is considered as the subject of discussion. The paper plants are large capital-oriented engineering system, comprising of subsystems, namely, chipping, feeding, pulping, washing, screening, bleaching, production of paper consisting of press unit, and collection, arranged in complex configuration. The present paper considers the most important functionary unit, namely, screening unit as a subject of discussion.

5.1. Screening System. The screening unit consists of four subsystems whose described are given as below:

(i) Filter (A): it works for removal of black liquor from the pulp. Its failure causes failure of the system.

(ii) Screen (B): it removes the knots and other undesirable materials from the pulp. Failure of the screen causes complete failure of the system.

(iii) Cleaner (C): it consists of three units in parallel. The failure of any one unit reduces the efficiency of the plant. Complete failure of the cleaner reduces the efficiency of the plant but the system remains operative. Manual operation is possible during the repair. Water is mixed here with the pulp by centrifugal action.

(iv) Decker (D): it reduces the blackness of the pulp. The failure of decker causes the complete failure of the system.

The reliability block diagram and its equivalent Petri net model are shown in Figures 3(a) and 3(b), respectively, where SSF represents the top place event of the screening unit system failure.

5.2. RAM Parameters Analysis. Under the information extraction phase, the data related to parameters of failure rate (β_i, θ_i) and repair time (τ_i) of the main component of the system are collected from the historical records and are

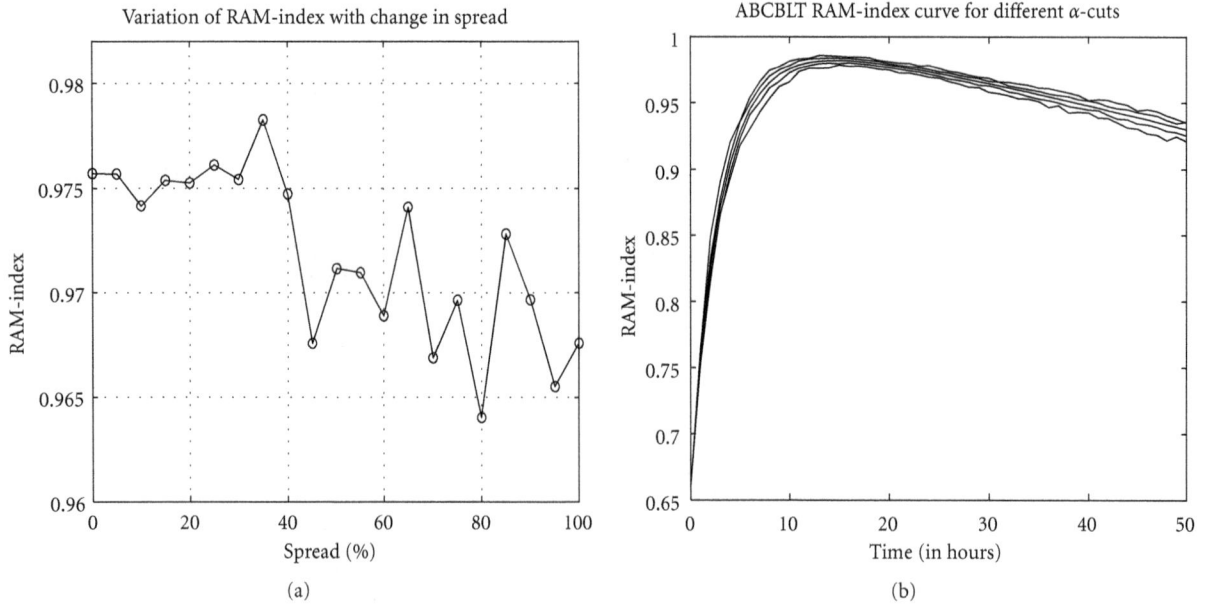

(a)

(b)

FIGURE 7: Variation of RAM-Index with (a) change in spread (b) different level of uncertainties.

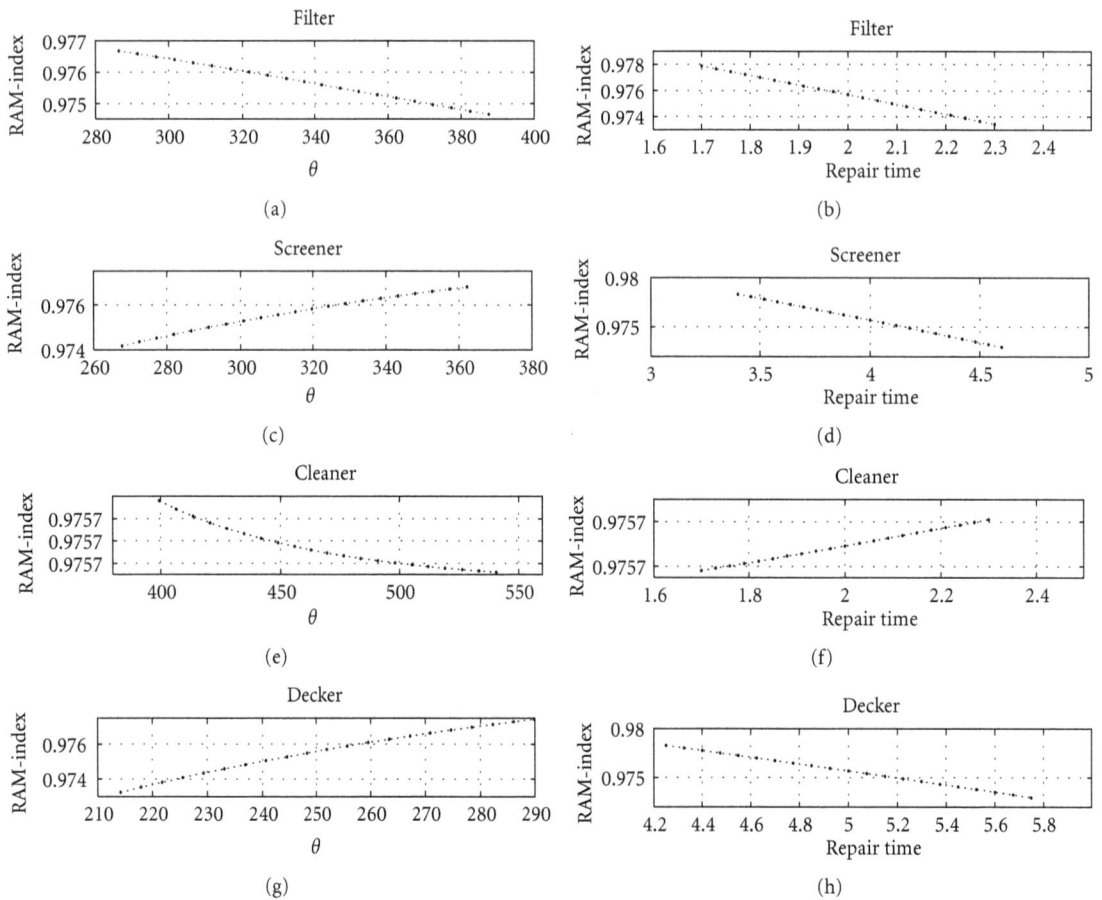

FIGURE 8: Effect of components parameters on RAM-Index when varies separately.

TABLE 7: Range of RAM-Index when parameters of components varies separately.

Component	Range of scale parameter θ (hrs)	RAM-Index	Range of repair time τ (hrs)	RAM-Index
Filter	286.45–387.55	Min: 0.97465822	1.70–2.30	Min: 0.97342489
		Max: 0.97667539		Max: 0.97787954
Screener	267.75–362.25	Min: 0.97417942	3.40–4.60	Min: 0.97294317
		Max: 0.97680809		Max: 0.97831576
Cleaner	399.50–540.50	Min: 0.97569885	1.70–2.30	Min: 0.97569885
		Max: 0.97569885		Max: 0.97569885
Decker	214.20–289.80	Min: 0.97324561	4.25–5.75	Min: 0.97298904
		Max: 0.97743319		Max: 0.97827127

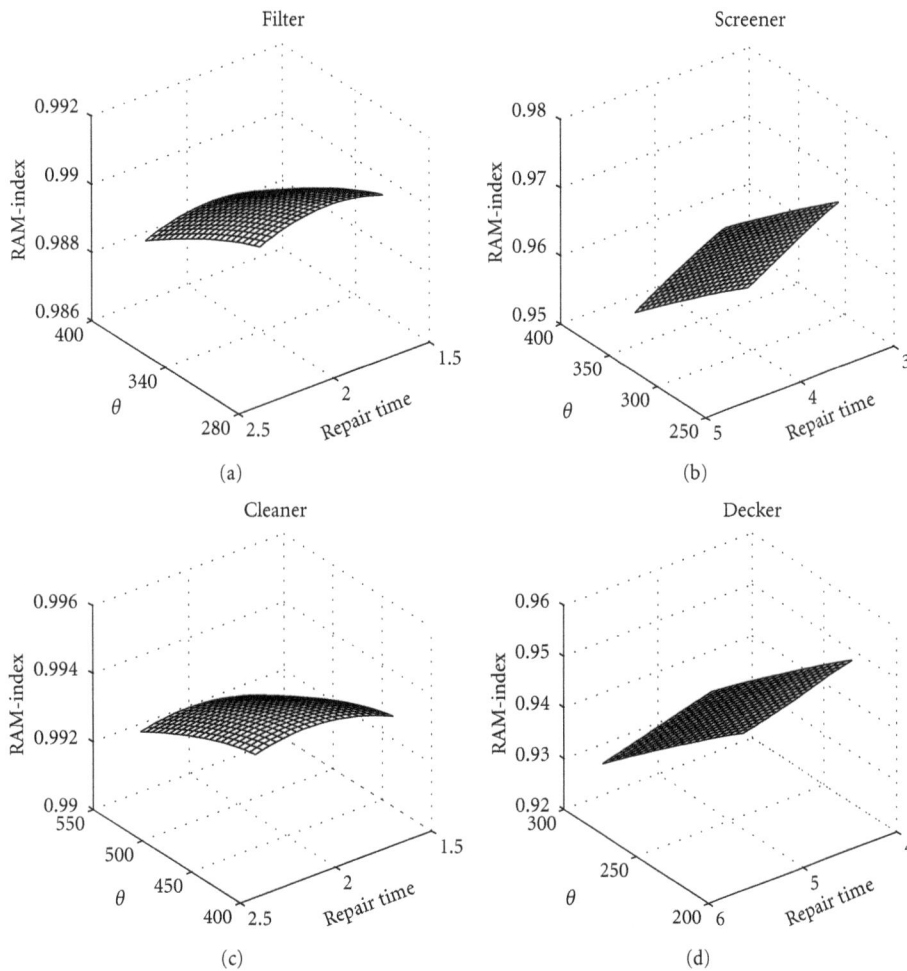

FIGURE 9: Effect of simultaneously varying components parameters on RAM-Index.

integrated with expertise of the system manager shown in Table 3.

As the collected data are taken from various resources which are out of data or contains large amount of uncertainty. So to handle these uncertainties and vagueness, the obtained (crisp) data are fuzzifier into fuzzy number with some known spread ±15% as suggested by the decision maker or system analysts. The minimal cut sets of the

system are obtained by using matrix method are {A}, {B}, {$C_1 C_2 C_3$}, and {D}. By using these sets and following the basic steps of ABCBLT technique, RAM parameters of the system are obtained in the form of fuzzy membership functions, at various membership grades for the mission time $t = 10$ (hrs) with left and right spread. The computed results by ABCBLT are depicted graphically in Figure 4 for ±15% spreads along with Lambda-Tau results. From the

TABLE 8: Range of RAM-Index when parameters of components varies simultaneously.

Component	Range of scale parameter θ (hrs)	Range of repair time τ (hrs)	RAM-Index
Filter	286.45–387.55	1.70–2.30	Min: 0.98675020 Max: 0.99197856
Screener	267.75–362.25	3.40–4.60	Min: 0.95034946 Max: 0.97417884
Cleaner	399.50–540.50	1.70–2.30	Min: 0.99011339 Max: 0.99565120
Decker	214.20–289.80	4.25–5.75	Min: 0.92456143 Max: 0.95706339

figure it has been concluded that the computed results have reduce region and have a smaller spread than the Lambda-Tau results.

The decrease in spread of the RAM parameters by ABCBLT technique from the Lambda-Tau technique have been calculated and shown in tabulated form in Table 4. From the Table 4, it has been concluded that the largest and smallest decrease in spread occurs corresponding to the reliability and maintainability respectively. On the other hand, the largest and smallest spreads occurs corresponding to availability and maintainability for ABCBLT technique which means prediction range of reliability parameters decreased. This suggests that decision makers (DMs) have smaller and more sensitive region to make more sound and effective decision in lesser time and hence ABCBLT is a superior than Lambda-Tau technique.

For defuzzification, center of gravity method [25] is used because it has the advantage of being whole membership function into account for this transformation. The crisp and defuzzified values of RAM parameters at ±15%, ±25%, and ±50% spreads are computed and compared with existing Lambda-Tau results which are shown in Table 5. It is clearly seen from the table that the result proposed by ABCBLT technique act as a bridge between Markovian process (crisp) and fuzzy Lambda-Tau (FLT) results. Also, variation in defuzzified values by ABCBLT technique are not so much as shown by results of Lambda-Tau technique, when the uncertainty level increases in the form of spread from ±15% to ±25%, and further ±50%. In addition to this, Table 5 reflects that the crisp values do not change irrespective of the spread chosen while the defuzzified values change with change of spreads. The change in the value of RAM parameters of ABCBLT results from the crisp and Lambda-Tau results have been computed and presented in Table 6. Based on these results the system analyst/decision maker may change their target goals rather comes from the traditional analysis. For an example, if plant personnel want to optimize reliability of the system using ABCBLT results then the new target of system reliability should be greater than 0.98249979 rather 0.98215233 comes from Lambda-Tau when uncertainty level taken as ±15% percent. Due to this and their reduced region of prediction, the value obtained through ABCBLT technique are conservative in nature which may be beneficial for system expert/analyst for future course

of action that is, now the maintenance will be based on the defuzzified values rather than crisp values.

At different α-cuts, (0, 0.5, 1), reliability, availability and maintainability curves for 0–50 (hrs) have been computed using ABCBLT technique and plotted in Figure 5 for depicting the behavior of the system with different levels of uncertainties. The behavior of these curves, using current conditions and uncertainties, shows that if current condition of equipments and subsystems are not changed then reliability of the systems will decrease rapidly while maintainability behave almost linearly after certain time for a long run period. The analysis suggests that to enhance the performance of these systems, current condition of equipments and subsystems should be changed according to effective maintenance program. But the problem is how to find the components or subsystems on which more attention should be given to save money, manpower and time for the effectiveness of the maintenance program. To overcome this problem a RAM analysis has been carried out using the proposed RAM-Index analysis.

5.3. RAM-Index Analysis. In order to analyze the effect of the system parameters on system performance, the RAM-Index analysis has been done. For this, the RAM-Index as given in (9) has been used. As the system analyst want to operated the system for a long run period for enhancing the production as well as productivity of the system. For this it is necessary that system should run failure free or have less range of uncertainties upto a desired levels. But failure is an unavoidable phenomenon, so in order to analyze the behavior of the system it is necessary that uncertainty levels in the analysis should be reduced upto a desired degree of accuracy. Hence fuzzy RAM-Index is computed by ABCBLT technique and their results are compared with Lambda-Tau results in Figure 6 at ±15% spread. It has been concluded that the uncertainties level by the proposed technique has a reduced region than Lambda-Tau. Moreover, to see the behavior of RAM index against different uncertainty (spread) levels, a plot between spread from 0 to 100 (in %) and RAM index has been plotted and shown in Figure 7(a). The variation of RAM-Index with a time range of 0–50(hrs) using ABCBLT technique for depicting the behavior of the system is shown in Figure 7(b) which shows that RAM index increases from 0 to 50(hrs) and then decreases. At time $t = 0$ hrs, the RAM-Index is 0.66 while it reaches to maximum in the range 0.978586–0.983811 at time $t = 16$ hrs and then decreases after that. This analysis shows that in order to improve the performance of the system, current conditions and equipment should be changed after time $t = 16$ (hrs).

In order to find the components, as per preferential order, on which more attention to be given for increasing the performance of the system, a sensitivity analysis on the RAM-Index has been done by varying the corresponding components failure rate and repair time parameters and fixing the other components parameter at the same time. The results thus obtained are shown graphically in Figure 8 which contains four subplots and each subplots has two subplots corresponding to failure and repair rate parameters of their component. The maximum and minimum values of the index during analysis has been obtained and given in Table 7. It has been observed that variation in failure and repair times of the filter, screener, and decker components significantly affect the RAM-Index of the system. For instance, an increase in scale parameter of filter components from 286.45 hrs to 387.55 hrs and the reduction in MTTR of the same from 2.30 hrs to 1.70 hrs reduce the system RAM-Index by approximately 0.529%. The effect of cleaners on system RAM-Index is found insignificant, because these unit have standby systems.

But in the real life modeling, the failure rate and repair time parameters affect simultaneously the system performance. For this, an analysis has been done on RAM-Index by varying simultaneously the parameters of failure and repair rate. The effects of individual component of the system on the system performance is noticed and shown in graphically in Figure 9 while their ranges corresponding to their components are tabulated in Table 8. On the basis of results, it can be analyzed that for improving the performance of the screening system, more attention should be given to the components as per the preferential order; decker, screener, filter, and cleaner.

6. Conclusion

In the present study an investigation has been done on the RAM analysis of the screening unit in paper industry by utilizing uncertain, vague, and limited data. The uncertainties in the collected or available data are removed with the help of fuzzy numbers. The development of fuzzy numbers from the available data and using fuzzy possibility theory can greatly increase the relevance of reliability study. RAM parameters of the system have been calculated by using ABCBLT technique and results are compared with Lambda-Tau results. The major advantage of this technique is that it optimizes the spread of the reliability indices upto a desired degree of accuracy which indicates the higher sensitivity zone and thus may be useful for the reliability engineers/experts to make more sound decisions. Plant personnel may be able to predict the system behavior more precisely and will plan future maintenance. To enhance the system performance, critical components of the system as per the preferential order have

been found by using proposed RAM-Index. Using RAM-Index, to improve the performance of the screening unit, more attention should be given in preferential order to the components; decker, screener, filter, and cleaner. Computed results will facilitate the management in reallocating the resources, making maintenance decisions, achieving long run availability of the system, and enhancing the overall productivity of the paper mill. These results will also help the concerned plant managers to plan and adapt suitable maintenance strategies for improving system performance and thereby reduce operational and maintenance costs.

References

[1] D. Kumar, *Analysis and optimization of systems availability in sugar, paper and fertilizer industries [Ph.D. thesis]*, University of Roorkee (Presently IIT Roorkee), Roorkee, India, 1991.

[2] C. Ebeling, *An Introduction to Reliability and Maintainability Engineering*, Tata McGraw-Hill, New York, NY, USA, 2001.

[3] H. Garg and S. P. Sharma, "RAM analysis of a coal crushing unit of a thermal power plant using fuzzy Lambda-Tau Methodology," in *Proceedings of the 1st International Conference on Emerging Trends in Mechanical Engineering (ICETME '11)*, pp. 795–802, Thapar University, Patiala, India.

[4] H. Garg and S. P. Sharma, "A two-phase approach for reliability and maintainability analysis of an industrial system," *International Journal of Reliability, Quality and Safety Engineering*, vol. 19, no. 3, 2012.

[5] A. Adamyan and D. He, "Analysis of sequential failures for assessment of reliability and safety of manufacturing systems," *Reliability Engineering and System Safety*, vol. 76, no. 3, pp. 227–236, 2002.

[6] A. Adamyan and D. He, "System failure analysis through counters of Petri net models," *Quality and Reliability Engineering International*, vol. 20, no. 4, pp. 317–335, 2004.

[7] H. Garg, "Reliability analysis of repairable systems using Petri nets and Vague Lambda-Tau methodology," *ISA Transactions*, http://dx.doi.org/10.1016/j.isatra.2012.06.009 . In press.

[8] P. S. Rajpal, K. S. Shishodia, and G. S. Sekhon, "An artificial neural network for modeling reliability, availability and maintainability of a repairable system," *Reliability Engineering and System Safety*, vol. 91, no. 7, pp. 809–819, 2006.

[9] Komal, S. P. Sharma, and D. Kumar, "RAM analysis of repairable industrial systems utilizing uncertain data," *Applied Soft Computing Journal*, vol. 10, no. 4, pp. 1208–1221, 2010.

[10] M. Rani, S. P. Sharma, and H. Garg, "Reliability analysis of press unit in paper plant using weibull fuzzy distribution function," in *Proceedings of the 16th Online World Conference on Soft computing in Industrial Application (WSC '11)*, December 2011, http://wsc16.cs.lboro.ac.uk/conference/sites/default/files/Paper_0.pdf.

[11] J. Knezevic and E. R. Odoom, "Reliability modelling of repairable systems using Petri nets and fuzzy Lambda-Tau methodology," *Reliability Engineering and System Safety*, vol. 73, no. 1, pp. 1–17, 2001.

[12] S. P. Sharma and H. Garg, "Behavioural analysis of urea decomposition system in a fertiliser plant," *International Journal of Industrial and Systems Engineering*, vol. 8, no. 3, pp. 271–297, 2011.

[13] H. Garg and S. P. Sharma, "Behavior analysis of synthesis unit in fertilizer plant," *International Journal of Quality and Reliability Management*, vol. 29, no. 2, pp. 217–232, 2012.

[14] D. L. Mon and C. H. Cheng, "Fuzzy system reliability analysis for components with different membership functions," *Fuzzy Sets and Systems*, vol. 64, no. 2, pp. 145–157, 1994.

[15] H. Garg and S. P. Sharma, "Stochastic behavior analysis of complex repairable industrial systems utilizing uncertain data," *ISA Transactions*, vol. 51, no. 6, pp. 752–762, 2012.

[16] S. M. Chen, "Fuzzy system reliability analysis using fuzzy number arithmetic operations," *Fuzzy Sets and Systems*, vol. 64, no. 1, pp. 31–38, 1994.

[17] H. Garg, S. P. Sharma, and M. Rani, "Behavior analysis of pulping unit in a paper mill with weibull fuzzy distribution function using ABCBLT technique," *International Journal of Applied Mathematics and Mechanics*, vol. 8, no. 4, pp. 86–96, 2012.

[18] D. Karaboga, "An idea based on honey bee swarm for numerical optimization," Tech. Rep. TR06, Computer Engineering Department, Engineering Faculty, Erciyes University, 2005.

[19] D. Karaboga and B. Basturk, "A powerful and efficient algorithm for numerical function optimization: artificial bee colony (ABC) algorithm," *Journal of Global Optimization*, vol. 39, no. 3, pp. 459–471, 2007.

[20] D. Karaboga and B. Basturk, "On the performance of artificial bee colony (ABC) algorithm," *Applied Soft Computing Journal*, vol. 8, no. 1, pp. 687–697, 2008.

[21] L. A. Zadeh, "Fuzzy sets," *Information and Control*, vol. 8, no. 3, pp. 338–353, 1965.

[22] W. Pedrycz, "Why triangular membership functions?" *Fuzzy Sets and Systems*, vol. 64, no. 1, pp. 21–30, 1994.

[23] X. Bai and S. Asgarpoor, "Fuzzy-based approaches to substation reliability evaluation," *Electric Power Systems Research*, vol. 69, no. 2-3, pp. 197–204, 2004.

[24] D. Karaboga and B. Akay, "A comparative study of Artificial Bee Colony algorithm," *Applied Mathematics and Computation*, vol. 214, no. 1, pp. 108–132, 2009.

[25] T. J. Ross, *Fuzzy Logic with Engineering Applications*, John Wiley & Sons, New York, NY, USA, 2nd edition, 2004.

5

Statistical Inferences and Applications of the Half Exponential Power Distribution

Wenhao Gui

Department of Mathematics and Statistics, University of Minnesota Duluth, Duluth, MN 55812, USA

Correspondence should be addressed to Wenhao Gui; guiwenhao@gmail.com

Academic Editor: Kai Yuan Cai

We investigate the statistical inferences and applications of the half exponential power distribution for the first time. The proposed model defined on the nonnegative reals extends the half normal distribution and is more flexible. The characterizations and properties involving moments and some measures based on moments of this distribution are derived. The inference aspects using methods of moment and maximum likelihood are presented. We also study the performance of the estimators using the Monte Carlo simulation. Finally, we illustrate it with two real applications.

1. Introduction

The well-known exponential power (EP) distribution or the generalized normal distribution has the following density function:

$$f(x) = \frac{p^{1-1/p}}{2\Gamma(1/p)} e^{-|x|^p/p}, \quad -\infty < x < \infty, \quad (1)$$

where $p > 0$ is the shape parameter. This family consists of a wide range of symmetric distributions and allows continuous variation from normality to nonnormality. It includes the normal distribution $Z \sim N(0,1)$ as the special case when $p = 2$ and the Laplace distribution when $p = 1$. Nadarajah [1] provided a comprehensive treatment of its mathematical properties.

Its tails can be more platykurtic ($p > 2$) or more leptokurtic ($p < 2$) than the normal distribution ($p = 2$). The distribution has been widely used in the Bayes analysis and robustness studies (see Box and Tiao [2], Genc [3], Goodman and Kotz [4], and Tiao and Lund [5].)

On the other hand, since the most popular models used to describe the lifetime process are defined on nonnegative measurements, which motivate us to take a positive truncation in the model (1) and develop a half exponential power (HEP) distribution. As far as we know, this model has not been previously studied although, we believe, it plays an important role in data analysis. The resulting nonnegative half

exponential power distribution generalizes the half normal (HN) distribution, and it is more flexible. In our work, we aim to investigate the statistical features of the nonnegative model and apply them to fit the lifetime data.

The rest of this paper is organized as follows: in Section 2, we present the new distribution and study its properties. Section 3 discusses the inference, moments, and maximum likelihood estimation for the parameters. In Section 4, we discuss a useful technique, a half normal plot with a simulated envelope, to assess the model adequacy. Simulation studies are performed in Section 5. Section 6 gives two illustrative examples and reports the results. Section 7 concludes our work.

2. The Half Exponential Power Distribution

2.1. The Density and Hazard Function

Definition 1. A random variable X has a half exponential power slash distribution if its density function with scale parameter $\sigma > 0$ takes

$$f(x) = \frac{p^{1-1/p}}{\sigma\Gamma(1/p)} e^{-x^p/p\sigma^p}, \quad x \geq 0, \quad (2)$$

where $\sigma > 0$ and $p > 0$. We denote it as $X \sim \text{HEP}(\sigma, p)$.

(a) Density function

(b) Hazard function

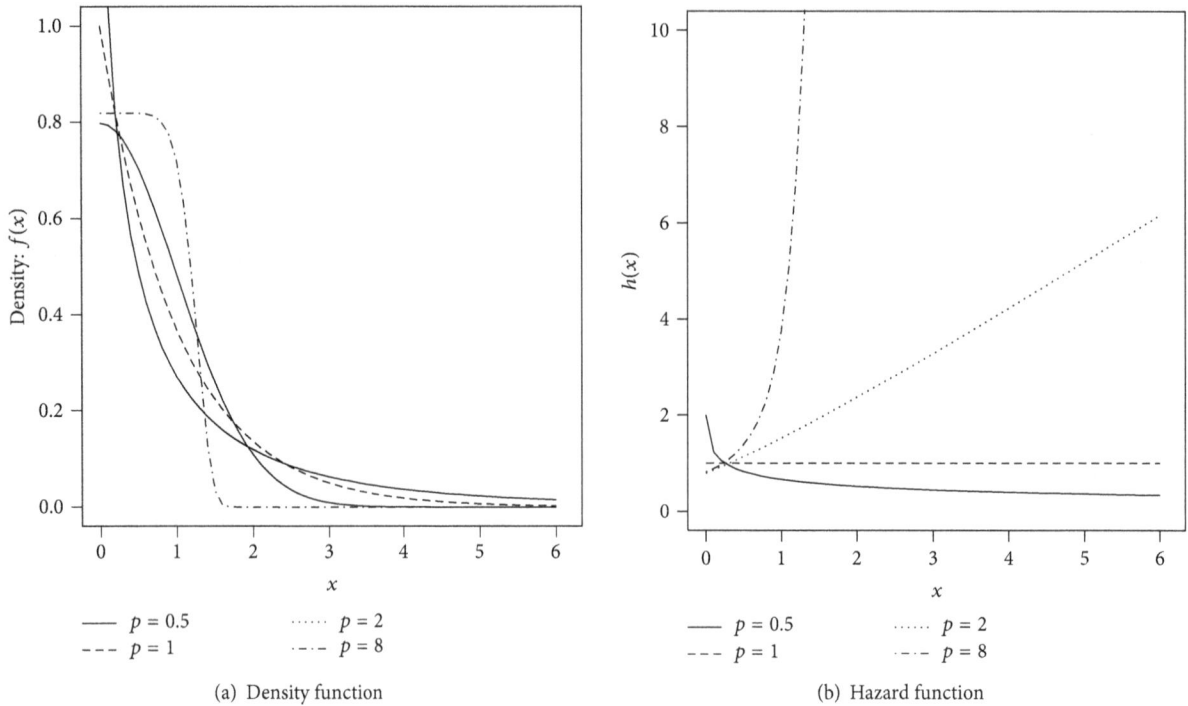

FIGURE 1: The density and hazard rate functions of HEP(σ, p) for $\sigma = 1$.

Figure 1(a) displays some plots of the density function of the half exponential power distribution with various parameters.

The cumulative distribution function of the half exponential power distribution $X \sim \text{HEP}(\sigma, p)$ is given as follows. For $x \geq 0$,

$$F(x) = \int_0^x f_X(u)\,du = \int_0^x \frac{p^{1-1/p}}{\sigma\Gamma(1/p)} e^{-u^p/p\sigma^p}\,du$$
$$= \frac{\gamma(1/p, x^p/p\sigma^p)}{\Gamma(1/p)}, \tag{3}$$

where $\gamma(,)$ is the lower incomplete gamma function, defined as $\gamma(s, x) = \int_0^x t^{s-1} e^{-t}\,dt$.

The hazard rate function (also known as the failure rate function) of the half exponential power distribution is given by, for $x \geq 0$,

$$h(x) = \frac{f(x)}{1 - F(x)} = \frac{p^{1-1/p}e^{-x^p/p\sigma^p}}{\sigma\left[\Gamma(1/p) - \gamma(1/p, x^p/p\sigma^p)\right]}. \tag{4}$$

Since $\Gamma(s) - \gamma(s, x) \sim x^{s-1} e^{-x}$, as $x \to \infty$, we obtain $h(x) \sim x^{p-1}/\sigma^p$. Therefore, the hazard rate function is increasing for $p \geq 1$ and decreasing for $0 < p < 1$. Figure 1(b) displays some plots of the hazard rate function of the half exponential power distribution with various parameters.

2.2. Moments and Measures Based on Moments

Proposition 2. Let $X \sim \text{HEP}(\sigma, p)$, for $k = 1, 2, 3, \ldots$; the kth noncentral moments are given by

$$\mu_k = \mathbb{E}X^k = \frac{p^{k/p}\sigma^k}{\Gamma(1/p)}\Gamma\left(\frac{k+1}{p}\right). \tag{5}$$

The following results are immediate consequences of (5).

Corollary 3. Let $X \sim \text{HEP}(\sigma, p)$. The mean and variance of X are given by

$$\mathbb{E}X = \frac{p^{1/p}\sigma}{\Gamma(1/p)}\Gamma\left(\frac{2}{p}\right),$$

$$\text{Var}(X) = \frac{p^{2/p}\sigma^2\left[\Gamma(1/p)\Gamma(3/p) - [\Gamma(2/p)]^2\right]}{[\Gamma(1/p)]^2}. \tag{6}$$

Corollary 4. Let $X \sim \text{HEP}(\sigma, p)$. The skewness and kurtosis coefficients of X are given by

$$\sqrt{\beta_1} = \frac{2[\Gamma(2/p)]^3 - 3\Gamma(1/p)\Gamma(2/p)\Gamma(3/p)}{\left(\Gamma(1/p)\Gamma(3/p) - [\Gamma(2/p)]^2\right)^{3/2}}$$
$$+ \frac{[\Gamma(1/p)]^2\Gamma(4/p)}{\left(\Gamma(1/p)\Gamma(3/p) - [\Gamma(2/p)]^2\right)^{3/2}},$$

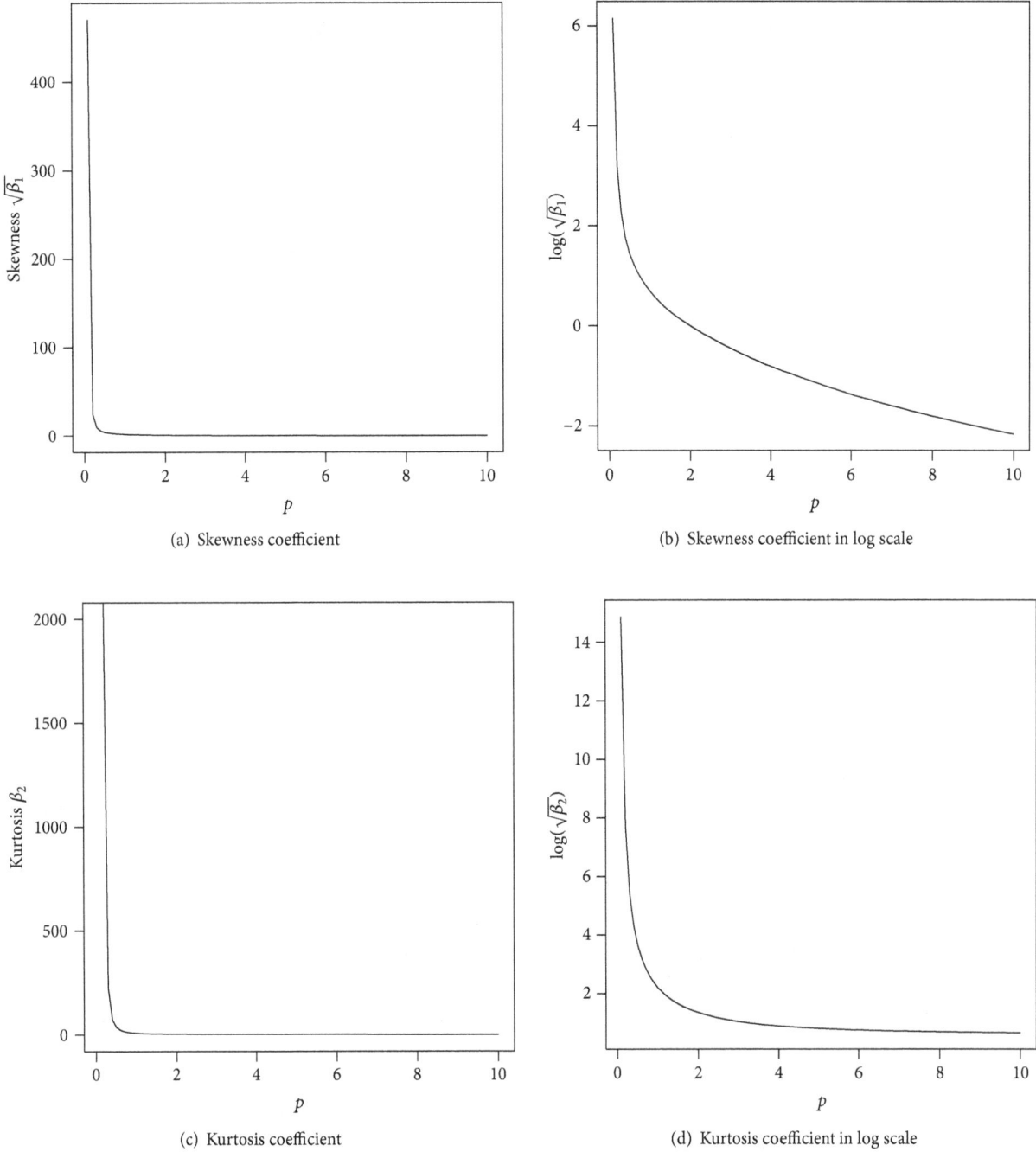

(a) Skewness coefficient

(b) Skewness coefficient in log scale

(c) Kurtosis coefficient

(d) Kurtosis coefficient in log scale

FIGURE 2: The plot for the skewness and kurtosis coefficients with various parameters.

$$\beta_2 = \frac{-3\left[\Gamma\left(2/p\right)\right]^4 + 6\Gamma\left(1/p\right)\left[\Gamma\left(2/p\right)\right]^2\Gamma\left(3/p\right)}{\left(\Gamma\left(1/p\right)\Gamma\left(3/p\right) - \left[\Gamma(2/p)\right]^2\right)^2}$$

$$- \frac{4\left[\Gamma\left(1/p\right)\right]^2\Gamma\left(2/p\right)\Gamma\left(4/p\right) + \left[\Gamma\left(1/p\right)\right]^3\Gamma\left(5/p\right)}{\left(\Gamma\left(1/p\right)\Gamma\left(3/p\right) - \left[\Gamma(2/p)\right]^2\right)^2}.$$

$$(7)$$

Figure 2 shows the skewness and kurtosis coefficients with various parameters for the HEP model.

3. Inference

3.1. *Moment Estimation.* Let $X_1, X_2, \ldots, X_n$ be a random sample from the distribution $\text{HEP}(\sigma, p)$. From (5), we have $\mathbb{E}X = (p^{1/p}\sigma/\Gamma(1/p))\Gamma(2/p)$ and

$\mathbb{E}X^2 = (p^{2/p}\sigma^2/\Gamma(1/p))\Gamma(3/p)$. Replacing $\mathbb{E}X$ and $\mathbb{E}X^2$ with the corresponding sample estimators, we obtain the moment equations

$$\overline{X} = \frac{1}{n}\sum_{i=1}^{n}X_i = \frac{p^{1/p}\sigma}{\Gamma(1/p)}\Gamma\left(\frac{2}{p}\right),$$

$$\overline{X^2} = \frac{1}{n}\sum_{i=1}^{n}X_i^2 = \frac{p^{2/p}\sigma^2}{\Gamma(1/p)}\Gamma\left(\frac{3}{p}\right).$$

$$(8)$$

The estimate $\widehat{p}$ is the solution to

$$\frac{\Gamma(1/p)\,\Gamma(3/p)}{[\Gamma(2/p)]^2} = \frac{\overline{X^2}}{\overline{X}^2},\qquad (9)$$

which can be solved numerically. And the estimate $\widehat{\sigma}$ is given by

$$\widehat{\sigma} = \frac{\overline{X}\Gamma(1/\widehat{p})}{\widehat{p}^{1/\widehat{p}}\,\Gamma(2/\widehat{p})}.\qquad (10)$$

It is clear that, for the special case when p is known, estimator $\widehat{\sigma}$ is unbiased and its mean squared error (MSE) is given by

$$\mathrm{MSE}(\widehat{\sigma}) = \frac{\sigma^2\left[\Gamma(1/p)\,\Gamma(3/p) - [\Gamma(2/p)]^2\right]}{n[\Gamma(2/p)]^2}.\qquad (11)$$

In the following proposition, we present the asymtotic property of the moment estimators.

Proposition 5. *Let $X_1, X_2, \ldots, X_n$ be a random sample of size n from the distribution $HEP(\sigma, p)$, and let $\boldsymbol{\theta} = (\sigma, p)$; then, if $\mu_6 = \mathbb{E}X^6 < \infty$ and $\widehat{\boldsymbol{\theta}}$ is the moment estimator of $\boldsymbol{\theta}$, one has*

$$\sqrt{n}\left(\widehat{\boldsymbol{\theta}} - \boldsymbol{\theta}\right) \xrightarrow{d} N_2\left(\mathbf{0}, \mathbf{H}^{-1}\Sigma\left[\mathbf{H}^{-1}\right]^T\right)\qquad (12)$$

as $n \to \infty$, where $\Sigma = (\{\mu_{i+j} - \mu_i\mu_j\}_{ij})$ and $\mathbf{H}$ is given by

$$\mathbf{H} = \mathbf{H}(\boldsymbol{\theta}) = \begin{pmatrix} \dfrac{\partial\mu_1}{\partial\sigma} & \dfrac{\partial\mu_1}{\partial p} \\ \dfrac{\partial\mu_2}{\partial\sigma} & \dfrac{\partial\mu_2}{\partial p} \end{pmatrix}\qquad (13)$$

whose entries are given by

$$\frac{\partial\mu_1}{\partial\sigma} = \frac{p^{1/p}\Gamma(2/p)}{\Gamma(1/p)},$$

$$\frac{\partial\mu_1}{\partial p} = -\frac{p^{-2+1/p}\sigma\Gamma(2/p)\left[-1 + \log p - \psi(1/p) + 2\psi(2/p)\right]}{\Gamma(1/p)},$$

$$\frac{\partial\mu_2}{\partial\sigma} = \frac{2p^{2/p}\sigma\Gamma(3/p)}{\Gamma(1/p)},$$

$$\frac{\partial\mu_2}{\partial p} = \frac{p^{-2+2/p}\sigma^2\Gamma(3/p)\left[-2 + 2\log p - \psi(1/p) + 3\psi(3/p)\right]}{\Gamma(1/p)},$$

$$(14)$$

where $\psi()$ is the digamma function defined as the logarithmic derivative of the gamma function, $\psi(x) = (d/dx)\log\Gamma(x) = \Gamma'(x)/\Gamma(x)$.

Remark 6. A consistent estimator for the asymptotic covariance matrix $\mathbf{H}^{-1}\Sigma[\mathbf{H}^{-1}]^T$ can be obtained by replacing parameters with their corresponding moment estimators.

3.2. Maximum Likelihood Estimation. In this section, we consider the maximum likelihood estimation about the parameter $\boldsymbol{\theta} = (\sigma, p)$ of the HEP model defined in (2). The log likelihood for a random sample $x_1, x_2, \ldots, x_n$ is

$$l(\boldsymbol{\theta}) = \log\prod_{i=1}^{n}f(x_i) = n\left(1 - \frac{1}{p}\right)\log p - n\log\sigma$$

$$- n\log\Gamma\left(\frac{1}{p}\right) - \frac{1}{p\sigma^p}\sum_{i=1}^{n}x_i^p.\qquad (15)$$

By taking the partial derivatives of the log-likelihood function with respect to σ and p, respectively, and equalizing the obtained expressions to zero, the following maximum likelihood estimating equations are obtained:

$$l_\sigma = -\frac{n}{\sigma} + \frac{1}{\sigma^{p+1}}\sum_{i=1}^{n}x_i^p = 0,$$

$$l_p = \frac{n(\log p + p - 1)}{p^2} + \frac{n\psi(1/p)}{p^2}$$

$$+ \frac{1 + p\log\sigma}{\sigma^p p^2}\sum_{i=1}^{n}x_i^p - \frac{1}{p\sigma^p}\sum_{i=1}^{n}x_i^p\log x_i = 0.$$

$$(16)$$

In general, there are no explicit solutions for the above maximum likelihood estimating equations. The estimates can be obtained by means of numerical procedures such as the Newton-Raphson method. The program R provides the nonlinear optimization routine *optim* for solving such problems.

For asymptotic inference of $\boldsymbol{\theta} = (\sigma, p)$, we need the Fisher information matrix $\mathbf{I}(\boldsymbol{\theta})$. It is known that its inverse is the asymptotic variance matrix of the maximum likelihood estimators. For the case of a single observation ($n = 1$), we take the second-order derivatives of the log-likelihood function in (15).

TABLE 1: Empirical means and SD for the moment estimators of σ and p.

σ	p	$n = 100$		$n = 150$		$n = 200$	
		$\hat{\sigma}$ (SD)	$\hat{p}$ (SD)	$\hat{\sigma}$ (SD)	$\hat{p}$ (SD)	$\hat{\sigma}$ (SD)	$\hat{p}$ (SD)
1	1	1.0116 (0.1274)	1.0643 (0.1949)	1.0099 (0.1077)	1.0450 (0.1675)	1.0084 (0.0935)	1.0380 (0.1426)
1	2	1.0046 (0.1014)	2.0544 (0.3443)	0.9989 (0.0816)	2.0369 (0.3167)	1.0034 (0.0745)	2.0484 (0.2869)
1	3	0.9972 (0.0844)	3.0454 (0.4233)	0.9998 (0.0714)	3.0375 (0.4089)	1.0044 (0.0640)	3.0547 (0.3970)
2	1	2.0365 (0.2499)	1.0660 (0.1959)	2.0390 (0.2099)	1.0559 (0.1635)	2.0233 (0.1872)	1.0443 (0.1505)
2	2	2.0090 (0.1983)	2.0726 (0.3453)	2.0111 (0.1710)	2.0541 (0.3117)	2.0014 (0.1424)	2.0372 (0.2814)
2	3	2.0033 (0.1660)	3.0516 (0.4338)	2.0013 (0.1392)	3.0344 (0.4054)	2.0116 (0.1275)	3.0607 (0.3974)

TABLE 2: Empirical means and SD for the MLE estimators of σ and p.

σ	p	$n = 100$		$n = 150$		$n = 200$	
		$\hat{\sigma}$ (SD)	$\hat{p}$ (SD)	$\hat{\sigma}$ (SD)	$\hat{p}$ (SD)	$\hat{\sigma}$ (SD)	$\hat{p}$ (SD)
1	1	1.0119 (0.1272)	1.0515 (0.2055)	1.0134 (0.1079)	1.0397 (0.1695)	1.0026 (0.0890)	1.0270 (0.1401)
1	2	1.0153 (0.1106)	2.2028 (0.6168)	1.0048 (0.0883)	2.0995 (0.4420)	1.0063 (0.0770)	2.0876 (0.3644)
1	3	1.0193 (0.1102)	3.4735 (1.3164)	1.0099 (0.0816)	3.2477 (0.7742)	1.0068 (0.0736)	3.1542 (0.6405)
2	1	2.0202 (0.2631)	1.0566 (0.2107)	2.0309 (0.2178)	1.0409 (0.1697)	2.0153 (0.1766)	1.0242 (0.1372)
2	2	2.0250 (0.2266)	2.1944 (0.6224)	2.0136 (0.1798)	2.1194 (0.4469)	2.0031 (0.1531)	2.0695 (0.3449)
2	3	2.0332 (0.2235)	3.4523 (1.4561)	2.0241 (0.1682)	3.2700 (0.8226)	2.0218 (0.1432)	3.2229 (0.7221)

Consider,

$$l_{\sigma\sigma} = \frac{1}{\sigma^2} - \frac{p+1}{\sigma^{p+2}} x^p,$$

$$l_{\sigma p} = \frac{1}{\sigma^{p+1}} x^p \left(\log x - \log \sigma \right),$$

$$
\begin{aligned}
l_{pp} = -\frac{1}{p^4 \sigma^p} \Big[&- 3p\sigma^p + p^2\sigma^p + 2px^p + 2p\sigma^p \log p \\
&+ 2p^2 x^p \log \sigma + p^3 x^p [\log \sigma]^2 \\
&- 2p^2 x^p \log x - 2p^3 x^p \log \sigma \log x \\
&+ p^3 x^p [\log x]^2 + 2p\sigma^p \psi\left(\frac{1}{p}\right) \\
&+ \sigma^p \psi'\left(\frac{1}{p}\right) \Big].
\end{aligned}
$$
(17)

Using the facts

$$\mathbb{E} x^p = \sigma^p,$$

$$\mathbb{E}\left(x^p \log x\right) = \frac{\sigma^p \left[p \log \sigma + \log p + \psi\left(1 + 1/p\right)\right]}{p},$$

$$\mathbb{E}\left(x^p [\log x]^2\right)$$

$$= \frac{\sigma^p \left[\left(p \log \sigma + \log p + \psi\left(1 + 1/p\right)\right)^2 + \psi'\left(1 + 1/p\right)\right]}{p^2},$$
(18)

TABLE 3: Summary of the plasma ferritin concentration measurements.

Sample size	Mean	Standard deviation	$\sqrt{b_1}$	b_2
202	76.88	47.50	1.28	4.42

we can obtain the elements of the Fisher information matrix:

$$I_{11} = -\mathbb{E} l_{\sigma\sigma} = \frac{p}{\sigma^2},$$

$$I_{12} = -\mathbb{E} l_{\sigma p} = \frac{\log p + \psi\left(1 + 1/p\right)}{\sigma p},$$

$$I_{21} = -\mathbb{E} l_{p\sigma} = \frac{\log p + \psi\left(1 + 1/p\right)}{\sigma p},$$
(19)

$$
\begin{aligned}
I_{22} = -\mathbb{E} l_{pp} = &\frac{-p - p^2 + p[\log p + \psi\left(1 + 1/p\right)]^2}{p^4} \\
&+ \frac{p\psi'\left(1 + 1/p\right) + \psi'\left(1/p\right)}{p^4}.
\end{aligned}
$$

Proposition 7. *Let $X_1, X_2, \ldots, X_n$ be a random sample of size n from the distribution $HEP(\sigma, p)$, let $\boldsymbol{\theta} = (\sigma, p)$, and $\hat{\boldsymbol{\theta}}$ is the maximum likelihood estimator of $\boldsymbol{\theta}$, one has*

$$\sqrt{n}\left(\hat{\boldsymbol{\theta}} - \boldsymbol{\theta}\right) \xrightarrow{d} N_2\left(\mathbf{0}, \mathbf{I}(\boldsymbol{\theta})^{-1}\right).$$
(20)

4. Assessment of Model Adequacy

In this section, we introduce a useful tool, a half normal plot with a simulated envelope which will be used to evaluate

TABLE 4: Maximum likelihood parameter estimates (with (SD)) of the HN and HEP models for the plasma ferritin concentration data.

Model	$\hat{\sigma}$	$\hat{p}$	Log lik.	AIC	BIC
HN	76.9436 (3.0588)	—	−1062.037	2126.074	2129.382
HEP	97.1311 (6.1496)	2.5109 (0.3318)	−1054.739	2113.478	2120.095

the HEP model in Section 6. The advantage of this technique is its ease of interpretation without knowing the distribution of the residuals.

Atkinson [6] proposed this diagnostic plot to detect potential outliers and influential observations in linear regression models. A simulated envelope is added to the plot to aid overall assessment, whereby the observed residuals are expected to lie within the boundary of the envelope if the presumed model has been correctly specified.

The method of simulated envelope and its corresponding transformations have been widely applied in many applications (see Flack and Flores [7], Ferrari and Cribari-Neto [8], da Silva Ferreira et al. [9], and so forth.) The simulated envelope technique compares the observed statistics with those of the data generated from the proposed model. Any sizeble departure of the observed residuals from the simulated quantities may be thought as evidence against the adequacy of the proposed model. Here is the procedure to produce the half normal plot with simulated envelopes.

(1) Fit the model to the observed data (sample size = n).

(2) Generate a sample of n observations based on the fitted model.

(3) Fit the model to the above generated sample and compute the ordered absolute values of the standard residuals.

(4) Repeat the above steps k times.

(5) Consider the n sets of the k-ordered statistics; calculate the average, minimum, and maximum values across each set.

(6) Plot these values together with the ordered residuals from the original data against the half normal scores $\Phi^{-1}((i+n-1/8)/(2n+1/2))$.

The minimum and maximum values of the k-ordered statistics constitute a simulated envelope to guide assessment of the model adequacy. Atkinson [6] suggested using $k = 19$ since there is a 5% chance to detect the largest residual being outside the boundary of the simulated envelope. Moreover, other types of residuals such as deviance or score residual may be used in the procedure. For example, da Silva Ferreira et al. [9] used the Mahalanobis distance to assess their models. The horizontal axis can also show other variables such as index.

5. Simulation Study

In this section, we conduct some simulations and study the properties of the estimators numerically.

We perform a simulation to illustrate the behaviors of the moment and MLE estimators for parameters $\theta = (\sigma, p)$,

TABLE 5: Summary of the life of fatigue fracture.

sample size	Mean	Standard deviation	$\sqrt{b_1}$	b_2
101	1.025	1.119	3.001	16.709

respectively. The simulation is conducted by the software R. We generate 1000 samples of size $n = 100$, $n = 150$, and $n = 200$ from the HEP(σ, p) distribution for fixed parameters σ and p.

The random numbers can be generated as follows. We first generate random numbers Y from an exponential power distribution with $\mu = 0, \sigma$, and p, the procedures can be found in Chiodi [10]; then we take the absolute value of the random numbers, $X = |Y|$. It follows that $X \sim \text{HEP}(\sigma, p)$.

The estimators are computed using the results in Section 3. The empirical means and standard deviations of the estimators are presented in Tables 1 and 2, respectively. The simulation studies show that the parameters are well estimated, and the estimates are asymptotically unbiased. The empirical MSEs decrease as sample size increases as expected. Further, MLEs are more efficient than moment estimators.

6. Real Data Illustration

In this section, we analyze two real datasets to fit with the proposed model. The applications demonstrate that the HEP model fits the data better than the HN model.

6.1. Application 1. The data are the plasma ferritin concentration measurements of 202 athletes collected at the Australian Institute of Sport. This dataset has been studied by several authors (see Azzalini and Dalla Valle [11], Cook and Weisberc [12], and Elal-Olivero et al. [13].)

The descriptive statistics for the dataset are shown in Table 3, where $\sqrt{b_1}$ and b_2 are the sample skewness and kurtosis coefficients. Notice that the dataset presents nonnegative measurements.

We fit the dataset with the half normal and the half exponential power distribution, respectively, using maximum likelihood method. The MLE estimators are computed using R, and the results are reported in Table 4. The usual Akaike information criterion (AIC) and Bayesian information criterion (BIC) to measure of the goodness of fit are also computed: $\text{AIC} = 2k - 2\log L$ and $\text{BIC} = k\log n - 2\log L$, where, k is the number of parameters in the distribution and L is the maximized value of the likelihood function. The results indicate that HEP model has the lower values for the AIC and BIC statistics, and thus it is a better model. Figures 3(a) and 3(b) display the fitted models using the MLE estimates.

TABLE 6: Maximum likelihood parameter estimates (with (SD)) of the HN and HEP models for the life of fatigue fracture data.

Model	$\hat{\sigma}$	$\hat{p}$	Log lik.	AIC	BIC
HN	1.5135 (0.1064)	—	−115.1666	232.3332	234.9483
HEP	0.9689 (0.1298)	0.8815 (0.1677)	−103.2537	210.5074	215.7376

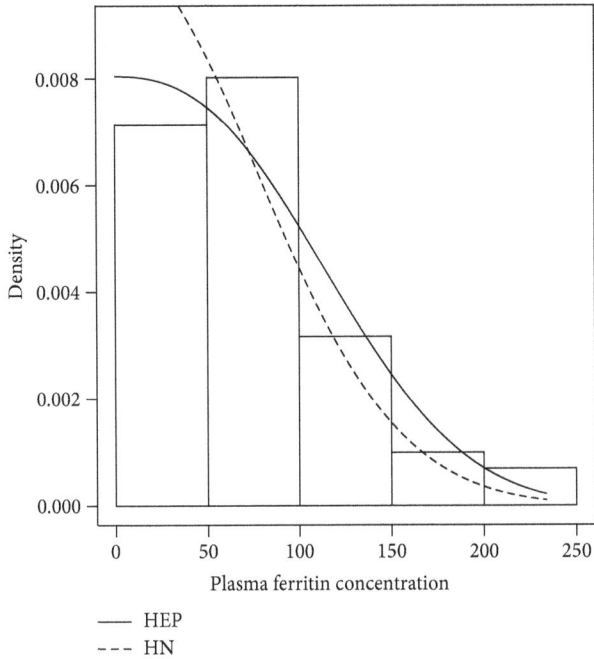

(a) Histogram and fitted curves

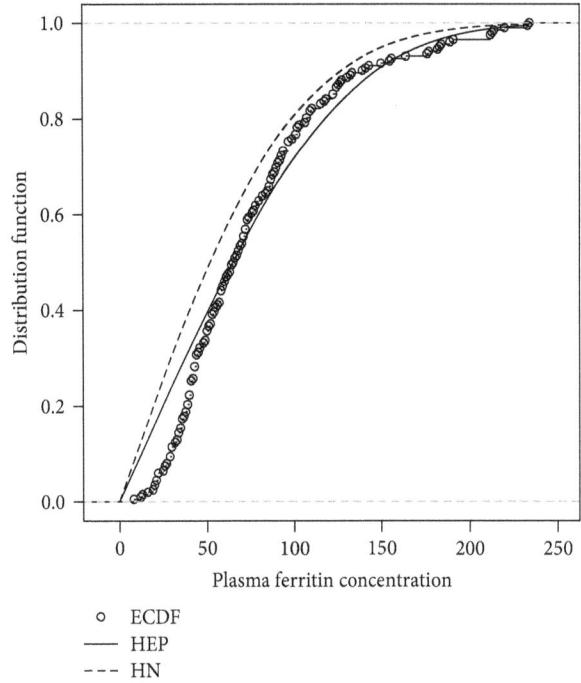

(b) Empirical and fitted CDF

FIGURE 3: Models fitted for the plasma ferritin concentration dataset.

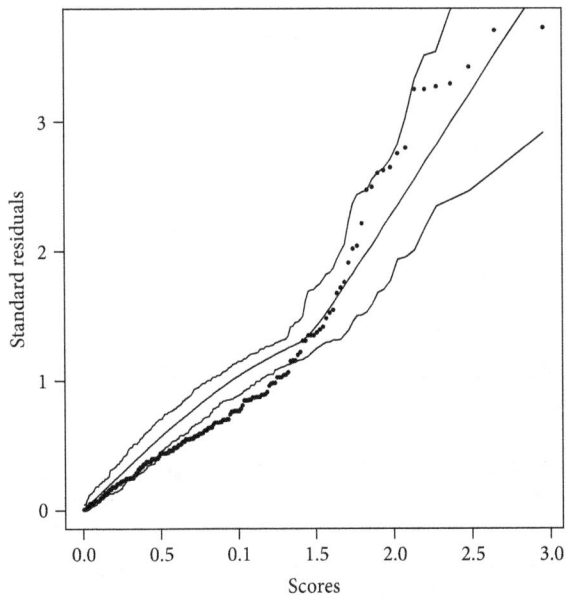

(a) Half normal

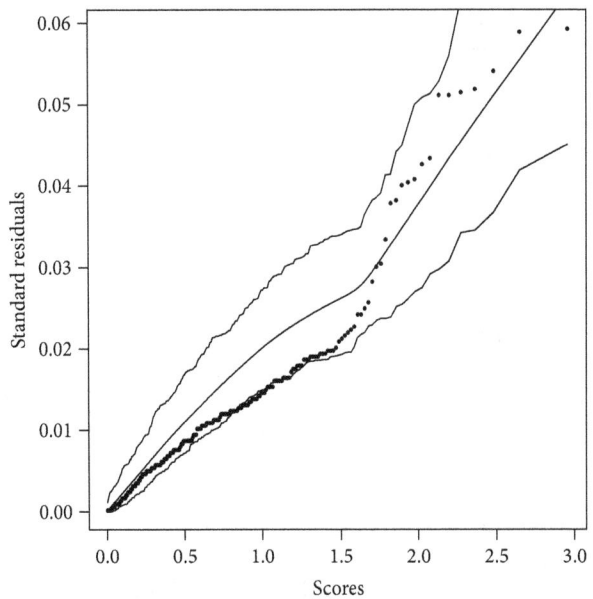

(b) Half exponential power

FIGURE 4: Simulated envelopes for on HN and HEP models.

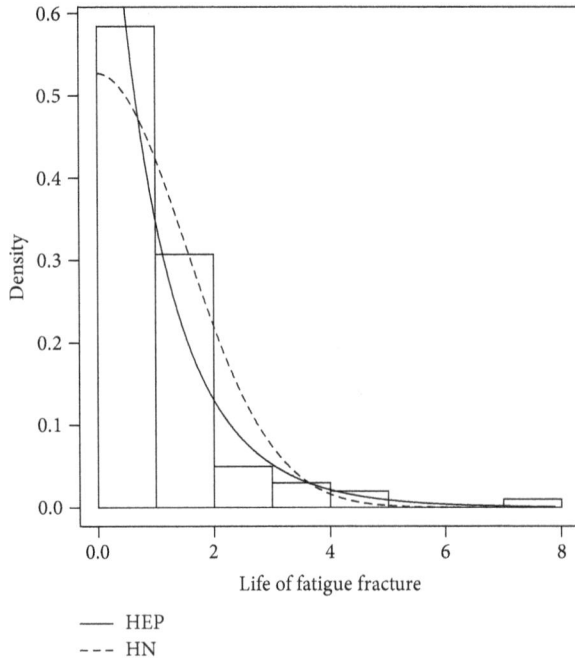

(a) Histogram and fitted curves

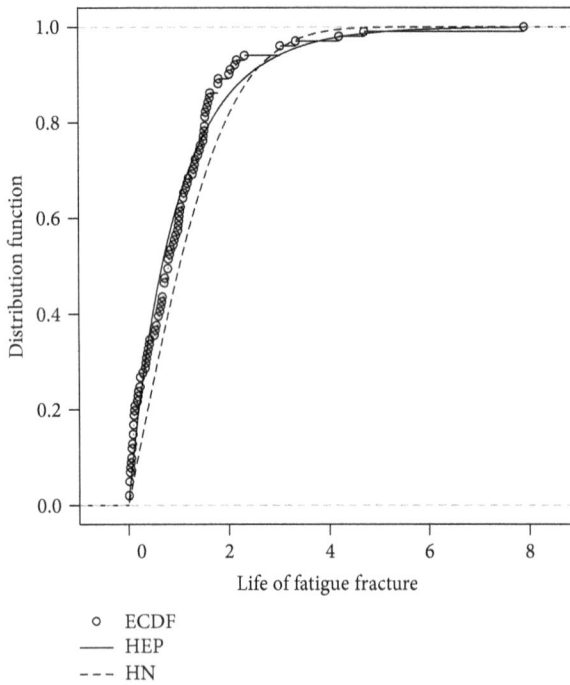

(b) Empirical and fitted CDF

FIGURE 5: Models fitted for the life of fatigue fracture dataset.

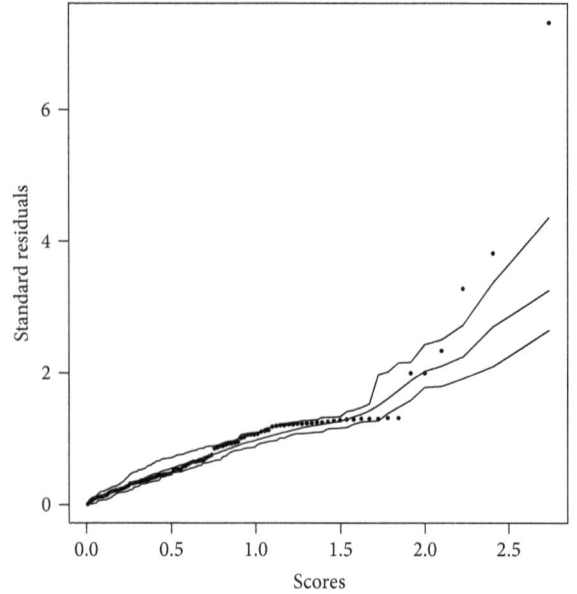

(a) Half normal

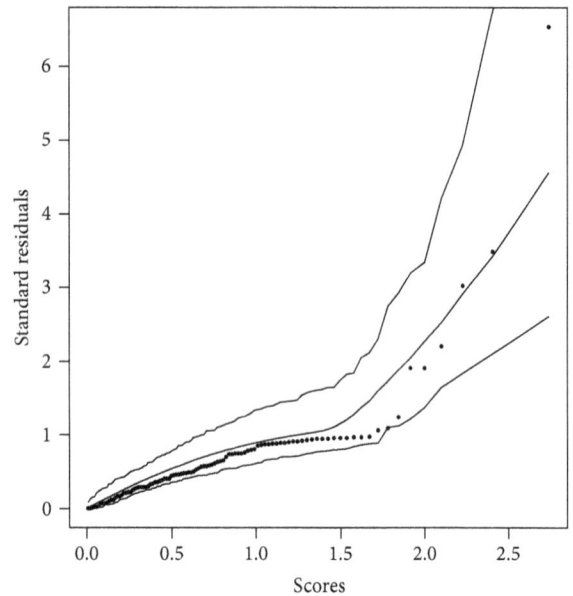

(b) Half exponential power

FIGURE 6: Simulated envelopes for on HN and HEP models.

The diagnostic procedure introduced in Section 4 is implemented for both models. The simulated envelope plots are shown in Figures 4(a) and 4(b). Most of the observed residuals are either near or outside the boundary of the envelope, indicating inadequacy of the fitted HN model. On the other hand, the observed residuals corresponding to the HEP model in Figure 4(b) are well within the simulated envelope, indicating that the HEP model provides a better fit to the data.

6.2. Application 2. We consider the stress-rupture dataset and the life of fatigue fracture of Kevlar 49/epoxy that are subject to the pressure at the 90% level. The dataset has been previously studied by Andrews and Herzberg [14], Barlow et al. [15], and Olmos et al. [16].

Table 5 summarizes the dataset. This dataset also shows nonnegative asymmetry. Same as before, we fit the dataset with the half normal and the half exponential power distribution, respectively, using maximum likelihood method.

The results are reported in Table 6. The AIC and BIC are presented as well, and the results show that HEP model fits better. Figures 5(a) and 5(b) display the fitted models using the MLE estimates.

The diagnostic procedure introduced in Section 4 is implemented for both models. The simulated envelope plots are shown in Figures 6(a) and 6(b). The observed residuals corresponding to the HEP model in Figure 6(b) are well within the simulated envelope, indicating that the HEP model provides a better fit to the data.

7. Concluding Remarks

In this paper, we have studied the half exponential power distribution $\text{HEP}(\sigma, p)$ in detail. This nonnegative distribution contains the half normal distribution as its special case. Probabilistic and inferential properties are studied. A simulation is conducted and demonstrates the good performance of the moment and maximum likelihood estimators. We apply the model to two real datasets, illustrating that the proposed model is appropriate and flexible in real applications. There are a number of possible extensions of the current work. Mixture modeling using the proposed distributions is the most natural extension. Other extensions of the current work include a generalization of the distribution to multivariate settings.

Appendix

Proofs of Propositions

Proof of Proposition 2. Consider,

$$
\begin{aligned}
\mathbb{E}X^k &= \int_0^\infty x^k \frac{p^{1-1/p}}{\sigma\Gamma(1/p)} e^{-x^p/p\sigma^p} dx \\
&= \frac{p^{1-1/p}}{\sigma\Gamma(1/p)} \int_0^\infty x^k e^{-x^p/p\sigma^p} dx \\
&= \frac{p^{1-1/p}}{\sigma\Gamma(1/p)} \sigma^{k+1} p^{(k+1)/p-1} \Gamma\left(\frac{k+1}{p}\right) \\
&= \frac{p^{k/p}\sigma^k}{\Gamma(1/p)} \Gamma\left(\frac{k+1}{p}\right).
\end{aligned}
\tag{A.1}
$$

$\square$

Proof of Proposition 5. This result follows directly by using standard large sample theory for moment estimators, as discussed in Sen and Singer [17]. $\square$

Proof of Proposition 7. It follows directly by using the large sample theory for maximum likelihood estimators and the Fisher information matrix given above. $\square$

References

[1] S. Nadarajah, "A generalized normal distribution," *Journal of Applied Statistics*, vol. 32, no. 7, pp. 685–694, 2005.

[2] G. Box and G. Tiao, "A further look at robustness via bayes's theorem," *Biometrika*, vol. 49, no. 3-4, pp. 419–432, 1962.

[3] A. I. Genç, "A generalization of the univariate slash by a scale-mixtured exponential power distribution," *Communications in Statistics*, vol. 36, no. 5, pp. 937–947, 2007.

[4] I. R. Goodman and S. Kotz, "Multivariate θ-generalized normal distributions," *Journal of Multivariate Analysis*, vol. 3, no. 2, pp. 204–219, 1973.

[5] G. Tiao and D. Lund, "The use of olumv estimators in inference robustness studies of the location parameter of a class of symmetric distributions," *Journal of the American Statistical Association*, vol. 65, pp. 370–386, 1970.

[6] A. Atkinson, *Plots, Transformations, and Regression: An Introduction to Graphical Methods of Diagnostic Regression Analysis*, Clarendon Press Oxford, 1985.

[7] V. F. Flack and R. A. Flores, "Using simulated envelopes in the evaluation of normal probability plots of regression residuals," *Technometrics*, vol. 31, no. 2, pp. 219–225, 1989.

[8] S. L. P. Ferrari and F. Cribari-Neto, "Beta regression for modelling rates and proportions," *Journal of Applied Statistics*, vol. 31, no. 7, pp. 799–815, 2004.

[9] C. da Silva Ferreira, H. Bolfarine, and V. H. Lachos, "Skew scale mixtures of normal distributions: properties and estimation," *Statistical Methodology*, vol. 8, no. 2, pp. 154–171, 2011.

[10] M. Chiodi, "Procedures for generating pseudo-random numbers from a normal distribution of order p $(P > 1)$," *Statistica Applicata*, vol. 1, pp. 7–26, 1986.

[11] A. Azzalini and A. Dalla Valle, "The multivariate skew-normal distribution," *Biometrika*, vol. 83, no. 4, pp. 715–726, 1996.

[12] R. Cook and S. Weisberc, "An introduction to regression graphic?" *Methods*, vol. 17, article 640, 1994.

[13] D. Elal-Olivero, J. F. Olivares-Pacheco, H. W. Gómez, and H. Bolfarine, "A new class of non negative distributions generated by symmetric distributions," *Communications in Statistics—Theory and Methods*, vol. 38, no. 7, pp. 993–1008, 2009.

[14] D. Andrews and A. Herzberg, *Data: A Collection of Problems from Many Fields for the Student and Research Worker*, vol. 18, Springer, New York, NY, USA, 1985.

[15] R. Barlow, R. Toland, and T. Freeman, "A bayesian analysis of the stress-rupture life of kevlar/epoxy spherical pressure vessels," in *Accelerated Life Testing and Experts Opinions in Reliability*, C. A. Clarotti and D. V. Lindley, Eds., 1988.

[16] N. M. Olmos, H. Varela, H. W. Gómez, and H. Bolfarine, "An extension of the half-normal distribution," *Statistical Papers*, pp. 1–12, 2011.

[17] P. Sen and J. M. Singer, *Large Sample Methods in Statistics: An Introduction with Applications*, Chapman and Hall/CRC, 1993.

6

Parameter Estimation Based on the Frèchet Progressive Type II Censored Data with Binomial Removals

Mohamed Mubarak[1,2]

[1] Mathematics Department, Faculty of Science, Minia University, El-Minia 61519, Egypt
[2] Mathematics Department, University College in Lieth, Umm Al-Qura University, Makkah 311, Saudi Arabia

Correspondence should be addressed to Mohamed Mubarak, mubarak_7061111@yahoo.com

Academic Editor: Chun-Ping Lin

This paper considers the estimation problem for the Frèchet distribution under progressive Type II censoring with random removals, where the number of units removed at each failure time has a binomial distribution. We use the maximum likelihood method to obtain the estimators of parameters and derive the sampling distributions of the estimators, and we also construct the confidence intervals for the parameters and percentile of the failure time distribution.

1. Introduction

Recently, the extreme value distribution is becoming increasingly important in engineering statistics as a suitable model to represent phenomena with usually large maximum observations. In engineering circles, this distribution is often called the Frèchet model. It is one of the pioneers of extreme value statistics. The Frèchet (extreme value type II) distribution is one of the probability distributions used to model extreme events. The generalization of the standard Frèchet distribution has been introduced by Nadarajah and Kotz [1] and Abd-Elfattah and omima [2]. There are over fifty applications ranging from accelerated life testing through to earthquakes, floods, rain fall, queues in supermarkets, sea currents, wind speeds, and track race records, see Kotz and Nadarajah [3]. Censoring arises in a life test when exact lifetimes are known for only a portion of test units and the remainder of the lifetimes are known only to exceed certain values under an experiment. There are several types of censored test. One of the most common censoring schemes is Type II censoring. In a Type II censoring, a total of n units is placed on test, but instead of continuing until all n units have failed, the test is terminated at the time of the mth ($1 \leq m \leq n$) unit failure. Type II censoring

with different failure time distribution has been studied by many authors including Mann et al. [4], Lawless [5], and Meeker and Escobar [6]. If an experiment desires to remove live units at points other than the final termination point of the life test, the above described scheme will not be of use to experimenter. Type II censoring does not allow for units to be lost or removed from the test at points other than the final termination point see Balakrishnan and Aggarwala [7, Chapter 1].

A generalization of Type II censoring is progressive Type II censoring. Under this scheme, n units are placed on test at time zero, and m failure are going to be observed. When the first failure is observed, r_1 of surviving units are randomly selected, removed, and so on. This experiment terminates at the time when the mth failures is observed and remaining $r_m = n - r_1 - r_2 - \cdots - r_{m-1} - m$ surviving units are all removed. The statistical inference on the parameters of failure time distribution under progressive Type II censoring has been studied by several authors [7–10]. Note that, in this scheme, $r_1,\ldots,r_m$ are all prefixed. However, in some practical situations, these numbers may occur at random [11]. In some reliability experiments, an experimenter may decide that it is inappropriate or too dangerous to carry on the testing on some of the tested units even though these units have not failed. In these cases, the pattern of removal at each failure is random. We assume that, any test unit being dropped out from the life test is independent of the others but with the same probability p. Then Tea et al. [12] indicated that the number of test units removed at each failure time has a binomial distribution.

In this paper, we will make inference on the parameters of three-parameter Frèchet distribution under progressive type II censoring with binomial removals. The maximum likelihood estimators (MLEs) for the parameters in an explicit and implicit form are obtained in Section 3. Section 4 discusses the sampling distribution of the MLEs and constructs the confidence intervals for the parameters. The estimator and confidence interval for the percentile of failure time distribution are also presented in Section 4.

2. Description of the Model

Let random variable X has a three-parameter Frèchet distribution. The probability density function and cumulative distribution function are

$$f(x) = \left(\frac{\alpha}{\beta}\right)\left(\frac{\beta}{x-\theta}\right)^{\alpha+1} \exp\left[-\left(\frac{\beta}{x-\theta}\right)^{\alpha}\right],$$
$$(x > \theta, \alpha, \beta > 0), \tag{1}$$

$$F(x) = \exp\left[-\left(\frac{\beta}{x-\theta}\right)^{\alpha}\right], \quad (x > \theta, \alpha, \beta > 0), \tag{2}$$

respectively, where α is continuous shape parameter, β is continuous scale parameter, and θ is continuous location parameter ($\theta = 0$ yields the two-parameter Frèchet distribution).

3. Estimation of Parameters

Suppose n independent units are placed on a test with the corresponding life times being identically distributed with probability density function $f_X(x)$ and cumulative distribution function $F_X(x)$. For simplicity of notation, let $X_1, X_2, \ldots, X_m$ denote a progressively Type II censored sample. Then, the joint probability density function of all m progressively Type II-censored order statistics is

$$f_{X_1,\ldots,X_m}(x_1,\ldots,x_m)$$
$$= k\prod_{i=1}^{m} f_X(x_i)[1 - F_X(x_i)]^{r_i}, \quad x_1 < \cdots < x_m, \tag{3}$$

where $k = n(n - r_1 - 1)\cdots(n - r_1 - \cdots - r_{m-1} - m + 1)$. Thus, for a progressive Type II with predetermined number

of removals $R = r$, the conditional likelihood and Log-likelihood functions, respectively, can be written as

$$L(\alpha,\beta,\theta \mid R = r)$$
$$= \frac{k\alpha^m}{\beta^m}\prod_{i=1}^{m}\left[\frac{\beta}{x_i - \theta}\right]^{(\alpha+1)}$$
$$\times \left[1 - \exp\left(-\left[\frac{\beta}{x_i - \theta}\right]^{\alpha}\right)\right]^{r_i} \exp\left(-\left[\frac{\beta}{x_i - \theta}\right]^{\alpha}\right), \tag{4}$$

$$L^*(\alpha,\beta,\theta \mid R = r)$$
$$= \ln k + m\ln\alpha + m\alpha\ln\beta - (\alpha+1)\sum_{i=1}^{m}\ln(x_i - \theta)$$
$$- \beta^\alpha\sum_{i=1}^{m}(x_i - \theta)^{-\alpha} + \sum_{i=1}^{m}r_i\ln\left[1 - \exp\left(-\left[\frac{\beta}{x_i - \theta}\right]^{\alpha}\right)\right], \tag{5}$$

where k is defined in (3). Now, suppose that an individual unit being removed from the life test is independent of the others but with the same probability p. Then, the number of units removed at each failure time follows a binomial distribution such that

$$P(R_1 = r_1) = \binom{n-m}{r_1} p^{r_1}(1-p)^{n-m-r_1}, \tag{6}$$

where $0 \leq r_1 \leq n$

$$P(R_i = r_i \mid R_{i-1} = r_{i-1}, \ldots, R_1 = r_1)$$
$$= \binom{n - m - \sum_{k=1}^{i-1}r_k}{r_i} \times p^{r_i}(1-p)^{n-m-\sum_{k=1}^{i-1}r_k}, \tag{7}$$

where $0 \leq r_i \leq n - m - \sum_{k=1}^{i}r_k$, $i = 2,\ldots,m-1$. Suppose further that R_i is independent of X_i. Then, the Log-likelihood function can be expressed as

$$L^*(\alpha,\beta,\theta) = L^*(\alpha,\beta,\theta \mid R = r)P(R = r), \tag{8}$$

where

$$P(R = r)$$
$$= P(R_{m-1} = r_{m-1} \mid R_{m-2} = r_{m-2}, \ldots, R_1 = r_1) \tag{9}$$
$$\cdots P(R_2 = r_2 \mid R_1 = r_1)P(R_1 = r_1),$$

$$P(R = r)$$
$$= \frac{(n-m)!}{\prod_{i=1}^{m-1}r_i!\left(n - m - \sum_{i=1}^{m-1}r_i\right)!} \tag{10}$$
$$\times p^{\sum_{i=1}^{m-1}r_i}(1-p)^{(m-1)(n-m)-\sum_{i=1}^{m-1}(m-i)r_i}.$$

Independently, the MLE of parameter p can be obtained by maximizing (10). Thus, we find immediately

$$\hat{p} = \frac{\sum_{i=1}^{m-1} r_i}{(m-1)(n-m) - \sum_{i=1}^{m-1}(m-i-1)r_i}. \quad (11)$$

Note that $P(R = r)$ does not depend on the parameters α, β, and θ, and hence the MLE's of the parameters can be derived from (5) by differentiating with respect to α, β, and θ and equating to zero, in this case we have

$$m - \sum_{i=1}^{m}\left[\frac{\beta}{x_i - \theta}\right]^{\alpha}$$

$$+ \sum_{i=1}^{m}\frac{r_i[\beta/(x_i - \theta)]^{\alpha}\exp\left(-[\beta/(x_i - \theta)]^{\alpha}\right)}{1 - \exp\left(-[\beta/(x_i - \theta)]^{\alpha}\right)} = 0,$$

$$\sum_{i=1}^{m}\left(\frac{1+\alpha}{\alpha}\left[\frac{\beta}{x_i - \theta}\right] - \left[\frac{\beta}{x_i - \theta}\right]^{(\alpha+1)}\right.$$

$$\left. + \frac{r_i[\beta/(x_i - \theta)]^{(\alpha+1)}\exp\left(-[\beta/(x_i - \theta)]^{\alpha}\right)}{1 - \exp\left(-[\beta/(x_i - \theta)]^{\alpha}\right)}\right) = 0,$$

$$\frac{m}{\alpha} + m\ln\beta$$

$$+ \sum_{i=1}^{m}\left(-\left[\frac{\beta}{x_i - \theta}\right]^{\alpha}\ln\left[\frac{\beta}{x_i - \theta}\right]\right.$$

$$\left. + \frac{r_i\ln[\beta/(x_i - \theta)]\exp\left(-[\beta/(x_i - \theta)]^{\alpha}\right)}{1 - \exp\left(-[\beta/(x_i - \theta)]^{\alpha}\right)}\right) = 0,$$

$$(12)$$

since (12) cannot be solved analytically for estimators $\hat{\alpha}$, $\hat{\beta}$, and $\hat{\theta}$. Hence, in this paper, the numerical solution is used to solve the problem.

4. Some Further Results

In this section, we are going to derive the sampling distributions of the MLE's and obtain the confidence intervals for the parameters [13–15]. In addition, we will obtain the MLE and confidence interval for the percentile of failure time distribution.

4.1. Distributions and Confidence Intervals for the Parameters. Let $X_1 < X_2 < \cdots < X_{m-1} < X_m$ denote a progressively Type II censored sample from a three-parameter Frèchet distribution with censored scheme $R = (r_1, \ldots, r_m)$. Let $Y_i = [(x_i - \theta)/\beta]^{-\alpha}$, $i = 1, \ldots, m$. It can be seen that, $Z_1 < Z_2 < \cdots < Z_m$ is a progressively Type II censored sample from

an exponential distribution with mean 1. Let us consider the following transformation

$$Z_1 = nY_1$$

$$Z_2 = (n - r_1 - 1)(Y_2 - Y_1)$$

$$\vdots \quad \vdots \quad (13)$$

$$Z_m = (n - r_1 - \cdots - r_{m-1} - m + 1)(Y_m - Y_{m-1}).$$

Thomas and Wilson [16] showed that the generalized spacings $Z_1, Z_2, \ldots, Z_m$ as defined in the previous equation (13) are all independent and identically distributed as standard exponential. $V = 2Z_1 = 2n[(x_i - \theta)/\beta]^{-\alpha}$ has a Chi-square distribution with 2 degrees of freedom. We can also write the numerator of $\hat{\beta}$ as the sum of $m - 1$ independent generalized spacings, that is, $2\sum_{i=1}^{m}(r_i+1)Y_i - nY_1 = 2\sum_{i=2}^{m}Z_i$. Therefore, we can find that, conditionally on a fixed set of $R = (r_1, \ldots, r_m)$, $U = 2\sum_{i=1}^{m}(r_i + 1)Y_i - nY_1$ has a Chi-square distribution with $2m - 2$ degrees of freedom. It is also easily seen that V and U are independent. Let

$$T_1 = \frac{U}{(m-1)V} = \frac{\sum_{i=1}^{m}(r_i+1)Y_i - nY_1}{n(m-1)Y_1},$$

$$T_2 = U + V = 2\sum_{i=1}^{m}(r_i+1)Y_i. \quad (14)$$

It is easy to show that, T_1 has an F distribution with $2m - 2$ and 2 degrees of freedom, and T_2 has a Chi-square distribution with $2m$ degrees of freedom. Furthermore, T_1 and T_2 are independent.

Theorem 1. *Suppose that $w_1 < w_2 < \cdots < w_m$. Let*

$$T_1(w) = \frac{\sum_{i=1}^{m}(r_i+1)(w_i - \theta)^{-\alpha} - n(w_1 - \theta)^{-\alpha}}{n(m-1)(w_1 - \theta)^{-\alpha}}. \quad (15)$$

Then, $T_1(w)$ is strictly increasing in w for any $w > 0$, furthermore, if $t > 0$, the equation $T_1(w) = t$ has unique solution for any $w > 0$.

Suppose that, X_i, $i = 1, \ldots, m$ are order statistics of a progressively Type II censored sample of size n from the Frèchet distribution, with censoring scheme $(r_1, \ldots, r_m)$ the a $100(1 - \phi)$ confidence interval for α is

$$\Psi\left(X_1, \ldots, X_m, F_{\phi/2}(2m-2, 2)\right)$$

$$< \alpha < \Psi\left(X_1, \ldots, X_m, F_{1-\phi/2}(2m-2, 2)\right). \quad (16)$$

Such that,

$$1 - \phi$$

$$= P\Big[F_{\phi/2}(2m - 2, 2) < T_1 < F_{1-\phi/2}(2m - 2, 2)\Big]$$

$$= P\Big[F_{\phi/2}(2m - 2, 2) \tag{17}$$

$$< \frac{\sum_{i=1}^m (r_i + 1)(x_i - \theta)^{-\alpha} - n(x_1 - \theta)^{-\alpha}}{n(m - 1)(x_1 - \theta)^{-\alpha}}$$

$$< F_{1-\phi/2}(2m - 2, 2)\Big].$$

Furthermore, the $100(1 - \phi)$ joint confidence region for α and β is determined by the following inequalities:

$$\Psi\Big(X_1, \ldots, X_m, F_{(1+\sqrt{1-\phi})/2}(2m - 2, 2)\Big),$$

$$< \alpha < \Psi\Big(X_1, \ldots, X_m, F_{(1-\sqrt{1-\phi})/2}(2m - 2, 2)\Big), \tag{18}$$

$$\frac{\chi^2_{(1+\sqrt{1+\phi})/2}(2m)}{\Big[2\sum_{i=1}^m (r_i + 1)(x_i - \theta)^\alpha\Big]^{1/\alpha}} < \beta < \frac{\chi^2_{(1+\sqrt{1-\phi})/2}(2m)}{\Big[2\sum_{i=1}^m (r_i + 1)(x_i - \theta)^\alpha\Big]^{1/\alpha}}. \tag{19}$$

Such that,

$$1 - \phi = \sqrt{1 - \phi}\sqrt{1 - \phi}$$

$$= \Big[PF_{(1+\sqrt{1-\phi})/2}(2m - 2, 2) < T_1$$

$$< F_{(1-\sqrt{1-\phi})/2}(2m - 2, 2)\Big] \tag{20}$$

$$\times P\Big[\chi^2_{(1+\sqrt{1-\phi})/2}(2m) < T_2 < \chi^2_{(1-\sqrt{1-\phi})/2}(2m)\Big].$$

4.2. Confidence Interval of x_F. In reliability analysis [17, 18], we are not only interested in making inference about parameter but also interested in deriving inference about percentiles of the failure time distribution. Let x_F be the 100_Fth percentile of the failure time distribution. One can obtain x_F by solving $F_X(x_F) = p$ where $F_X(\cdot)$ is as given in (2). Then it easy to see that $x_F = \theta - \beta(\log p)^{-1/\alpha}$. Consequently, the MLE of x_F is given by $x_F = \hat{\theta} - \hat{\beta}(\log p)^{-1/\hat{\alpha}}$. Confidence interval for x_F can be derived by using the pivotal quantity $U_F = [(x_F - \hat{\theta})/\hat{\beta}]^{-\hat{\alpha}}$ if $(u_{f,1-\phi})^{-\alpha}$ is the $(1 - \phi)$th percentile of U_F, then

$$1 - \phi = P\Big[U_F \le u_{F,1-\phi}\Big] = P\Big[x_F \ge \hat{\theta} - \hat{\beta}u_{F,1-\phi}\Big]. \tag{21}$$

Hence, $\hat{\theta} - \hat{\beta}u_{F,1-\phi}$ is a lower confidence limit for x_F with confidence coefficient $1 - \phi$. We can rewrite (21) as

$$1 - \phi = P\Big[\frac{V}{2n} + \lambda U \le -\ln p\Big], \tag{22}$$

such that, $V \sim \chi^2(2)$, $U \sim \chi^2(2m - 2)$, and $\lambda = -u_{F,1-\phi}/2m$. In addition, V and U are independent, which is discussed in the previous subsection. Hence, we need to find the value of λ by using solving (22). There are several methods for computing λ, see Engelhardt and Bain [19].

5. Conclusions

We develop some results on a three-parameter Frèchet distribution when progressive Type II censoring with binomial removals is performed. We derive the MLEs and confidence for the parameters. The MLE and confidence interval for the percentiles of failure time distribution are obtained. In practice, it is often useful to have an idea of the duration of a life test. Therefore, it is important to compute the expected time required to complete a life test. In the case of progressively type II-censored sampling plan with binomial removals, one can obtain this information by calculating the expectation of the mth order statistic. In fact, we believe that the value of removal probability p is very important when we compare the expected test times of progressive Type II censoring with binomial removals and complete sampling plan. In addition, the removal probability p may not be fixed for each stage. Such belief is not discussed in this paper and we will be investigated in the future.

List of Symbols

$f(x)$: Probability density function
$F(x)$: Cumulative distribution function
α, β, θ: Three-parameter Frèchet distribution
R: Number of removals
p: Removal probability
χ^2: Chi-square
F: F-distribution
$100(1 - \phi)$: Confidence interval
X_F: 100-th percentile of the failure time distribution.

References

[1] S. Nadarajah and S. Kotz, "The Exponentiated Frechet Distribution," Interstat Electronic Journal, 2003, http://interstat.statjournals.net/YEAR/2003/articles/.

[2] A. M. Abd-Elfattah and A. M. Omima, "Estimation of the unknown parameters of the generalized Frechet distribution," *Journal of Applied Sciences Research*, vol. 5, no. 10, pp. 1398–1408, 2009.

[3] S. Kotz and S. Nadarajah, *Extreme Value Distributions: Theory and Applications*, Imperial College Press, London, UK, 2000.

[4] N. R. Mann, R. E. Schafer, and N. D. Singpurwalla, *Methods for Statistical Analysis of Reliability and Life Data*, John Wiley & Sons, New York, NY, USA, 1974.

[5] J. F. Lawless, "Statistical Methods and Methods for Lifetime Data, John Wiley & Sons, New York, NY, USA, 1982.

[6] W. Q. Meeker and L. A. Escobar, *Statistical Methods for Reliability Data*, John Wiley & Sons, New York, NY, USA, 1998.

[7] N. Balakrishnan and R. Aggarwala, *Progressive Censoring—Theory, Moethods and Applications*, Birkhäuser, Boston, Mass, USA, 2000.

[8] A. C. Cohen, "Progressively censored samples in life testing," *Technometrics*, vol. 5, pp. 327–339, 1976.

[9] N. R. Mann, "Best liner invariant estimation for Weibull parameters under progressive censoring," *Technometrics*, vol. 13, pp. 521–533, 1971.

[10] R. Viveros and N. Balakrishnan, "Interval estemation of parameters of life from progressivve censoring data," *Technometrics*, vol. 36, pp. 84–91, 1994.

[11] H. K. Yuen and S. K. Tse, "Parameters estimation for weibull distributed lifetimes under progressive censoring with random removals," *Journal of Statistical Computation and Simulation*, vol. 55, no. 1-2, pp. 57–71, 1996.

[12] S. K. Tae, C. Yang, and H. K. Yuen, "Statistical analysis of Weibull distributed lifetime data under Type II prograssive censoring with binomial removals," *Journal of Applied Statistics*, vol. 27, pp. 1033–1043, 2000.

[13] K. Alakuş, "Confidence intervals estimation for survival function in weibull proportional hazards regression based on censored survival time data," *Scientific Research and Essays*, vol. 5, no. 13, pp. 1589–1594, 2010.

[14] M. Maswadah, "Conditional confidence interval estimation for the inverse weibull distribution based on censored generalized order statistics," *Journal of Statistical Computation and Simulation*, vol. 73, no. 12, pp. 887–898, 2003.

[15] J. A. Griggs and Y. Zhang, "Determining the confidence intervals of Weibull parameters estimated using a more precise probability estimator," *Journal of Materials Science Letters*, vol. 22, no. 24, pp. 1771–1773, 2003.

[16] D. R. Thomas and W. M. Wilson, "Linear order statistic estimation for the two parameter Weibull and extreme value distribution from type -II progressively censored samples," *Technometrics*, vol. 14, pp. 679–691, 1972.

[17] M. Han, "Estimation of failure probability and its applications in lifetime data analysis," *International Journal of Quality, Statistics, and Reliability*, vol. 2011, Article ID 719534, 6 pages, 2011.

[18] S. Loehnert, "About statistical analysis of qualitative survey data," *International Journal of Quality, Statistics, and Reliability*, vol. 2010, Article ID 849043, 12 pages, 2010.

[19] M. Engelhardt and L. J. Bain, "Tolerance limits and confidence limits on reliability for the two parameter exponential distribution," *Technometrics*, vol. 20, pp. 37–39, 1978.

7

On Bayesian Analysis of Burr Type VII Distribution under Different Censoring Schemes

Navid Feroze[1] and Muhammad Aslam[2]

[1] Department of Mathematics and Statistics, AIOU, Islamabad 44000, Pakistan
[2] Department of Statistics, Quaid-i-Azam University, Islamabad 44000, Pakistan

Correspondence should be addressed to Navid Feroze, navidferoz@hotmail.com

Academic Editor: Hongzhong Huang

This paper includes the Bayesian analysis of Burr type VII distribution. Three censoring schemes, namely, left censoring, singly type II censoring, and doubly type II censoring have been used for posterior estimation. The results of different censoring schemes have been compared with those under complete samples. The comparative study among the performance of different censoring schemes has also been made. Two noninformative (uniform and Jeffreys) priors have been assumed to derive the posterior distributions under each case. The performance of Bayes estimators has been compared in terms of posterior risks under a simulation study.

1. Introduction

Burr [1] introduced twelve forms of the Burr distribution. However, most of the authors have considered the estimation of Burr type XII distribution. The Burr type VII distribution has rarely received any attention. Wahed [2] presented Bayes estimators for the parameters of Burr type XII distribution under the symmetric squared error loss function and the asymmetric linear exponential loss function based on a simple prior distribution. As the estimator turns out to be ratios of integrals, different approximation techniques have been used to obtain approximate Bayes estimators. Real life example has been used to demonstrate the application of Burr type XII distribution. Dasgupta [3] discussed that under certain conditions, the distribution of Burr can be shown to follow an extreme value distribution. Hence, a result on extremal process based on stationary sequence has been proved. Some data sets have been analyzed, and applications of the results have been indicated. Makhdoom and Jafari [4] obtained Bayesian estimators for the shape parameter of the Burr Type XII distribution using grouped and ungrouped data and also consider relationship between them. Bayes point and interval estimators have been derived. Squared error and precautionary loss functions have been

considered for the posterior analysis. Monte Carlo simulation has been used to compare the performance of different estimators. Panahi and Asadi [5] considered the statistical inferences based on a Type-II hybrid censored sample from a Burr type XII distribution. As the maximum likelihood estimators cannot be obtained in closed form, a simple fixed point type algorithm has been proposed to compute the maximum likelihood estimators. The approximate confidence intervals for the parameters based on the s-normal approximation to the asymptotic distribution of MLE have been constructed. Bayes estimates of the unknown parameters have also been obtained under the Linex loss function using two approximations. Monte Carlo simulations have been performed to observe the behavior of the proposed methods.

The probability density function of Burr type VII distribution is

$$f(x) = \theta \, 2^{-\theta} \text{sech}^2(x)\{1 + \tanh(x)\}^{\theta-1},$$
$$-\infty < x < \infty, \ \theta > 0. \tag{1}$$

And the cumulative distribution function of the distribution is

$$F(x) = 2^{-\theta}\{1 + \tanh(x)\}^{\theta}. \tag{2}$$

The probability density function and cumulative distribution function of the Burr type VII distribution can be, respectively, presented as

$$f(x) = \theta \, \text{sech}^2(x) e^{-\ln\{1+\tanh(x)\}} e^{-\theta \ln\{(1+\tanh(x))/2\}^{-1}}, \tag{3}$$

$$F(x) = e^{-\theta \ln\{(1+\tanh(x))/2\}^{-1}}.$$

Under inverse transformation method of simulation the random numbers can be generated from the distribution by using the following formula, where U is uniformly distributed random variable:

$$x = \tanh^{-1}\left\{\left(\frac{u}{2^{-\theta}}\right)^{1/\theta} - 1\right\}, \tag{4}$$

where $\tanh^{-1}(x) = (1/2)\ln((1+x)/(1-x))$.

2. Posterior Analysis under Complete Data

In this section, the Bayes estimators along with posterior risks have been derived under the assumption of uniform and Jeffreys priors using complete data. Squared error loss function (SELF) and precautionary loss function (PLF) have been used for estimation. In order to derive the Bayes estimators and corresponding risks, the first step is to obtain the likelihood function. The likelihood function for a sample of size "n" observation is

$$L(\theta \mid \underline{x}) \propto \theta^n e^{-\theta T}, \tag{5}$$

where $T = \sum_{i=1}^{n} \ln\{(1 + \tanh(x_i))/2\}^{-1}$.

The uniform prior is assumed to be

$$p(\theta) \propto 1; \quad \theta > 0. \tag{6}$$

The posterior distribution under the assumption of uniform prior is

$$p(\theta \mid \underline{x}) = \frac{[T]^{n+1}}{\Gamma(n+1)} \theta^n e^{-\theta T}, \quad \theta > 0. \tag{7}$$

The Bayes estimators and corresponding posterior risks under squared error loss function (SELF) can be derived by using the following formulae:

$$\theta_{\text{SELF}} = E(\theta \mid \underline{x}),$$

$$\rho(\theta_{\text{SELF}}) = E(\theta^2 \mid \underline{x}) - \{E(\theta \mid \underline{x})\}^2. \tag{8}$$

Bayes estimator and posterior risk under uniform prior using SELF are

$$\theta_{\text{SELF}} = \frac{n+1}{T}, \qquad \rho(\theta_{\text{SELF}}) = \frac{n+1}{T^2}. \tag{9}$$

Similarly, The Bayes estimators and corresponding posterior risks under precautionary loss function (PLF) can be derived by using the following formulae:

$$\theta_{\text{PLF}} = \{E(\theta^2 \mid \underline{x})\}^{1/2},$$

$$\rho(\theta_{\text{PLF}}) = 2\left[\{E(\theta^2 \mid \underline{x})\}^{1/2} - E(\theta \mid \underline{x})\right]. \tag{10}$$

The derivations of Bayes estimators and associated posterior risks using precautionary loss function have not been presented in the paper. These can easily be derived from the concerned posterior distributions.

The Jeffreys prior has been derived to be

$$p(\theta) \propto \frac{1}{\theta}, \quad \theta > 0. \tag{11}$$

The posterior distribution under the assumption of Jeffreys prior is

$$p(\theta \mid \underline{x}) = \frac{T^n}{\Gamma(n)} \theta^{n-1} e^{-\theta T}, \quad \theta > 0. \tag{12}$$

Bayes estimator and risk under Jeffreys prior using SELF are

$$\theta_{\text{SELF}} = \frac{n}{T}, \qquad \rho(\theta_{\text{SELF}}) = \frac{n}{T^2}. \tag{13}$$

3. Posterior Analysis under Left Censored Samples

Let $X_{r+1}, \ldots, X_n$ be last $n - r$ order statistics from a sample of size n from Burr type VII distribution. Then the likelihood function for the $X_{r+1}, \ldots, X_n$ observations is

$$L(\theta \mid \underline{x}) \propto \{F(x_{r+1})\}^r \prod_{i=r+1}^{n} f(x_i), \tag{14}$$

$$L(\theta \mid \underline{x}) \propto \theta^{n-r} e^{-\theta \xi_{1r}},$$

where

$$\xi_{1r} = r \ln\left\{\frac{1 + \tanh(x_{r+1})}{2}\right\}^{-1} + \sum_{i=r+1}^{n} \ln\left\{\frac{1 + \tanh(x_i)}{2}\right\}^{-1}. \tag{15}$$

The posterior distribution under uniform prior is

$$p(\theta \mid \underline{x}) = \frac{(\xi_{1r})^{n-r+1}}{\Gamma(n-r+1)} \theta^{n-r} e^{-\theta \xi_{1r}}. \tag{16}$$

The Bayes estimator and posterior risk under uniform prior using SELF are

$$\theta_{\text{SELF}} = \frac{n - r + 1}{\xi_{1r}}, \qquad \rho(\theta_{\text{SELF}}) = \frac{n - r + 1}{(\xi_{1r})^2}. \qquad (17)$$

The posterior distribution under Jeffreys prior is

$$p(\theta \mid \underline{x}) = \frac{(\xi_{1r})^{n-r}}{\Gamma(n - r)} \theta^{n-r-1} e^{-\theta \xi_{1r}}. \qquad (18)$$

The Bayes estimator and posterior risk under Jeffreys prior using SELF are

$$\theta_{\text{SELF}} = \frac{n - r}{\xi_{1r}}, \qquad \rho(\theta_{SELF}) = \frac{n - r}{(\xi_{1r})^2}. \qquad (19)$$

4. Posterior Analysis under Singly Type II Censored Samples

Suppose "n" items are put on a life-testing experiment, and only first "m" failure times have been observed, that is, $x_1 < x_2 \cdots < x_m$ and remaining "$n - m$" items are still working. Under the assumptions that the lifetimes of the items are independently and identically distributed Burr type VII random variable, the likelihood function of the observed data, is

$$L(\theta \mid \underline{x}) \propto \left[\prod_{i=1}^{m} f(x_i) \right] [1 - F(x_m)]^{n-m}, \qquad (20)$$

$$L(\theta \mid \underline{x}) \propto \sum_{j=0}^{n-m} (-1)^j \binom{n - m}{j} \theta^m e^{-\theta \psi_{1j}},$$

where

$$\psi_{1j} = \sum_{i=1}^{m} \ln \left\{ \frac{1 + \tanh(x_i)}{2} \right\}^{-1} + j \ln \left\{ \frac{1 + \tanh(x_m)}{2} \right\}^{-1}. \qquad (21)$$

The posterior distribution under uniform prior is

$$p(\theta \mid \underline{x}) = \frac{1}{C_1} \sum_{j=0}^{n-m} (-1)^j \binom{n - m}{j} \theta^m e^{-\theta \psi_{1j}}, \qquad (22)$$

where $C_1 = \sum_{j=0}^{n-m} (-1)^j \binom{n-m}{j} (\Gamma(m+1)/(\psi_{1j})^{m+1})$.

The Bayes estimator and posterior risk under uniform prior using SELF are

$$\theta_{\text{SELF}} = \frac{1}{C_1} \sum_{j=0}^{n-m} (-1)^j \binom{n - m}{j} \frac{\Gamma(m + 2)}{(\psi_{1j})^{m+2}},$$

$$\rho(\theta_{\text{SELF}}) = \frac{1}{C_1} \sum_{j=0}^{n-m} (-1)^j \binom{n - m}{j} \frac{\Gamma(m + 3)}{(\psi_{1j})^{m+3}} \qquad (23)$$

$$- \left\{ \frac{1}{C_1} \sum_{j=0}^{n-m} (-1)^j \binom{n - m}{j} \frac{\Gamma(m + 2)}{(\psi_{1j})^{m+2}} \right\}^2.$$

The posterior distribution under Jeffreys prior is

$$p(\theta \mid \underline{x}) = \frac{1}{C_2} \sum_{j=0}^{n-m} (-1)^j \binom{n - m}{j} \theta^{m-1} e^{-\theta \psi_{1j}}, \qquad (24)$$

where $C_2 = \sum_{j=0}^{n-m} (-1)^j \binom{n-m}{j} (\Gamma(m)/(\psi_{1j})^m)$.

The Bayes estimator and posterior risk under Jeffreys prior using SELF are

$$\theta_{\text{SELF}} = \frac{1}{C_1} \sum_{j=0}^{n-m} (-1)^j \binom{n - m}{j} \frac{\Gamma(m + 1)}{(\psi_{1j})^{m+1}},$$

$$\rho(\theta_{\text{SELF}}) = \frac{1}{C_1} \sum_{j=0}^{n-m} (-1)^j \binom{n - m}{j} \frac{\Gamma(m + 2)}{(\psi_{1j})^{m+2}}$$

$$- \left\{ \frac{1}{C_1} \sum_{j=0}^{n-m} (-1)^j \binom{n - m}{j} \frac{\Gamma(m + 1)}{(\psi_{1j})^{m+1}} \right\}^2. \qquad (25)$$

5. Posterior Analysis under Doubly Type II Censored Samples

Consider a random sample of size "n" from an Burr type X distribution, and let $x_r, \ldots, x_s$ be the ordered observations remaining when the "$r - 1$" smallest observations and the "$n - s$" largest observations have been censored. The likelihood function for θ given the type II doubly censored sample $\underline{x} = (x_r, \ldots, x_s)$ is

$$L(\theta \mid \underline{x}) \propto [F(x_r \mid \theta)]^{r-1} [1 - F(x_s \mid \theta)]^{n-s} \prod_{i=r}^{s} f(x_i \mid \theta),$$

$$L(\theta \mid \underline{x}) \propto \sum_{j=0}^{n-s} (-1)^j \binom{n - s}{j} \theta^k e^{-\theta \psi_{2j}},$$

$$\psi_{2j} = (r - 1) \ln \left\{ \frac{1 + \tanh(x_r)}{2} \right\}^{-1}$$

$$+ \sum_{i=r}^{s} \ln \left\{ \frac{1 + \tanh(x_i)}{2} \right\}^{-1} + j \ln \left\{ \frac{1 + \tanh(x_s)}{2} \right\}^{-1}, \qquad (26)$$

and $k = s - r + 1$.

The posterior distribution under uniform prior is

$$p(\theta \mid \underline{x}) = \frac{1}{C_3} \sum_{j=0}^{n-s} (-1)^j \binom{n-s}{j} \theta^k e^{-\theta \psi_{2j}}, \qquad (27)$$

where $C_3 = \sum_{j=0}^{n-s} (-1)^j \binom{n-s}{j} (\Gamma(k+1)/(\psi_{2j})^{k+1})$.

The Bayes estimator and posterior risk under uniform prior using SELF are

$$\theta_{\text{SELF}} = \frac{1}{C_3} \sum_{j=0}^{n-s} (-1)^j \binom{n-s}{j} \frac{\Gamma(k+2)}{(\psi_{2j})^{k+2}},$$

$$\rho(\theta_{\text{SELF}}) = \frac{1}{C_3} \sum_{j=0}^{n-s} (-1)^j \binom{n-s}{j} \frac{\Gamma(k+3)}{(\psi_{2j})^{k+3}}$$
$$- \left\{ \frac{1}{C_3} \sum_{j=0}^{n-s} (-1)^j \binom{n-s}{j} \frac{\Gamma(k+2)}{(\psi_{2j})^{k+2}} \right\}^2.$$
$$(28)$$

The posterior distribution under Jeffreys prior is

$$p(\theta \mid \underline{x}) = \frac{1}{C_4} \sum_{j=0}^{n-s} (-1)^j \binom{n-s}{j} \theta^{k-1} e^{-\theta \psi_{2j}}, \qquad (29)$$

where $C_4 = \sum_{j=0}^{n-s} (-1)^j \binom{n-s}{j} (\Gamma(k)/(\psi_{2j})^k)$.

The Bayes estimator and posterior risk under Jeffreys prior using SELF are

$$\theta_{\text{SELF}} = \frac{1}{C_4} \sum_{j=0}^{n-s} (-1)^j \binom{n-s}{j} \frac{\Gamma(k+1)}{(\psi_{2j})^{k+1}},$$

$$\rho(\theta_{\text{SELF}}) = \frac{1}{C_4} \sum_{j=0}^{n-s} (-1)^j \binom{n-s}{j} \frac{\Gamma(k+2)}{(\psi_{2j})^{k+2}}$$
$$- \left\{ \frac{1}{C_4} \sum_{j=0}^{n-s} (-1)^j \binom{n-s}{j} \frac{\Gamma(k+1)}{(\psi_{2j})^{k+1}} \right\}^2.$$
$$(30)$$

6. Simulation Study

Simulation study is a useful technique to assess and compare the performance of different estimators numerically. A simulation study often reflects the patterns of the real life data. Here, the inverse transformation technique of simulation has been used for $n = 50, 70, 100$, and 150 under the parametric space $\theta \in (3, 6)$. As a single sample cannot fully describe the behaviors and properties of the estimators, the samples have been replicated 1000 times, and an average of the results has been presented. The magnitude of posterior risks, associated with each estimator, has been underlined in the tables.

TABLE 1: Bayes estimators and risks under uniform prior using SELF.

Sample size	Uniform prior ($\theta = 3$)			
	CD	LC	S2C	D2C
50	3.0503	3.2268	3.3793	2.9098
	0.1824	0.1829	0.2278	0.1852
70	3.0108	3.1851	3.3382	2.8617
	0.1277	0.1280	0.1596	0.1288
100	3.0388	3.2147	3.3712	2.8854
	0.0914	0.0917	0.1144	0.0921
150	3.0082	3.1823	3.3260	2.8514
	0.0599	0.0601	0.0744	0.0601

CD: complete data, LC: left censored data, S2C: singly type II censored data, D2C: doubly type II censored data.

TABLE 2: Bayes estimators and risks under uniform prior using PLF.

Sample size	Uniform prior ($\theta = 3$)			
	CD	LC	S2C	D2C
50	3.0800	3.2583	3.4128	2.9414
	0.0595	0.0630	0.0671	0.0633
70	3.0319	3.2074	3.3620	2.8841
	0.0423	0.0447	0.0476	0.0448
100	3.0538	3.2305	3.3881	2.9013
	0.0300	0.0318	0.0339	0.0318
150	3.0181	3.1928	3.3372	2.8620
	0.0199	0.0210	0.0223	0.0211

CD: complete data, LC: left censored data, S2C: singly type II censored data, D2C: doubly type II censored data.

TABLE 3: Bayes estimators and risks under Jeffreys prior using SELF.

Sample size	Jeffreys prior ($\theta = 3$)			
	CD	LC	S2C	D2C
50	2.9905	3.1636	3.3119	2.8461
	0.1789	0.1793	0.2231	0.1811
70	2.9684	3.1402	3.2904	2.8167
	0.1259	0.1262	0.1572	0.1267
100	3.0087	3.1828	3.3372	2.8535
	0.0905	0.0908	0.1132	0.0910
150	2.9883	3.1613	3.3037	2.8303
	0.0595	0.0597	0.0738	0.0597

CD: complete data, LC: left censored data, S2C: singly type II censored data, D2C: doubly type II censored data.

TABLE 4: Bayes estimators and risks under Jeffreys prior using PLF.

Sample size	Jeffreys prior ($\theta = 3$)			
	CD	LC	S2C	D2C
50	3.0800	3.2583	3.3454	2.8778
	0.1791	0.1895	0.0670	0.0633
70	3.0319	3.2074	3.3142	2.8391
	0.1271	0.1344	0.0476	0.0448
100	3.0538	3.2305	3.3542	2.8694
	0.0902	0.0954	0.0338	0.0318
150	3.0181	3.1928	3.3149	2.8409
	0.0597	0.0632	0.0223	0.0211

CD: complete data, LC: left censored data, S2C: singly type II censored data, D2C: doubly type II censored data.

TABLE 5: Bayes estimators and risks under uniform prior using SELF.

Sample size	Uniform prior ($\theta = 6$)			
	CD	LC	S2C	D2C
50	5.9775	6.3235	6.6528	5.6995
	0.7006	0.7025	0.8830	0.7109
70	5.9814	6.3276	6.6728	5.7147
	0.5039	0.5053	0.6381	0.5138
100	6.0442	6.3940	6.6872	5.7463
	0.3617	0.3627	0.4501	0.3651
150	6.0154	6.3636	6.6541	5.7169
	0.2396	0.2403	0.2979	0.2418

CD: complete data, LC: left censored data, S2C: singly type II censored data, D2C: doubly type II censored data.

TABLE 6: Bayes estimators and risks under uniform prior using PLF.

Sample size	Uniform prior ($\theta = 6$)			
	CD	LC	S2C	D2C
50	6.0358	6.3852	6.7189	5.7615
	0.1166	0.1234	0.1321	0.1241
70	6.0233	6.3720	6.7205	5.7595
	0.0839	0.0888	0.0953	0.0896
100	6.0740	6.4256	6.7208	5.7780
	0.0597	0.0632	0.0671	0.0634
150	6.0353	6.3847	6.6764	5.7380
	0.0398	0.0421	0.0447	0.0422

CD: complete data, LC: left censored data, S2C: singly type II censored data, D2C: doubly type II censored data.

TABLE 7: Bayes estimators and risks under Jeffreys prior using SELF.

Sample size	Jeffreys prior ($\theta = 6$)			
	CD	LC	S2C	D2C
50	5.8603	6.1995	6.5202	5.5748
	0.6869	0.6887	0.8648	0.6952
70	5.8971	6.2385	6.5773	5.6249
	0.4968	0.4981	0.6286	0.5056
100	5.9843	6.3307	6.6200	5.6828
	0.3581	0.3591	0.4454	0.3611
150	5.9756	6.3215	6.6093	5.6746
	0.2381	0.2387	0.2963	0.2400

CD: complete data, LC: left censored data, S2C: singly type II censored data, D2C: doubly type II censored data.

TABLE 8: Bayes estimators and risks under Jeffreys prior using PLF.

Sample size	Jeffreys prior ($\theta = 6$)			
	CD	LC	S2C	D2C
50	6.0358	6.3852	6.5862	5.6368
	0.3510	0.3714	0.1320	0.1240
70	6.0233	6.3720	6.6249	5.6696
	0.2524	0.2671	0.0952	0.0895
100	6.0740	6.4256	6.6535	5.7145
	0.1794	0.1898	0.0671	0.0634
150	6.0353	6.3847	6.6316	5.6957
	0.1194	0.1264	0.0448	0.0422

CD: complete data, LC: left censored data, S2C: singly type II censored data, D2C: doubly type II censored data.

The results of simulation study, presented in Tables 1–8, are obtained under complete and 20% censored samples. It is immediate from the above study that the magnitude of risk decreases with the increase in the sample size irrespective of nature (complete and censored) of the samples. The estimated value of the parameter converges to the true value very rapidly under complete data; however, in case of censored data the convergence is not that good. While, increase in the true parametric value and the censoring rate imposes a negative impact on the convergence and performance of the estimates. It can also be observed that the performance, in terms of convergence and posterior risk, of precautionary loss function is better than squared error loss function. It is because that the posterior distributions are asymmetric. It can also be assessed that the magnitudes of risks associated with estimates under Jeffreys prior are lesser as compared to those under uniform prior. In comparison of censoring schemes it is found that the left censored samples provide the best results. The estimates under doubly censored samples stand at second position performance wise.

7. Conclusions and Recommendations

From the findings of simulation study it can be concluded that in order to estimate the parameter of Burr type VII distribution under a Bayesian framework using censored data, the use of Jeffreys prior, precautionary loss function, and left censored samples can be preferred. The study can further be extended by using some other loss functions; informative priors and mixture of two are more components of Burr type VII distribution. The work on the mixture of two components of Burr type VII distribution is continued.

References

[1] W. I. Burr, "Cumulative frequency distribution," *Annals of Mathematical Statistics*, vol. 13, pp. 215–232, 1942.

[2] A. S. Wahed, "Bayesian inference using Burr model under asymmetric loss function: an application to Carcinoma survival data," *Journal of Statistical Research*, vol. 40, no. 1, pp. 45–57, 2006.

[3] R. Dasgupta, "On the distribution of burr with applications," *Sankhya B*, vol. 73, pp. 1–19, 2011.

[4] I. Makhdoom and A. Jafari, "Bayesian estimations on the Burr type XII distribution using grouped and un-grouped data," *Australian Journal of Basic and Applied Sciences*, vol. 5, no. 6, pp. 1525–1531, 2011.

[5] H. Panahi and S. Asadi, "Analysis of the type-II hybrid censored Burr type XII distribution under LINEX loss function," *Applied Mathematical Sciences*, vol. 5, no. 79, pp. 3929–3942, 2011.

8

A Nonparametric Scheme for Monitoring a Process Output with a Block Effect

Saad T. Bakir

College of Business Administration, Alabama State University, 915 South Jackson Street, Montgomery, AL 36104, USA

Correspondence should be addressed to Saad T. Bakir, bakir00@yahoo.com

Academic Editor: Shuen L. Jeng

This paper proposes a distribution-free (or nonparametric) control scheme to monitor a process output that contains two special causes of variation called "block or batch" effects and "treatment or position" effects. The scheme properties (control limits, false alarm rate, and in-control average run length) stay the same under any assumed continuous probability distribution. For moderate sample sizes, these properties can be computed exactly from available tables without the need to estimate the mean or variance of the process. The proposed monitoring scheme requires ranking the observations within blocks and using the method of analysis of means by ranks. The paper includes an illustrative example concerning the grinding process of silicon wafers used in integrated circuits production.

1. Introduction

In statistical process control, there are instances in which the process output contains block effects component in addition to the treatment effects component that is to be controlled. In manufacturing integrated circuits on silicon wafers, Roes and Does ([1, Table 1]) reported data bearing two effects: the batch (or block) effect and the position (or treatment) effect of the wafer under the grinder. To take account of both effects, they proposed a Shewhart-type control chart that is based on a two-way analysis of variance (ANOVA) model for controlling the process treatment mean and for controlling certain linear contrasts of the wafer positions. In manufacturing paper or plastic films, Palm and DeAmico [2] reported that the raw manufacturing material was formed into a continuous sheet (web) which would then be wound up into rolls at the end of the production line. Because of nonuniformities in the product along the cross-direction perpendicular to the direction of the travel of the web, samples were taken from each roll at different cross-directional positions (e.g., the middle, the front or operator side, and the back or motor side.) Palm and DeAmico [2] modeled the data as a two-way ANOVA setup where rolls served as blocks and cross-directional positions served as treatments. To account for the block effects, they suggested performing periodic cross-directional studies that are based on the analysis of main effects (ANOME) when monitoring the silicon oxide on computer chips, Yashchin [3] reported that from each lot, a sample of wafers is selected and then several measurements are made on each of the selected wafers. Based on a nested random effects model, he developed a cumulative sum (CUSUM) control chart for monitoring the process variance components. Earlier, Woodall and Thomas [4] considered monitoring variance components in a nested random effects model situation.

All the above monitoring schemes are distribution based (or parametric) that assume a normal distribution for the process output. The normality assumption, however, may not be valid for some manufacturing data; citations are given in Bakir [5], Chakraborti and Graham [6], and Chakraborti et al. [7].

The purpose of this paper is to develop a distribution-free (or nonparametric) control scheme for monitoring a process output that is under the influence of two special causes of variation called "block or batch" effects and "treatment or position" effects. While accounting for the block component, the proposed scheme monitors and detects differences in the treatment components of the process output. Our approach involves ranking the raw measurements of the process output within blocks and then using the method of analysis of means by ranks (ANOMR) as developed in Bakir [8]. The proposed scheme has the following merits.

(a) The scheme takes account of the variability in the data that is due to block effects.

(b) The scheme is distribution-free because its properties (control limits, false alarm rate, and in-control average run length) stay the same under any continuous probability distribution.

(c) For moderate sample sizes, the scheme properties can be computed exactly without the need to estimate the mean or the standard deviation of the process. For large sample sizes, large sample approximations are available.

2. A Nonparametric Monitoring Scheme

The purpose of the monitoring scheme is to monitor a process output that contains two special causes of variation called "block or batch" effects and "treatment or position" effects. Examples of such output were cited in the Introduction.

2.1. Sampling Plan. At each time instance we take a sample of n blocks (e.g., rolls of plastic film or batches), then we make k measurements (e.g., at fixed positions on the roll or the batch) on the process output within each of the n blocks. Thus, a total of nk observations x_{tij} ($i = 1, 2, \ldots, n; j = 1, 2, \ldots, k$) is gathered at each sampling time instance t, $t = 1, 2, \ldots$.

2.2. Model and Assumptions. For such x_{tij} data, an appropriate model is the following two-way ANOVA model:

$$X_{tij} = \mu + \beta_{i(t)} + \tau_{tj} + \varepsilon_{tij} \quad \begin{array}{l} \text{for } t = 1, 2, \ldots; \\ i = 1, 2, \ldots, n; \ j = 1, 2, \ldots, k. \end{array} \quad (1)$$

In model (1), at each sampling instance t, $t = 1, 2, \ldots$, we have that X_{tij} is the measurement on the jth treatment in the ith block, μ is an unknown constant representing the process overall mean, $\beta_{i(t)}$ is an unknown constant representing the effect of the ith block nested within time, τ_{tj} is an unknown constant representing the fixed effect of the jth treatment, and ε_{tij} is a random variable having a continuous probability distribution with zero median; it represents common cause variation in the (tij)th observation. The $\varepsilon'_{tij}s$ ($t = 1, 2, \ldots, i = 1, 2, \ldots, n$, and $j = 1, 2, \ldots, k$) are assumed mutually independent.

2.3. Quality Characteristic to Be Monitored. The purpose of the monitoring scheme is to detect differences among the treatment components while accounting for the block effects. This translates to the process being in control if the hypothesis, H_0 (called the in-control hypothesis), holds true for all t, $t = 1, 2, \ldots$

$$H_0 : \tau_{t1} = \tau_{t2} = \cdots = \tau_{tk}. \quad (2)$$

The out-of-control hypothesis is

$$H_1 : \text{the } \tau's \text{ are not all equal.} \quad (3)$$

2.4. Charting Statistic. At each sampling instance t, $t = 1, 2, \ldots$ and for $i = 1, 2, \ldots, n$, let R_{tij} denote the rank of X_{tij} within the ith block $\{X_{ti1}, X_{ti2}, \ldots, X_{tik}\}$. Formally,

$$R_{tij} = 1 + \sum_{s=1}^{j-1} I\left(X_{tis} < X_{tij}\right), \quad (4)$$

where $I(\cdot)$ is the indicator function.

Calculate the treatment rank totals

$$R_{t \cdot j} = \sum_{i=1}^{n} R_{tij} \quad \text{for } t = 1, 2, \ldots, \ j = 1, 2, \ldots, k. \quad (5)$$

It can be verified that under the in-control hypothesis H_0 in (2), the rank totals, $R_{t.j}$, have mean μ and variance σ^2 given by

$$\mu = \frac{n(k+1)}{2}, \qquad \sigma^2 = \frac{n\ (k+1)\ (k-1)}{12}. \quad (6)$$

Define the treatment rank deviations

$$D_{t \cdot j} = R_{t \cdot j} - \mu \quad \text{for } t = 1, 2, \ldots, \ j = 1, 2, \ldots, k. \quad (7)$$

We will use the $D_{t \cdot j}$'s as charting statistics for a follow-up monitoring scheme of the process. Define the maximum absolute rank deviations

$$D_t = \max_{1 \leq j \leq k} \left| D_{t \cdot j} \right| \quad \text{for } t = 1, 2, \ldots. \quad (8)$$

We will use D_t as a charting statistic for an initial monitoring scheme.

2.5. The In-Control Probability Distribution of the Charting Statistic, D_t. At each sampling instance t, $t = 1, 2, \ldots$ and under the in-control hypothesis H_0 in (2), the $k!$ possible rank configurations $(R_{ti1}, R_{ti2}, \ldots, R_{tik})$ are equally likely within each of the n blocks. Since the blocks are independent, there are $(k!)^{n-1}$ equally likely rank configurations at each sampling time instance. Using these facts, Bakir ([8, Table A.1]) tabulates the exact in-control distribution of D_t for the cases when $k = 3$ and $n = 3(1)9$; $k = 4$ and $n = 3(1)6$; and when $k = 5$ and $n = 3, 4$. Bakir's table gives the values for α and $d_{\alpha;k,n}$ satisfying

$$\Pr(D_t \geq d_{\alpha;k,n} \mid H_0) = \alpha, \quad (9)$$

D_t is a nonparametric test statistic because its distribution (under H_0) is the same for any parent continuous process distribution. The large sample in-control distribution of D_t is discussed in Bakir [8] and its upper percentile points are given in Bakir ([9, Table IV]).

2.6. Signaling Rule and Control Limits

2.6.1. Initial Monitoring. First, we start an initial monitoring scheme by plotting the maximum absolute rank deviation, D_t, against time $t = 1, 2, \ldots$. For a desired prechosen probability of false signal, α, the scheme signals at the first sampling time instance t, $t = 1, 2, \ldots$ for which

$$D_t \geq d_{\alpha;k,n}. \quad (10)$$

TABLE 1: Values of UCL, FAR, and ARL_0.

Number of blocks, n	Number of treatments, k								
	$k=3$			$k=4$			$k=5$		
	UCL	FAR	ARL_0	UCL	FAR	ARL_0	UCL	FAR	ARL_0
$n=3$	3	.1944	5.3	4.5	.1181	8.5	5	.2800	3.6
							6	.0075	133.3
$n=4$	4	.0694	14.4	5	.1418	7.1	6	.2120	4.70
				6	.0307	32.6	7	.0769	13.0
							8	.0159	62.9
$n=5$	4	.1242	8.1	5.5	.1452	6.9			
	5	.0239	41.8	6.5	.0451	22.2			
				7.5	.0078	128.2			
$n=6$	4	.1840	5.4	6	.1443	6.9			
	5	.0521	9.2	7	.0518	19.3			
	6	.0081	23.5	8	.0135	74.1			
				9	.0019	526.3			
$n=7$	4	.0854	11.7						
	5	.0207	37.0						
	6	.0027	370.4						
$n=8$	5	.1197	8.4						
	6	.0375	26.7						
	7	.0080	125.0						
	8	.0009							
$n=9$	5	.1540	6.5						
	6	.0570	17.5						
	7	.0158	63.3						
	8	.0030	333.3						
	9	.0003	3333.3						

The initial monitoring scheme has one control limit given by

$$\text{UCL} = d_{\alpha;k,n}, \qquad (11)$$

where $d_{\alpha;k,n}$ is defined in (9).

A signal by this initial monitoring only indicates an out-of-control condition; it does not, however, pinpoint the treatment components that triggered the signal. To achieve this objective, we need to perform a follow-up monitoring.

2.6.2. Follow-Up Monitoring. In the follow-up monitoring scheme we plot the individual rank deviations $D_{t \cdot j}$ ($j = 1, 2, \ldots, k$), in (7), across time. A signal is given at the first time t, $t = 1, 2, \ldots$ for which

$$D_{t \cdot j} \geq d_{\alpha;k,n} \quad \text{or} \quad D_{t \cdot j} \leq -d_{\alpha;k,n},$$
$$\text{for at least one } j = 1, 2, \ldots, k. \qquad (12)$$

The upper and lower control limits for this follow-up monitoring scheme, respectively, are

$$\text{UCL} = d_{\alpha;k,n}, \qquad \text{LCL} = -d_{\alpha;k,n}. \qquad (13)$$

The signaling rules in (10) and (12) are equivalent in the sense that one signals if and only if the other signals too. For large values of n, the value of $d_{\alpha;k,n}$ may be replaced in all

the formulae that appear in this section by $\sigma \omega(\alpha; k)$, where $\omega(\alpha; k)$ are the upper percentile points of the large sample distribution of D_t given in Bakir ([9, Table IV.])

3. The Average Run Length

Let T be a discrete random variable denoting the time (number of sampling time instances) required until the monitoring scheme's first signal; that is, when $D_t \geq d_{\alpha;k,n}$. T is called the run length of the scheme and its expected value, $E(T)$, is called the average run length (ARL). Since the random variables $D_t, t = 1, 2, \ldots$ are independent, T follows a geometric distribution with parameter (probability of a signal), p:

$$p = \Pr(D_t \geq d_{\alpha;k,n}). \qquad (14)$$

By properties of the geometric distribution, we get

$$\text{ARL} = E(T) = \frac{1}{p}. \qquad (15)$$

The monitoring scheme is nonparametric because its in-control (under H_0) run length distribution is geometric with the same parameter p for any parent continuous process distribution. When the process is in control, let ARL_0 denote

ARL and let FAR denote the false alarm rate (probability of a false signal). Then

$$\text{FAR} = \Pr(D_t \geq d_{\alpha;k,n} \mid H_0) = \alpha, \qquad \text{ARL}_0 = \frac{1}{\alpha}. \quad (16)$$

It is clear that ARL_0 and FAR do not depend on the functional form of the probability distribution of the process. Table 1 gives the exact values of FAR, ARL_0, and the corresponding UCL for some values of n and k. Table 1 was generated by using results in Bakir ([8, Table A.1]). For large values of n, one may use the large sample distribution of D_t to compute ARL_0.

4. Illustrative Example

We apply the proposed monitoring scheme to data relating to the grinding process of silicon wafers used in integrated circuits production. The wafers are positioned in a circular arrangement under a grindstone to achieve a certain desired thickness. Thickness measurements (μm) are taken on wafers seated on five fixed positions (labeled P1, P2, P3, P4, and P5) under the grinder. The P1 and P2 positions fall on the outer circle, P3 and P4 in the middle, and P5 in the inner circle. Thickness data of 30 successive batches of wafers are given in Roes and Does ([1, Table 1]). To apply our scheme, we break the data into ten ($t = 10$) time (sampling) instances with three ($n = 3$) batches (blocks) at each time. Measurements are made at the five ($k = 5$) positions (treatments) within each block.

Formulas in (6) give the following values for the mean and variance of the position (treatment) rank totals, $R_{t,j}$:

$$\mu = \frac{3(5+1)}{2} = 9.0, \qquad \sigma^2 = \frac{3(5+1)(5-1)}{12} = 6.0. \quad (17)$$

Table 2 displays the thickness data, assignments of the appropriate ranks, and further computations of the rank totals for the five positions (P1, P2, ..., P5). Using Table 1 with $n = 3$, $k = 5$, and FAR of $\alpha = 0.0075$ (equivalent to $\text{ARL}_0 = 133$), we read the upper control limit: UCL = 6. Figure 1 displays the resulting chart for the initial monitoring scheme and Figures 2 and 3 show the resulting charts for the follow-up monitoring scheme.

Figure 1 shows that, except for the fifth sampling time instance, the successive values of the charting statistic, D_t, fall right on the upper control limit. Thus, the initial monitoring scheme triggers an out-of-control signal indicating significant variation among the wafers' positions. However, Figure 1 does not pinpoint which position or positions caused the signal. We proceed now to perform the follow-up monitoring scheme as shown in Figures 2 and 3.

Figure 2 is a multiline chart showing the individual position profiles across time. This follow-up scheme signals at each time except for the fifth when all positions become within the control limits. Moreover, Figure 2 pinpoints the positions that caused the signal; namely, P1 and P4. Positions 2, 3, and 5 are within the control limits at all sampling times.

The follow-up scheme in Figure 3 is a single-line chart showing the individual position profiles across time. Figure 3

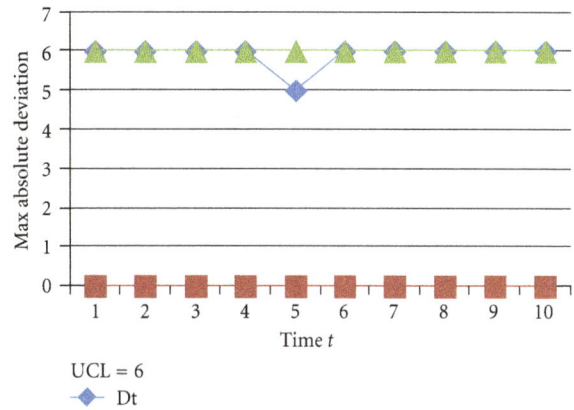
FIGURE 1: Initial Monitoring Chart.

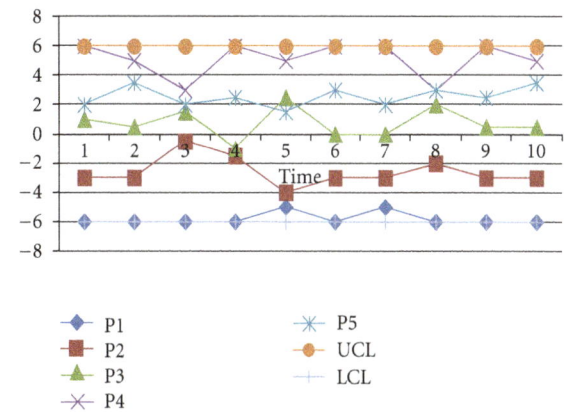
FIGURE 2: Follow-up Multi-line Monitoring Chart.

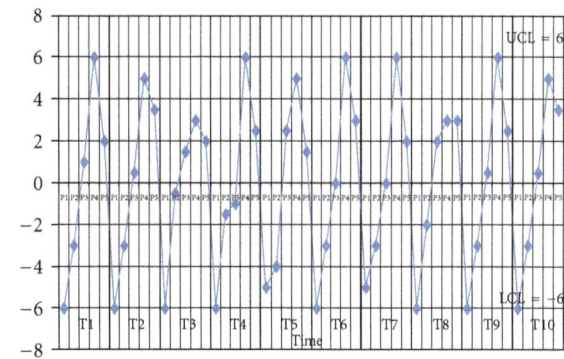
FIGURE 3: Follow-up Single-line Monitoring Chart.

leads to the same conclusions as in Figure 2. Charting the follow-up monitoring scheme by either Figure 2 or Figure 3 is a matter of choice by the practitioner. It is interesting to note that Figure 3 resembles the chart of the famous Red Bead Experiment of Deming ([10], Figure 38 page 347). In the Red Bead Experiment, the numbers of defectives produced by each one of six workers are plotted over four days. The workers correspond to the wafer positions in our current example.

TABLE 2: Thickness (μm) of wafers. P indicates position of wafer.

Time t	Block i	P1	Rank	P2	Rank	P3	Rank	P4	Rank	P5	Rank
1	1	240	1	243	2	250	4	253	5	248	3
1	2	238	1	242	2	245	3	251	5	247	4
1	3	239	1	242	2	246	3	250	5	248	4
	$R1.j$		3		6		10		15		11
	$D1.j$		−6		−3		1		6		2
2	1	235	1	237	2	246	3.5	249	5	246	3.5
2	2	240	1	241	2	246	3	247	4	249	5
2	3	240	1	243	2	244	3	248	5	245	4
	$R2.j$		3		6		9.5		14		12.5
	$D2.j$		−6		−3		0.5		5		3.5
3	1	240	1	243	2	244	3	249	5	246	4
3	2	245	1	250	4.5	250	4.5	247	2	248	3
3	3	238	1	240	2	245	3	248	5	246	4
	$R3.j$		3		8.5		10.5		12		11
	$D3.j$		−6		−0.5		1.5		3		2
4	1	240	1	242	2	246	3	249	5	248	4
4	2	240	1	243	2	246	3	250	5	248	4
4	3	241	1	245	3.5	243	2	247	5	245	3.5
	$R4.j$		3		7.5		8		15		11.5
	$D4.j$		−6		−1.5		−1		6		2.5
5	1	247	2	245	1	255	5	250	4	249	3
5	2	237	1	239	2	243	3	247	5	246	4
5	3	242	1	244	2	245	3.5	248	5	245	3.5
	$R5.j$		4		5		11.5		14		10.5
	$D5.j$		−5		−4		2.5		5		1.5
6	1	237	1	239	2	242	3	247	5	245	4
6	2	242	1	244	2	246	3	251	5	248	4
6	3	243	1	245	2	247	3	252	5	249	4
	$R6.j$		3		6		9		15		12
	$D6.j$		−6		−3		0		6		3
7	1	243	1	245	2	248	3	251	5	250	4
7	2	244	1	246	3	246	3	250	5	246	3
7	3	241	2	239	1	244	3	250	5	246	4
	$R7.j$		4		6		9		15		11
	$D7.j$		−5		−3		0		6		2
8	1	242	1	245	2	248	3	251	5	249	4
8	2	242	1	245	3	248	5	243	2	246	4
8	3	241	1	244	2	245	3	249	5	247	4
	$R8.j$		3		7		11		12		12
	$D8.j$		−6		−2		2		3		3
9	1	236	1	239	2	241	3	246	5	242	4
9	2	243	1	246	2	247	3.5	252	5	247	3.5
9	3	241	1	243	2	245	3	248	5	246	4
	$R9.j$		3		6		9.5		15		11.5
	$D9.j$		−6		−3		0.5		6		2.5
10	1	239	1	240	2	242	3	243	4	244	5
10	2	239	1	240	2	250	3.5	252	5	250	3.5
10	3	241	1	243	2	249	3	255	5	253	4
	$R10.j$		3		6		9.5		14		12.5
	$D10.j$		−6		−3		0.5		5		3.5

5. Concluding Remarks

In this paper, we propose a distribution-free (or nonparametric) scheme for monitoring treatment components while accounting for block effects in a process output. The procedure is based on a nonparametric version of the analysis of means method. Being distribution-free, the FAR and the ARL_0, of the procedure stay the same under any assumed continuous distribution for the process. For small sample sizes and small number of treatments, the control limits, FAR and ARL_0, can be computed exactly from available tables without the need to estimate the mean or variance of the process. The paper includes an example concerning the grinding process of silicon wafers used in integrated circuits production.

References

[1] K. C. B. Roes and R. J. M. M. Does, "Shewhart-type charts in nonstandard situations," *Technometrics*, vol. 37, pp. 15–24, 1995.

[2] A. C. Palm and R. L. DeAmico, "Shewhart-type charts in nonstandard situations: discussion," *Technometrics*, vol. 37, pp. 26–29, 1995.

[3] E. Yashchin, "Monitoring variance components," *Technometrics*, vol. 36, no. 4, pp. 379–393, 1994.

[4] W. H. Woodall and E. V. Thomas, "Statistical process control with several components of common cause variability," unpublished paper presented at the Joint Statistical Meetings, Atlanta, Ga, USA, 1991.

[5] S. T. Bakir, "A distribution-free Shewhart quality control chart based on signed-ranks," *Quality Engineering*, vol. 16, no. 4, pp. 613–623, 2004.

[6] S. Chakraborti and M. A. Graham, "Nonparametric control charts," in *Encyclopedia of Quality and Reliability*, John Wiley, New York, NY, USA, 2007.

[7] S. Chakraborti, S. W. Human, and M. A. Graham, "Nonparametric (distribution-free) quality control charts," in *Handbook of Methods and Applications of Statistics: Engineering, Quality Control, and Physical Sciences*, N. Balakrishnan, Ed., pp. 298–329, John Wiley & Sons, New York, NY, USA, 2011.

[8] S. T. Bakir, "Analysis of means using ranks for the randomized complete block design," *Communications in Statistics—Simulation*, vol. 23, no. 2, pp. 547–568, 1994.

[9] S. T. Bakir, "Analysis of means using ranks," *Communications in Statistics—Simulation*, vol. 18, pp. 757–776, 1989.

[10] W. E. Deming, *Out of the Crisis*, MIT, Cambridge, Mass, USA, 1986.

9

On Stress-Strength Reliability with a Time-Dependent Strength

Serkan Eryilmaz

Department of Industrial Engineering, Atilim University, Incek, 06836 Ankara, Turkey

Correspondence should be addressed to Serkan Eryilmaz; seryilmaz@atilim.edu.tr

Academic Editor: Shey-Huei Sheu

The study of stress-strength reliability in a time-dependent context needs to model at least one of the stress or strength quantities as dynamic. We study the stress-strength reliability for the case in which the strength of the system is decreasing in time and the stress remains fixed over time; that is, the strength of the system is modeled as a stochastic process and the stress is considered to be a usual random variable. We present stochastic ordering results among the lifetimes of the systems which have the same strength but are subjected to different stresses. Multicomponent form of the aforementioned stress-strength interference is also considered. We illustrate the results for the special case when the strength is modeled by a Weibull process.

1. Introduction

Stress-strength models are of special importance in reliability literature and engineering applications. A technical system or unit may be subjected to randomly occurring environmental stresses such as pressure, temperature, and humidity and the survival of the system heavily depends on its resistance. In the simplest form of the stress-strength model, a failure occurs when the strength (or resistance) of the unit drops below the stress. In this case the reliability R is defined as the probability that the unit's strength is greater than the stress, that is, $R = P\{X < Y\}$, where Y is the random strength of the unit and X is the random stress placed on it. This reliability has been widely studied under various distributional assumptions on X and Y. (See, e.g., Johnson [1] and Kotz et al. [2] for an extensive and lucid review of the topic.)

In the aforementioned simplest form, stress and strength quantities are considered to be both static. Dynamic modeling of stress-strength interference might offer more realistic applications to real-life reliability studies than static modeling and it enables us to investigate the time-dependent (dynamic) reliability properties of the system. Let $X(t)$ and $Y(t)$ denote the stress that the system is experiencing and strength of the system at time t, respectively. Then the lifetime of the system is represented as

$$T = \inf\{s \geq 0 : X(s) > Y(s)\}. \tag{1}$$

The most important characteristic in reliability analysis is the reliability function of a system which is defined as the probability of surviving at time t, that is,

$$R(t) = P\{T > t\}. \tag{2}$$

This function is also known as the survival function in the reliability literature and its exact formulation is of special importance in engineering applications.

The reliability function for the lifetime random variable given in (1) is

$$R(t) = P\{X(s) < Y(s), \ 0 \leq s \leq t\}. \tag{3}$$

The function given by (3) has been investigated in several papers. Whitmore [3] computed the function $R(t)$ under the assumptions that $X(t)$ and $Y(t)$ are independent Brownian motions. Ebrahimi [4] studied the properties of $R(t)$ assuming the strength of the system $Y(t)$ is decreasing in time.

In this paper, we study $R(t)$ assuming (i) $Y(t)$ is decreasing in time, that is, $P\{Y(t_2) \leq Y(t_1)\} = 1$ for all $t_1 < t_2$ and (ii) $X(t) = X$; that is, stress remains fixed over time (static). The first assumption is common in reality because the system's strength could degrade due to aging. In Section 2, we provide some stochastic ordering results among the lifetimes of the systems which have the same strength but are subjected to different stresses. In Section 3, stress-strength interference is considered for multicomponent systems. Finally, in Section 4 some results are presented for the special case when the strength is modeled by a Weibull process.

2. Reliability and Ordering Properties

Under the assumptions (i), (ii) X and $Y(t)$ are independent and the reliability function can be formulated as

$$
R(t) = P\{X < Y(t)\} = \int_{-\infty}^{\infty} F(x)\, dG_t(x)
$$

$$
= \int_{-\infty}^{\infty} \overline{G}_t(x)\, dF(x), \tag{4}
$$

where $F(x) = P\{X \le x\}$, $G_t(x) = P\{Y(t) \le x\}$ and $\overline{G}_t(x) = 1 - G_t(x)$.

The following example illustrates the computation of reliability function for the stochastic strength process given with its analytical form. That is, the strength aging deterioration process is expressed as a function of time, and a random variable.

Example 1. Let $Y(t)$ be defined by

$$
Y(t) = e^{-C \cdot t}, \quad t \ge 0, \tag{5}
$$

where C follows Pareto distribution with c.d.f. $F_C(x) = 1 - (\mu/x)^2$, $x > \mu$, $\mu > 0$. Then the c.d.f. of $Y(t)$ is

$$
G_t(x) = \frac{\mu^2 t^2}{(\ln x)^2}, \quad 0 < x < e^{-\mu t}. \tag{6}
$$

Let X have a c.d.f. $F(x) = x^\theta$, $0 < x < 1$, $\theta > 0$. Then using (4) we have

$$
R(t) = \int_0^{e^{-\mu t}} \left(1 - \frac{\mu^2 t^2}{(\ln x)^2}\right) \theta x^{\theta-1}\, dx
$$

$$
= e^{-\mu t \theta} - t^2 \mu^2 \theta \int_0^{e^{-\mu t}} \frac{x^{\theta-1}}{(\ln x)^2}\, dx, \quad t \ge 0. \tag{7}
$$

Differentiating $1 - R(t)$ w.r.t. t using the rule of differentiation under the integral sign, the p.d.f. of T is found as

$$
f(t) = 2\mu^2 \theta t \int_0^{e^{-\mu t}} \frac{x^{\theta-1}}{(\ln x)^2}\, dx. \tag{8}
$$

Using (8) in (7) we obtain

$$
R(t) = e^{-\mu t \theta} - \frac{1}{2} t f(t). \tag{9}
$$

Integrating both sides of the last equation over $(0, \infty)$ we get

$$
E(T) = \frac{2}{3\mu\theta} \tag{10}
$$

which is the MTTF of the system.

The process defined by (5) can be considered in a more general form given by

$$
Y(t) = e^{-C \cdot S(t)}, \quad t \ge 0, \tag{11}
$$

where $S(0) = 0$ and $S(t)$ is a nondecreasing function. In this case the reliability function and MTTF of the system are found to be

$$
R(t) = e^{-\mu \theta S(t)} - S^2(t) \mu^2 \theta \int_0^{e^{-\mu S(t)}} \frac{x^{\theta-1}}{(\ln x)^2}\, dx,
$$

$$
\text{MTTF} = \int_0^{\infty} e^{-\mu \theta S(t)}\, dt - \frac{1}{2} E\left[\frac{S(T)}{S'(T)}\right]. \tag{12}
$$

For the system defined in Example 1 it can be easily seen that an increase in θ leads to a decrease in MTTF of the system. Since $EX = \theta/(\theta + 1)$, the larger θ the harsher the stress and hence the smaller the reliability. This can be theoretically established using the concept of stochastic ordering as in the following lines.

We investigate the behaviour of the lifetime of the system under different stresses in terms of stochastic ordering. In this context, we consider the following stochastic orderings between the lifetimes. Let X and Y be two random variables having cumulative distributions F and G, densities f and g, hazard rates r_F and r_G, and reversed hazard rates h_F and h_G, respectively. Note that the hazard and reversed hazard rates are defined, respectively, as

$$
r_F(t) = \frac{f(t)}{\overline{F}(t)}, \qquad h_F(t) = \frac{f(t)}{F(t)}. \tag{13}
$$

Usual Stochastic Ordering. X is said to be smaller than Y in usual stochastic ordering if $P\{X > x\} \le P\{Y > x\}$ for all x, or equivalently $E[\Psi(X)] \le E[\Psi(Y)]$ for all increasing functions Ψ for which the expectations exist. This relation is denoted by $X \overset{st}{\le} Y$.

Hazard Rate Ordering. X is said to be smaller than Y in hazard rate ordering if $r_F(x) \ge r_G(x)$ for all x or $\overline{G}(x)/\overline{F}(x)$ is a nondecreasing function of x and we write $X \overset{hr}{\le} Y$.

Reversed Hazard Rate Ordering. X is said to be smaller than Y in the reversed hazard rate order if $h_F(x) \le h_G(x)$ for all x or $G(x)/F(x)$ is a nondecreasing function of x and we write $X \overset{rh}{\le} Y$.

Likelihood Ratio Ordering. X is said to be smaller than Y in likelihood ratio ordering (written as $X \overset{lr}{\le} Y$) if $g(x)/f(x)$ is nondecreasing function of x.

We have the following implications among these orderings:

$$
X \overset{lr}{\le} Y \implies \left(\begin{array}{c} X \overset{hr}{\le} Y \\ X \overset{rh}{\le} Y \end{array} \right) \implies X \overset{st}{\le} Y. \tag{14}
$$

For more details on stochastic orderings refer to Shaked and Shanthikumar [5].

The following concepts will be useful for the next section.

Definition 2. A function $k(x, y)$ is said to be *sign regular of order 2* (SR$_2$) if $\varepsilon_1 k(x, y) \geq 0$ and

$$\varepsilon_2 \begin{vmatrix} k(x_1, y_1) & k(x_1, y_2) \\ k(x_2, y_1) & k(x_2, y_2) \end{vmatrix} \geq 0, \tag{15}$$

whenever $x_1 < x_2$, $y_1 < y_2$, and $\varepsilon_i \in \{-1, 1\}$ for $i = 1, 2$.

If the conditions given in Definition 2 hold with $\varepsilon_1 = 1$ and $\varepsilon_2 = 1$ then k is said to be *totally positive of order 2* (TP$_2$); and k is said to be *reverse regular of order 2* (RR$_2$) if they hold with $\varepsilon_1 = 1$ and $\varepsilon_2 = -1$.

Proposition 3. *Let T_i denote the lifetime of the system whose stress-strength pair is $(X_i, Y(t))$, $i = 1, 2$. Then*

(a) *if $X_1 \overset{st}{\leq} X_2$ then $T_1 \overset{st}{\geq} T_2$;*

(b) *if $X_1 \overset{rh}{\leq} X_2$ and $g(x, t) = (\partial/\partial x) G_t(x)$ is RR$_2$ in (x, t) then $T_1 \overset{hr}{\geq} T_2$;*

(c) *if $X_1 \overset{lr}{\leq} X_2$ and $h(x, t) = (\partial/\partial t) G_t(x)$ is RR$_2$ in (x, t) then $T_1 \overset{lr}{\geq} T_2$.*

Proof. The proof of (a) immediately follows because $P\{T_i \leq t\} = E[G_t(X_i)]$, $i = 1, 2$, and $G_t(x)$ is increasing in x. The proofs of (b) and (c) can be obtained as an application of Theorems 1.B.14, 1.B.52, and 1.C.17 in Shaked and Shanthikumar [5]. These results are obtained using basic composition formula of Karlin [6]. $\square$

Example 4. Consider the process defined in Example 1 with $F_C(x) = 1 - e^{-x}$, $x > 0$ and let X_i have a c.d.f. $F_i(x) = x^{\theta_i}$, $0 < x < 1$, $i = 1, 2$. In this case

$$G_t(x) = x^{1/t}, \quad 0 < x < 1,$$

$$h(x, t) = \frac{\partial}{\partial t} G_t(x) = -\frac{1}{t^2} x^{1/t} \ln x, \tag{16}$$

$$g(x, t) = \frac{\partial}{\partial x} G_t(x) = \frac{1}{t} x^{1/t-1}.$$

For any $x_1 < x_2$ and $t_1 < t_2$

$$h(x_1, t_1) h(x_2, t_2) - h(x_1, t_2) h(x_2, t_1)$$

$$= \frac{1}{t_1^2 t_2^2} \ln x_1 \ln x_2 \left[x_1^{1/t_1} x_2^{1/t_2} - x_2^{1/t_1} x_1^{1/t_2} \right] \leq 0 \tag{17}$$

which implies the RR$_2$ property of $h(x, t)$. If $\theta_1 \leq \theta_2$ then $X_1 \overset{lr}{\leq} X_2$. Thus by the Proposition 3 we have $T_1 \overset{lr}{\geq} T_2$. Similarly, $g(x, t)$ is also RR$_2$ in (x, t). If $\theta_1 \leq \theta_2$ then $X_1 \overset{rh}{\leq} X_2$ and hence by the Proposition 3 we have $T_1 \overset{hr}{\geq} T_2$.

3. Multicomponent Setup

In the previous sections we analyzed stress-strength reliability for a single component system. Most of the engineering systems consist of several components and the components might have different statistical properties. Multicomponent stress-strength reliability in a static form has been studied in various papers including Bhattacharyya and Johnson [7], Chandra and Owen [8], Johnson [1], Pandey et al. [9], Eryilmaz [10], and Eryilmaz [11].

Assume that a system consists of n components and the deteriorating strength of the ith component at time t is denoted by the process $Y_i(t)$, $i = 1, 2, \ldots, n$. The components are subjected to a common random stress X. If T_i denotes the lifetime of the ith component then the joint survival function of $T_1, T_2, \ldots, T_n$ is given by

$$R(t_1, t_2, \ldots, t_n)$$

$$= P\{T_1 > t_1, T_2 > t_2, \ldots, T_n > t_n\}$$

$$= P\{X < Y_1(t_1), X < Y_2(t_2), \ldots, X < Y_n(t_n)\}. \tag{18}$$

If the components are independent then we have

$$R(t_1, t_2, \ldots, t_n) = \int \prod_{i=1}^{n} P\{Y_i(t_i) > x\} dF(x). \tag{19}$$

From (19) it follows that the lifetimes of the components are dependent even if the strengths of them are independent. This positive dependence among the lifetimes arises from common environmental stress characterized by X. There are many types of positive dependence. The likelihood ratio (or TP$_2$) dependence as the strongest notion of positive dependence is defined as follows. Let T_1, T_2 have the joint probability density $f(t_1, t_2)$. Then recall from Definition 2 that $f(t_1, t_2)$ is TP$_2$ if

$$\begin{vmatrix} f\left(t_1^{(1)}, t_2^{(1)}\right) & f\left(t_1^{(1)}, t_2^{(2)}\right) \\ f\left(t_1^{(2)}, t_2^{(1)}\right) & f\left(t_1^{(2)}, t_2^{(2)}\right) \end{vmatrix} \geq 0, \tag{20}$$

for $t_1^{(1)} < t_1^{(2)}$ and $t_2^{(1)} < t_2^{(2)}$. The random variables T_1 and T_2 are said to be likelihood ratio (or TP$_2$) dependent if their joint density is TP$_2$.

The following result can be proved using the basic composition formula of Karlin [6] together with

$$f(t_1, t_2) = \int h^{(1)}(x, t_1) h^{(2)}(x, t_2) dF(x), \tag{21}$$

where $h^{(i)}(x, t) = (\partial/\partial t) G_{t,i}(x)$ and $G_{t,i}(x) = P\{Y_i(t) \leq x\}$, $i = 1, 2$.

Proposition 5. *If $h^{(1)}(x, t_1)$ is TP$_2$ (RR$_2$) in (t_1, x) and $h^{(2)}(x, t_2)$ is TP$_2$ (RR$_2$) in (x, t_2), then T_1 and T_2 are likelihood ratio dependent.*

Example 6. Let $Y_i(t) = e^{-C_i t}$, $t \geq 0$, $i = 1, 2$ and C_i be an exponential random variable with c.d.f. $F_{C_i}(x) = 1 - e^{-\mu_i x}$, $x > 0$. Also assume that the common random stress X has c.d.f. $F(x) = x^\theta$, $0 < x < 1$. In this case the joint survival function of T_1 and T_2 is found to be

$$R(t_1, t_2) = 1 - \frac{\theta t_1}{\mu_1 + \theta t_1} - \frac{\theta t_2}{\mu_2 + \theta t_2}$$

$$+ \frac{\theta}{\mu_1/t_1 + \mu_2/t_2 + \theta}, \quad t_1, t_2 \geq 0. \quad (22)$$

Since

$$h^{(i)}(x, t) = -\frac{\mu_i}{t^2} x^{\mu_i/t} \ln x \quad (23)$$

is RR_2 for $i = 1, 2$, T_1 and T_2 are likelihood ratio dependent.

Consider a system ϕ with n components which has two possible states; $\phi = 1$ if the system is functioning and $\phi = 0$ if the system has failed. Since the state of the system is determined by the states of its components we can write $\phi = \phi(x_1, \ldots, x_n)$, where $x_i = 1$ if the ith component is functioning and $x_i = 0$ if it has failed. The function ϕ is called structure function. A system with structure function ϕ is coherent if ϕ is nondecreasing in each argument, and each component is relevant to the performance of the system. If the components' lifetimes are denoted by $T_1, \ldots, T_n$, then $T = \phi(T_1, T_2, \ldots, T_n)$ represents the lifetime of the system.

Let $T_1, T_2, \ldots, T_n$ denote the i.i.d. lifetime random variables with continuous distribution. Samaniego [12] introduced the signature of a coherent system $T = \phi(T_1, T_2, \ldots, T_n)$ as the vector $\mathbf{p} = (p_1, p_2, \ldots, p_n)$, where

$$p_i = P\{\phi(T_1, T_2, \ldots, T_n) = T_{(i)}\}, \quad i = 1, \ldots, n, \quad (24)$$

where $T_{(i)}$ denotes the smallest ith in $T_1, T_2, \ldots, T_n$, showing that

$$p_i = \# \text{ of orderings for which the } i\text{th failure causes}$$
$$\text{system failure} \times (n!)^{-1}. \quad (25)$$

A general formula for the reliability function of any coherent structure consisting of n components can be given by using the concept of "signature" if the components are independent and identical. Samaniego [12] (see also [13]) showed that the reliability function of a coherent system $T = \phi(T_1, T_2, \ldots, T_n)$ can be represented as

$$P\{T > t\} = \sum_{i=1}^{n} p_i P\{T_{(i)} > t\}. \quad (26)$$

Navarro et al. [14] (see also Navarro and Rychlik [15]) proved that the representation (26) also holds whenever $(T_1, T_2, \ldots, T_n)$ has an absolutely continuous exchangeable joint distribution. The following theorem provides the reliability function of any coherent structure consisting of n components.

Theorem 7. Let T_i denote the lifetime of the ith component whose strength is $Y_i(t)$, $i = 1, 2, \ldots, n$, that is, $T_i = \inf\{s : s \geq 0, X > Y_i(s)\}$. If ϕ denotes the structure function of the coherent

system with lifetime T, that is, $T = \phi(T_1, T_2, \ldots, T_n)$ and $Y_1(t), Y_2(t), \ldots, Y_n(t)$ are i.i.d. with c.d.f. $G_t(x) = P\{Y_i(t) \leq x\}$, $i = 1, 2, \ldots, n$ then

$$P\{T > t\} = \sum_{i=1}^{n} p_i P\{T_{(i)} > t\}, \quad (27)$$

where

$$P\{T_{(i)} > t\} = \sum_{j=n-i+1}^{n} \sum_{m=0}^{n-j} (-1)^m \binom{n}{j} \binom{n-j}{m} E(\overline{G}_t(X))^{j+m}. \quad (28)$$

Proof. Under the assumption that $Y_1(t), Y_2(t), \ldots, Y_n(t)$ are i.i.d. the joint survival function of $T_1, T_2, \ldots, T_n$ is

$$R(t_1, t_2, \ldots, t_n) = \int \prod_{i=1}^{n} \overline{G}_{t_i}(x) \, dF(x). \quad (29)$$

The function given by (33) is a mixture of independent n-variate d.f.'s with equal marginals; that is, the random vector $(T_1, T_2, \ldots, T_n)$ is positive dependent by mixture (PDM). PDM d.f.'s are exchangeable. (See, e.g., Shaked [16] for the concept of PDM and associated exchangeability). Since the representation (26) also holds for exchangeable lifetimes we get (27) with

$$P\{T_{(i)} > t\} = \sum_{j=n-i+1}^{n} \binom{n}{j} P\{X < Y_1(t), \ldots, X < Y_j(t),$$

$$X \geq Y_{j+1}(t), \ldots, X \geq Y_n(t)\}. \quad (30)$$

The usage of inclusion-exclusion principle for the probability inside the sum gives

$$P\{T_{(i)} > t\} = \sum_{j=n-i+1}^{n} \binom{n}{j} \sum_{m=0}^{n-j} (-1)^m \binom{n-j}{m}$$

$$\times P\{X < Y_1(t), \ldots, X < Y_{j+m}(t)\}. \quad (31)$$

The proof is now completed by conditioning on X. $\qquad \square$

4. Weibull Stress-Strength Model

In this section we study the stress-strength reliability for the Weibull process which can be used to model the decreasing strength of a unit. Chiodo and Mazzanti [17] studied stress-strength reliability and its estimation for aged power system components subjected to voltage surges using Weibull process.

Let $Y(t)$ be a Weibull process whose one-dimensional distribution is

$$G_t(x) = P\{Y(t) \leq x\} = 1 - \exp\left\{-\left(\frac{x}{\alpha(t)}\right)^\beta\right\}, \quad x > 0, \quad (32)$$

where the shape parameter β is assumed to be time independent and the intensity function $\alpha(t)$ is decreasing in time with $\alpha(0) = \infty$. Similarly, assume that the stress random variable X has a Weibull distribution with c.d.f.

$$F(x) = 1 - \exp\left\{-\left(\frac{x}{\theta}\right)^{\beta}\right\}, \quad x > 0, \ \theta, \beta > 0. \tag{33}$$

Under these assumptions the reliability function is found to be

$$R(t) = \frac{1}{1 + (\theta/\alpha(t))^{\beta}}, \quad t > 0. \tag{34}$$

The following results can be obtained from Proposition 3.

Corollary 8. *Let T_i denote the lifetime of the system whose stress-strength pair is $(X_i, Y(t))$, $i = 1, 2$, where X_i has a Weibull distribution with scale parameter θ_i and shape parameter β and $Y(t)$ is a Weibull process whose distribution is given by (32). Then, since $g(x,t) = (\partial/\partial x)G_t(x) = (\beta/\alpha(t))(x/\alpha(t))^{\beta-1}e^{-(x/\alpha(t))^{\beta}}$ is RR_2 in (x,t), if $\theta_1 \leq \theta_2$ then $T_1 \overset{hr}{\geq} T_2$.*

Corollary 9. *Under the same assumptions of Corollary 8, the function*

$$h(x,t) = \frac{\partial}{\partial t}G_t(x) = -\beta\frac{x^{\beta}}{\alpha^{\beta+1}(t)}\alpha'(t)e^{-(x/\alpha(t))^{\beta}} \tag{35}$$

is RR_2 in (x,t). Indeed, because $\alpha(t)$ is decreasing, $\alpha'(t) < 0$, and hence $h(x,t) \geq 0$. For $x_1 < x_2$ and $t_1 < t_2$,

$$h(x_1,t_1)h(x_2,t_2) - h(x_1,t_2)h(x_2,t_1) \leq 0 \tag{36}$$

which implies that $h(x,t)$ is RR_2 in (x,t). Therefore, from Proposition 3, if $\theta_1 \leq \theta_2$ then $T_1 \overset{lr}{\geq} T_2$.

Remark 10. *If $\alpha(t) = 1/t$ then*

$$R(t) = \frac{1}{1 + (\theta t)^{\beta}}, \quad t > 0 \tag{37}$$

which is the survival function of the log-logistic distribution. That is, the lifetime of the system has log-logistic distribution with scale parameter $\lambda = 1/\theta$ and shape parameter β. If $\beta > 1$ then the MTTF of the system is found to be

$$E(T) = \frac{\pi/\theta\beta}{\sin(\pi/\beta)}. \tag{38}$$

Theorem 11. *Let $Y_i(t)$ be a Weibull process with intensity $\alpha_i(t)$ associated with ith component, $i = 1, \ldots, n$. Assume that the common random stress X has a Weibull distribution with c.d.f. given by (33). Then the joint survival function of $T_1, \ldots, T_n$ is*

$$R(t_1,\ldots,t_n) = \frac{1}{1 + (\theta/\alpha_1(t_1))^{\beta} + \cdots + (\theta/\alpha_n(t_n))^{\beta}}, \tag{39}$$

and the survival copula associated with (39) belongs to the Clayton family and is given by

$$C(u_1,\ldots,u_n) = \frac{1}{1/u_1 + \cdots + 1/u_n - n + 1}. \tag{40}$$

Proof. Using (19) one can write

$$R(t_1,\ldots,t_n)$$
$$= P\{X < Y_1(t_1), \ldots, X < Y_n(t_n)\}$$
$$= \int_0^{\infty} e^{-(x/\alpha_1(t_1))^{\beta}} \cdots e^{-(x/\alpha_n(t_n))^{\beta}} \frac{\beta}{\theta}\left(\frac{x}{\theta}\right)^{\beta-1} e^{-(x/\theta)^{\beta}} dx$$
$$= \frac{1}{1 + (\theta/\alpha_1(t_1))^{\beta} + \cdots + (\theta/\alpha_n(t_n))^{\beta}}. \tag{41}$$

By the definition of survival copula (see, e.g., Nelsen [18, page 32])

$$C(u_1,\ldots,u_n)$$
$$= R\left(R_1^{-1}(t_1),\ldots,R_n^{-1}(t_n)\right)$$
$$= R\left(\alpha_1^{-1}\left(\theta\left(\frac{1}{u_1}-1\right)^{-1/\beta}\right),\ldots,\alpha_n^{-1}\left(\theta\left(\frac{1}{u_n}-1\right)^{-1/\beta}\right)\right)$$
$$= \left(u_1^{-1} + \cdots + u_n^{-1} - n + 1\right)^{-1} \tag{42}$$

which is known to be a Clayton copula (see, e.g., Nelsen [18, page 152]). Thus the proof is completed. $\square$

Remark 12. For $\alpha_i(t) = 1/t$, $i = 1, \ldots, n$, (39) becomes the survival function of the multivariate log-logistic distribution generated by the Clayton family of copulas.

Example 13. Consider the system consisting of n components whose deteriorating strengths are modeled by a Weibull process with the common intensity function $\alpha(t)$. Suppose that these components are subjected to a common random stress X which has c.d.f. given by (33). Then, the lifetimes of the components are exchangeable and we have

$$E\left(\overline{G}_t(X)\right)^{j+m} = \frac{1}{1 + (j+m)(\theta/\alpha(t))^{\beta}}. \tag{43}$$

If, for example, a system has a 2-out-of-3 structure, that is, the system functions if and only if at least two components function, then since $\mathbf{p} = (0,1,0)$ using Theorem 7, the reliability of the system is found to be

$$P\{T_{(2)} > t\} = \frac{3}{1 + 2(\theta/\alpha(t))^{\beta}} - \frac{2}{1 + 3(\theta/\alpha(t))^{\beta}}. \tag{44}$$

Acknowledgment

The author thanks the anonymous referee for his/her helpful comments and suggestions.

References

[1] R. A. Johnson, "Stress-strength models for reliability," in *Handbook of Statistics*, P. R. Krishnaiah and C. R. Rao, Eds., vol. 7, pp. 27–54, Elsevier, Amsterdam, North-Holland, 1988.

[2] S. Kotz, Y. Lumelskii, and M. Pensky, *The Stress-Strength Model and its Generalizations*, World Scientific, River Edge, NJ, USA, 2003.

[3] G. A. Whitmore, "On the reliability of stochastic systems: a comment," *Statistics & Probability Letters*, vol. 10, no. 1, pp. 65–67, 1990.

[4] N. Ebrahimi, "Two suggestions of how to define a stochastic stress-strength system," *Statistics & Probability Letters*, vol. 3, no. 6, pp. 295–297, 1985.

[5] M. Shaked and J. G. Shanthikumar, *Stochastic Orders*, Springer Series in Statistics, Springer, New York, NY, USA, 2007.

[6] S. Karlin, *Total Positivity. Vol. I*, Stanford University Press, Stanford, Calif, USA, 1968.

[7] G. K. Bhattacharyya and R. A. Johnson, "Estimation of reliability in a multicomponent stress-strength model," *Journal of the American Statistical Association*, vol. 69, pp. 966–970, 1974.

[8] S. Chandra and D. B. Owen, "On estimating the reliability of a component subject to several different stresses (strengths)," *Naval Research Logistics Quarterly*, vol. 22, pp. 31–39, 1975.

[9] M. Pandey, M. B. Uddin, and J. Ferdous, "Reliability estimation of an s-out-of-k system with non-identical component strengths: the Weibull case," *Reliability Engineering and System Safety*, vol. 36, no. 2, pp. 109–116, 1992.

[10] S. Eryilmaz, "Consecutive k-out-of n: G system in stress-strength setup," *Communications in Statistics. Simulation and Computation*, vol. 37, no. 3–5, pp. 579–589, 2008.

[11] S. Eryilmaz, "Multivariate stress-strength reliability model and its evaluation for coherent structures," *Journal of Multivariate Analysis*, vol. 99, no. 9, pp. 1878–1887, 2008.

[12] F. Samaniego, "On closure of the IFR class under formation of coherent systems," *IEEE Transactions on Reliability*, vol. 34, no. 1, pp. 69–72, 1985.

[13] S. Kochar, H. Mukerjee, and F. J. Samaniego, "The "signature" of a coherent system and its application to comparisons among systems," *Naval Research Logistics*, vol. 46, no. 5, pp. 507–523, 1999.

[14] J. Navarro, J. M. Ruiz, and C. J. Sandoval, "A note on comparisons among coherent systems with dependent components using signatures," *Statistics & Probability Letters*, vol. 72, no. 2, pp. 179–185, 2005.

[15] J. Navarro and T. Rychlik, "Reliability and expectation bounds for coherent systems with exchangeable components," *Journal of Multivariate Analysis*, vol. 98, no. 1, pp. 102–113, 2007.

[16] M. Shaked, "A concept of positive dependence for exchangeable random variables," *The Annals of Statistics*, vol. 5, no. 3, pp. 505–515, 1977.

[17] E. Chiodo and G. Mazzanti, "Bayesian reliability estimation based on a weibull stress-strength model for aged power system components subjected to voltage surges," *IEEE Transactions on Dielectrics and Electrical Insulation*, vol. 13, no. 1, pp. 146–159, 2006.

[18] R. B. Nelsen, *An Introduction to Copulas*, Springer Series in Statistics, Springer, New York, NY, USA, 2nd edition, 2006.

10

On Parameters Estimation of Lomax Distribution under General Progressive Censoring

Bander Al-Zahrani[1] and Mashail Al-Sobhi[2]

[1] *Department of Statistics, Faculty of Sciences, King Abdulaziz University, Jeddah 21589, Saudi Arabia*
[2] *Department of Mathematics, Umm Al-Qura University, Makkah, Saudi Arabia*

Correspondence should be addressed to Bander Al-Zahrani; bmalzahrani@kau.edu.sa

Academic Editor: Antoine Grall

We consider the estimation problem of the probability $S = P(Y < X)$ for Lomax distribution based on general progressive censored data. The maximum likelihood estimator and Bayes estimators are obtained using the symmetric and asymmetric balanced loss functions. The Markov chain Monte Carlo (MCMC) methods are used to accomplish some complex calculations. Comparisons are made between Bayesian and maximum likelihood estimators via Monte Carlo simulation study.

1. Introduction

The Lomax distribution, also called "Pareto type II" distribution is a particular case of the generalized Pareto distribution (GPD). The Lomax distribution has been used in the literature in a number of ways. For example, it has been extensively used for reliability modelling and life testing; see, for example, Balkema and de Haan [1]. It also has been used as an alternative to the exponential distribution when the data are heavy tailed; see Bryson [2]. Ahsanullah [3] studied the record values of Lomax distribution. Balakrishnan and Ahsanullah [4] introduced some recurrence relations between the moments of record values from Lomax distribution. The order statistics from nonidentical right-truncated Lomax random variables have been studied by Childs et al. [5]. Also, the Lomax model has been studied, from a Bayesian point of view, by many authors; see, for example, Arnold et al. [6] and El-Din et al. [7]. Howlader and Hossain [8] presented Bayesian estimation of the survival function of the Lomax distribution. Ghitany et al. [9] considered Marshall-Olkin approach and extended Lomax distribution. Cramer and Schmiedt [10] considered progressively type-II censored competing risks data from Lomax distribution. The Lomax distribution has applications in economics, actuarial modelling, queuing problems and biological sciences; for details, we refer to Johnson et al. [11].

A positive random variable X is said to have the Lomax distribution, abbreviated as $X \sim L(\alpha, \beta)$, if it has the probability density function (pdf)

$$f(x; \alpha, \beta) = \alpha\beta(1 + \beta x)^{-(\alpha+1)}, \quad x > 0, \ \alpha, \beta > 0. \quad (1)$$

Here, α and β are the shape and the scale parameters, respectively. The survival function (sf) associated with (1) is

$$\overline{F}(x; \alpha, \beta) = (1 + \beta x)^{-\alpha}, \quad x > 0. \quad (2)$$

Further probabilistic properties of this distribution are given, for example, in Arnold [12].

This paper is concerned with the problem of estimating $S = P(Y < X)$ for Lomax based on general progressive censored data. The reliability of a component during a given period of time is defined as the probability that its strength X exceeds the stress Y, and symbolically we write $S = P(Y < X)$. We assume X and Y to be independent, and each follows a Lomax distribution. A good overview on estimating S can be found in the monograph of Kotz et al. [13]. Later, the problem of estimating S attracted the attention of many authors; for example, see Baklizi [14], Raqab et al. [15], Kundu and Raqab [16], and Panahi and Asadi [17], and references cited therein.

The rest of the paper is organized as follows. In Section 2, we give a brief overview of the general progressive censoring.

The maximum likelihood estimators (MLEs) are obtained in Section 3. In Section 4, we obtain Bayes estimators using the symmetric and asymmetric balanced loss functions. In Section 5, the MCMC methods are used to accomplish some complex calculations, and, therefore, comparisons are made between Bayesian and maximum likelihood estimators via Monte Carlo simulation study.

2. General Progressive Censoring

We refer to the paper of Soliman et al. [18, page 452], for introducing the general progressive censoring as follows. Consider a general type-II progressive censoring scheme, proposed by Balakrishnan and Sandhu [19]. This scheme of censoring can be explained as follows: at time $X_0 \equiv 0$, n randomly selected components were placed on a life test. The failure times of the first r components to fail, $X_1, \ldots, X_r$, were not observed. At the time of the $(r + 1)$th failure, $X_{r+1:n}$, R_{r+1} number of surviving components are removed from the test randomly and so on; at the time of the $(r + i)$th observed failure, $X_{r+i:n}$, R_{r+i} number of surviving components are removed from the test randomly; and finally, at the time of the mth failure, the remaining $R_m = n - m - R_{r+1} - R_{r+2} - \cdots - R_{m-1}$ are removed from the test. Suppose that $X_{r+1:m:n} \leq X_{r+2:m:n} \leq \cdots \leq X_{m:m:n}$ are the lifetimes of the completely observed components to fail and that $R_{r+1}, R_{r+2}, \ldots, R_m$ are the number of components removed from the test at these failure times, respectively. The R_i's, m, and r are prespecified integers such that $0 \leq r < m \leq n$, $0 \leq R_i \leq n-i$ for $i = r+1, \ldots, m-1$, and $R_m = n - m - \sum_{i=r+1}^{m-1} R_i$. The resulting $(m - r)$ ordered values $X_{r+1:m:n}, X_{r+2:m:n}, \ldots, X_{m:m:n}$ are appropriately referred to as general progressively type-II censored order statistics.

Also referring to Soliman et al. [18], it should be noted that (i) if $R_i = 0$, for $i = r+1, \ldots, m-1$, and $R_m = n-m$, the general progressively type-II censoring scheme is reduced to the case of type-II doubly censored sample. (ii) If $r = 0$, this scheme is reduced to the progressive type-II right censoring. (iii) If $r = 0$ and $R_i = 0$, for $i = r+1, \ldots, m-1$ so that $R_m = n - m$, the general progressively type-II censoring scheme is reduced to conventional type-II one-stage right censoring, where just the first m usual order statistics are observed. (iv) If $r = 0$ and $R_i = 0$, for $i = r+1, \ldots, m$, so that $m = n$, the general progressively type-II censoring scheme is reduced to the case of no censoring (complete sample case), where all n usual order statistics are observed. In this scheme $R_1, R_2, \ldots, R_m$ are all prefixed. Saraoglu et al. [20] discussed two examples showing the motivation behind the developments of the stress-strength models under censored samples. For more details, see Balakrishnan and Aggarwala [21].

Suppose that n randomly selected components from the $L(\alpha, \beta)$ are put in test. Further, let $X_{r+1:m:n}, X_{r+2:m:n}, \ldots, X_{m:m:n}$ denote a general progressively type-II censored sample from that population, with $(R_{r+1}, R_{r+2}, \ldots, R_m)$ being the progressive censoring scheme. For simplicity of notation, we will use x_i instead of $x_{i:m:n}$, and then $\mathbf{x} = (x_{r+1}, \ldots, x_m)$ is the observed general progressive censored sample. The likelihood function for the parameters α and β is then

$$\mathscr{L}(\alpha, \beta \mid \mathbf{x}) = c^* \left\{ 1 - \overline{F}(x_{r+1}) \right\}^r \prod_{i=r+1}^{m} f(x_i) \left[\overline{F}(x_i) \right]^{R_i}, \quad (3)$$

where

$$c^* = \binom{n}{r}(n - r) \prod_{j=r+2}^{m} \left[n - \sum_{i=r+1}^{j-1} R_i - j + 1 \right], \quad (4)$$

and the functions $f(x)$ and $\overline{F}(x)$ are given, respectively, by (1) and (2). Substituting (1) and (2) into (3), the likelihood function is

$$\mathscr{L}(\alpha, \beta \mid \mathbf{x}) = c^* \left[1 - (1 + \beta x_{r+1})^{-\alpha} \right]^r (\alpha\beta)^{m-r}$$
$$\times \prod_{i=r+1}^{m} (1 + \beta x_i)^{-\alpha(1+R_i)-1}. \quad (5)$$

Using the binomial expansion, r is a positive integer, one can rewrite the likelihood function as follows:

$$\mathscr{L}(\alpha, \beta \mid \mathbf{x}) = c^* (\alpha\beta)^{m-r} e^{-u} \sum_{s=0}^{r} C_{r,s} e^{-\alpha T_s}, \quad (6)$$

where

$$u = u(\mathbf{x}; \beta) = \sum_{i=r+1}^{m} \ln(1 + \beta x_i),$$

$$T_s = T_s(\mathbf{x}; \beta) = s \ln(1 + \beta x_{r+1})$$
$$- \sum_{i=r+1}^{m} (1 + R_i) \ln(1 + \beta x_i), \quad (7)$$

$$C_{r,s} = \binom{r}{s}(-1)^s.$$

We focus our attention on the estimation of the probability $S = P(Y < X)$, where X and Y are two independent random variables each is $L(\alpha_j, \beta_j)$, $j = 1, 2$ distributed, and the data obtained from both distributions are general progressively type-II censored. Here, X and Y are typically modeled as independent. The probability $P(Y < X)$ has been widely studied under different approaches and distributional assumptions on X and Y. The case where X and Y are dependent has been considered by Nandi and Aich [22], Barbiero [23], and Rubio and Steel [24]. We investigate properties of S when the common scale parameter $\beta_1 = \beta_2 = \beta$ is known. Then, it can be shown that

$$S = P(Y < X) = \iint_{0<y<x} f(x, y)\,dx\,dy = \frac{\alpha_1}{\alpha_1 + \alpha_2}. \quad (8)$$

Here, $f(x, y)$ is the joint pdf of X and Y, $f(x, y) = f_1(x)f_2(y)$ by the independence. The general case, when $\beta_1 \neq \beta_2$, can be studied in a similar manner. We obtain that

$$S = P(Y < X)$$
$$= \alpha_1\beta_1 \int_0^\infty (1 + \beta_1 x)^{-(\alpha_1+1)}(1 + \beta_2 x)^{-\alpha_2}\,dx. \quad (9)$$

3. Maximum Likelihood Estimation of S

Suppose that $X_1, X_2, \ldots, X_{n_1}$ and $Y_1, Y_2, \ldots, Y_{n_2}$ are two independent random samples of size n_1 and n_2 from $L(\alpha_1, \beta_1)$ and $L(\alpha_2, \beta_2)$ distributions, respectively. The log-likelihood function, with ignoring constants, is given by

$$\ln \mathscr{L}(\Theta \mid \mathbf{x}, \mathbf{y}) = \sum_{j=1}^{2} (m_j - r_j)(\ln \alpha_j + \ln \beta_j)$$

$$+ r_1 \ln \left[1 - (1 + \beta_1 x_{r_1+1})^{-\alpha_1} \right]$$

$$- \sum_{i=r_1+1}^{m_1} (\alpha_1 (1 + R_{1i}) + 1) \ln (1 + \beta_1 x_i)$$

$$+ r_2 \ln \left[1 - (1 + \beta_2 y_{r_2+1})^{-\alpha_2} \right]$$

$$- \sum_{i=r_2+1}^{m_2} (\alpha_2 (1 + R_{2i}) + 1) \ln (1 + \beta_2 y_i), \tag{10}$$

where $\Theta = (\alpha_1, \alpha_2, \beta_1, \beta_2)$. The MLEs of (α_1, β_1), say $(\widehat{\alpha}_{1\mathrm{ML}}, \widehat{\beta}_{1\mathrm{ML}})$, are obtained as the solution of the system of equations

$$\frac{\partial \ln \mathscr{L}}{\partial \alpha_1} = \frac{m_1 - r_1}{\alpha_1} + \frac{r_1 \ln (1 + \beta_1 x_{r_1+1})}{(1 + \beta_1 x_{r_1+1})^{\alpha_1} - 1}$$

$$- \sum_{i=r_1+1}^{m_1} (1 + R_{1i}) \ln (1 + \beta_1 x_i) = 0,$$

$$\tag{11}$$

$$\frac{\partial \ln \mathscr{L}}{\partial \beta_1} = \frac{m_1 - r_1}{\beta_1} + \frac{\alpha_1 r_1 x_{r_1+1}(1 + \beta_1 x_{r_1+1})^{-1}}{(1 + \beta_1 x_{r_1+1})^{\alpha_1} - 1}$$

$$- \sum_{i=r_1+1}^{m_1} (\alpha_1 (1 + R_{1i}) + 1) \frac{x_i}{1 + \beta_1 x_i} = 0.$$

In a similar way, we can obtain the MLEs of (α_2, β_2): say $(\widehat{\alpha}_{2\mathrm{ML}}, \widehat{\beta}_{2\mathrm{ML}})$.

The corresponding "ML plug-in estimation" of S, when $\beta_1 = \beta_2 = \beta$, is obtained by replacing α_1 and α_2 by its MLEs, $\widehat{\alpha}_{1\mathrm{ML}}, \widehat{\alpha}_{2\mathrm{ML}}$, and substituting them into relation (8) which yields

$$\widehat{S}_{\mathrm{ML}} = \frac{\widehat{\alpha}_{1\mathrm{ML}}}{\widehat{\alpha}_{1\mathrm{ML}} + \widehat{\alpha}_{2\mathrm{ML}}}. \tag{12}$$

For the general case, $\beta_1 \neq \beta_2$, the corresponding "ML plug-in estimation" of S is obtained by replacing $\alpha_1, \beta_1, \alpha_2,$ and β_2 by its MLEs, $\widehat{\alpha}_{1\mathrm{ML}}, \widehat{\beta}_{1\mathrm{ML}}, \widehat{\alpha}_{2\mathrm{ML}},$ and $\widehat{\beta}_{2\mathrm{ML}}$ and substituting them into relation (9) which results in obtaining $\widehat{S}_{\mathrm{ML}}$ as

$$\widehat{S}_{\mathrm{ML}} = \widehat{\alpha}_{1\mathrm{ML}} \widehat{\beta}_{1\mathrm{ML}}$$

$$\times \int_0^\infty (1 + \widehat{\beta}_{1\mathrm{ML}} x)^{-(\widehat{\alpha}_{1\mathrm{ML}}+1)} (1 + \widehat{\beta}_{2\mathrm{ML}} x)^{-\widehat{\alpha}_{2\mathrm{ML}}} dx. \tag{13}$$

4. Bayes Estimation

In this section, Bayesian estimation for the probability S in the stress-strength model involving Lomax distribution is obtained. The estimation is based on balanced loss function (BLF) which is introduced by Zellner [25]. We will use an extended class of BLF introduced by Jozani et al. [26]. It is of the following form:

$$L_{\rho,\omega,\delta_0}^{q}(\zeta(\theta),\delta) = \omega q(\theta)\rho(\delta_0,\delta) + (1-\omega)q(\theta)\rho(\zeta(\theta),\delta), \tag{14}$$

where $q(\cdot)$ is a suitable positive weight function and $\rho(\zeta(\theta),\delta)$ is an arbitrary loss function when estimating $\zeta(\theta)$ by δ. The parameter δ_0 is a chosen priori estimate of $\zeta(\theta)$, obtained for instance from the criterion of maximum likelihood, least squares, or unbiasedness among others. An intuitive interpretation of the BLFs is given by Ahmadi et al. [27] who argue that they give a general Bayesian connection between the case of $\omega > 0$, and $\omega = 0$ where $0 \leq \omega < 1$. By choosing $\rho(\zeta(\theta),\delta) = (\delta - \zeta(\theta))^2$ and $q(\theta) = 1$, the BLF is reduced to the balanced squared error loss (BSEL) function, used by Ahmadi et al. [27], in the form

$$L_{\omega,\delta_0}(\zeta(\theta),\delta) = \omega(\delta - \delta_0)^2 + (1-\omega)\rho(\delta - \zeta(\theta))^2. \tag{15}$$

The corresponding Bayes estimate of the function $\zeta(\theta)$ is given by

$$\delta_{\omega,\zeta,\delta_0}(\mathbf{x}) = \omega\delta_0 + (1-\omega)E(\zeta(\theta)\mid\mathbf{x}). \tag{16}$$

Also, by choosing $\rho(\zeta(\theta),\delta) = e^{a(\delta-\zeta(\theta))} - a(\delta-\zeta(\theta)) - 1$ and $q(\theta) = 1$, we get the balanced LINEX, abbreviated as BLINEX, loss function written as

$$L_{a,\omega,\delta_0}(\zeta(\theta),\delta) = \omega\left[e^{a(\delta-\delta_0)} - a(\delta-\delta_0) - 1\right]$$

$$+ (1-\omega)\left[e^{a(\delta-\zeta(\theta))} - a(\delta-\zeta(\theta)) - 1\right]. \tag{17}$$

In this case, the Bayes estimate of $\zeta(\theta)$ takes the form

$$\delta_{a,\omega,\zeta,\delta_0}(\mathbf{x}) = -\frac{1}{a} \ln \left[\omega e^{-a\delta_0} + (1-\omega)E\left(e^{-a\zeta(\theta)}\mid\mathbf{x}\right) \right], \tag{18}$$

where $a \neq 0$ is the shape parameter of BLINEX loss function.

4.1. Bayes Estimation When $\beta_1 = \beta_2 = \beta$. Assuming that $\beta_1 = \beta_2 = \beta$ (β is known) and α_1, α_2 are random variable each having gamma prior with some parameters, we can write

$$\pi(\alpha_j) = \frac{\delta_j^{\nu_j}}{\Gamma(\nu_j)} \alpha_j^{\nu_j-1} e^{-\delta_j \alpha_j}, \quad \alpha_j > 0, \ (\nu_j, \delta_j > 0), \ j = 1, 2. \tag{19}$$

Since $\alpha_j, j = 1, 2$ are independent, by combining the likelihood function with the priors pdf, the joint posterior density function of α_1 and α_2 is given by

$$\pi^*(\alpha_1, \alpha_2 \mid \mathbf{x}, \mathbf{y}) = J \prod_{j=1}^{2} \alpha_j^{A_j^*-1} e^{-\delta_j \alpha_j - u_j^*} \sum_{s=0}^{r_j} C_{r_j,s} e^{-\alpha_j T_{js}^*}, \tag{20}$$

where $A_j^* = m_j - r_j + \nu_j$, $T_{1s}^* = T_s(\mathbf{x}; \beta)$, $T_{2s}^* = T_s(\mathbf{y}; \beta)$, $u_1^* = u(\mathbf{x}; \beta)$, $u_2^* = u(\mathbf{y}; \beta)$, and

$$J^{-1} = \prod_{j=1}^{2} \int_0^\infty \alpha_j^{A_j^*-1} e^{-\delta_j \alpha_j - u_j^*} \sum_{s=0}^{r_j} C_{r_j,s} e^{-\alpha_j T_{js}^*} d\alpha_j. \quad (21)$$

Under the BSEL function and using (12) and (20), the proposed "Bayesian estimators" of S are actually Bayesian "plug-in" estimators as they are obtained by replacing $(\alpha_1, \beta_1, \alpha_2, \beta_2)$ with its Bayesian estimator $(\widehat{\alpha}_1, \widehat{\beta}_1, \widehat{\alpha}_2, \widehat{\beta}_2)$

$$\widehat{S}_{\text{BS}} = \omega \widehat{S}_{\text{ML}} + (1 - \omega) \int_0^\infty \int_0^\infty S \pi^*(\alpha_1, \alpha_2 \mid \mathbf{x}, \mathbf{y}) \, d\alpha_1 \, d\alpha_2. \quad (22)$$

With the same argument, we can obtain Bayes estimator under the BLINEX loss function. It is obtained as follows:

$$\widehat{S}_{\text{BL}} = -\frac{1}{a} \ln \left[\omega e^{-a\widehat{S}_{\text{ML}}} + (1 - \omega) \right. $$
$$\left. \times \int_0^\infty \int_0^\infty e^{-aS} \pi^*(\alpha_1, \alpha_2 \mid \mathbf{x}, \mathbf{y}) \, d\alpha_1 \, d\alpha_2 \right], \quad (23)$$

where $\widehat{S}_{\text{ML}}$ is the ML "plug-in" estimate of S as given by (12).

4.2. Bayes Estimation when $\beta_1 \neq \beta_2$. We assume that α_j and β_j, $j = 1, 2$, are random variables each having gamma prior with some parameters; that is,

$$\pi(\alpha_j \mid \beta_j) = \frac{b_j^{a_j}}{\Gamma(a_j)} \alpha_j^{a_j-1} e^{-b_j \alpha_j},$$

$$\alpha_j > 0, \quad (a_j, b_j > 0), \quad j = 1, 2,$$

$$\pi(\beta_j) = \frac{d_j^{c_j}}{\Gamma(c_j)} \beta_j^{c_j-1} e^{-d_j \beta_j}, \quad (24)$$

$$\beta_j > 0, \quad (c_j, d_j > 0), \quad j = 1, 2.$$

Since α_j and β_j are independent, then the joint density function of (α_j, β_j) is given by

$$\pi(\alpha_j, \beta_j) = \frac{b_j^{a_j} d_j^{c_j}}{\Gamma(a_j) \Gamma(c_j)} \alpha_j^{a_j-1} \beta_j^{c_j-1} e^{-b_j \alpha_j - d_j \beta_j}, \quad j = 1, 2. \quad (25)$$

Combining the likelihood function with the priors pdf yields the posterior density function of all parameters $\Theta = (\alpha_1, \alpha_2, \beta_1, \beta_2)$ as follows:

$$\pi^*(\Theta \mid \mathbf{x}, \mathbf{y}) = K \prod_{j=1}^{2} \alpha_j^{A_j-1} \beta_j^{B_j-1} e^{-u_j - d_j \beta_j}$$
$$\times \sum_{s=0}^{r_j} C_{r_j,s} e^{-\alpha_j (T_{js} + b_j)}, \quad (26)$$

where $A_j = m_j - r_j + a_j$, $B_j = m_j - r_j + c_j$, $T_{1s} = T_s(\mathbf{x}; \beta_1)$, $T_{2s} = T_s(\mathbf{y}; \beta_2)$, $u_1 = u(\mathbf{x}; \beta_1)$, $u_2 = u(\mathbf{y}; \beta_2)$, and

$$K^{-1} = \prod_{j=1}^{2} \int_0^\infty \int_0^\infty \alpha_j^{A_j-1} \beta_j^{B_j-1} e^{-u_j - d_j \beta_j}$$
$$\times \sum_{s=0}^{r_j} C_{rj,s} e^{-\alpha_j (T_{js} + b_j)} d\alpha_j \, d\beta_j. \quad (27)$$

Under the BSEL function, and by using (13) and (26), the Bayesian "plug-in" estimate of S is given by

$$\widehat{S}_{\text{BS}} = \omega \widehat{S}_{\text{ML}} + (1 - \omega) \int_\Theta S \pi^*(\Theta \mid \mathbf{x}, \mathbf{y}) \, d\Theta. \quad (28)$$

Also, based on the BLINEX loss function, the Bayes estimate of S is obtained by using (13) and is written as

$$\widehat{S}_{\text{BL}} = -\frac{1}{a} \ln \left[\omega e^{-a\widehat{S}_{\text{ML}}} + (1 - \omega) \int_\Theta e^{-aS} \pi^*(\Theta \mid \mathbf{x}, \mathbf{y}) \, d\Theta \right], \quad (29)$$

where $\widehat{S}_{\text{ML}}$ is the ML "plug-in" estimate of S as given by (13).

It may be noted, from (22), (23), (28), and (29), that the Bayes estimates of S contain integrals that cannot be obtained in simple closed form, and numerical techniques must be used for computations. We, therefore, propose to consider MCMC methods.

5. MCMC Algorithm for Bayesian Estimation

The MCMC algorithm is conducted to compare the Bayes estimates of S. We consider the Metropolis-Hastings algorithm to generate samples from the conditional posterior distributions, and then we compute the Bayes estimates. For more details about the MCMC methods, see, for example, Robert and Casella [28], Upadhyaya and Gupta [29], and Jaheen and Al Harbi [30]. The Metropolis-Hastings algorithm generates samples from an arbitrary proposal distribution.

5.1. The Case When $\beta_1 = \beta_2 = \beta$. The conditional posteriors distributions of the parameters α_j, $j = 1, 2$, can be computed and written, respectively, as

$$\pi^*(\alpha_1 \alpha_2, \mathbf{x}, \mathbf{y}) \propto J \alpha_1^{A_1^*-1} e^{-\delta_1 \alpha_1} \sum_{s=0}^{r_1} C_{r_1,s} e^{-\alpha_1 T_{1s}^*},$$

$$\pi^*(\alpha_2 \mid \alpha_1, \mathbf{x}, \mathbf{y}) \propto J \alpha_2^{A_2^*-1} e^{-\delta_2 \alpha_2} \sum_{s=0}^{r_2} C_{r_2,s} e^{-\alpha_2 T_{2s}^*}. \quad (30)$$

5.2. The Case When $\beta_1 \neq \beta_2$. The conditional posteriors distributions of the parameters α_j and β_j, $j = 1, 2$, can be computed and written, respectively, by

$$\pi^* \left(\alpha_1 \Theta - \alpha_1, \mathbf{x}, \mathbf{y} \right) \propto \alpha_1^{A_1 - 1} \sum_{s=0}^{r_1} C_{r_1, s} e^{-\alpha_1 (T_{1s} + b_1)},$$

$$\pi^* \left(\beta_1 \Theta - \beta_1, \mathbf{x}, \mathbf{y} \right) \propto \beta_1^{B_1 - 1} e^{-u_1 - d_1 \beta_1} \sum_{s=0}^{r_1} C_{r_1, s} e^{-\alpha_1 (T_{1s} + b_1)},$$

$$\pi^* \left(\alpha_2 \Theta - \alpha_2, \mathbf{x}, \mathbf{y} \right) \propto \alpha_2^{A_2 - 1} \sum_{s=0}^{r_2} C_{r_2, s} e^{-\alpha_2 (T_{2s} + b_2)},$$

$$\pi^* \left(\beta_2 \mid \Theta - \beta_2, \mathbf{x}, \mathbf{y} \right) \propto \beta_2^{B_2 - 1} e^{-u_2 - d_2 \beta_2} \sum_{s=0}^{r_2} C_{r_2, s} e^{-\alpha_2 (T_{2s} + b_2)}.$$

$$(31)$$

The following MCMC procedure is proposed to compute Bayes estimators for $S = P(Y < X)$.

Step 1. Start with initial guess of α_j and β_j; say α_{j0} and β_{j0}, $j = 1, 2$.

Step 2. Set $i = 1$.

Step 3. Generate α_{1i} from $\pi^*(\alpha_1 \mid \Theta - \alpha_1, \mathbf{x}, \mathbf{y})$ and β_{1i} from $\pi^*(\beta_1 \mid \Theta - \beta_1, \mathbf{x}, \mathbf{y})$.

Step 4. Generate α_{2i} from $\pi^*(\alpha_2 \mid \Theta - \alpha_2, \mathbf{x}, \mathbf{y})$ and β_{2i} from $\pi^*(\beta_2 \mid \Theta - \beta_2, \mathbf{x}, \mathbf{y})$.

Step 5. Compute $S(\alpha_i, \beta_i) = \alpha_{1i} \beta_{1i} \int_0^\infty (1 + \beta_{1i} x)^{-(\alpha_{1i} + 1)} (1 + \beta_{2i} x)^{-\alpha_{2i}} dx$.

Step 6. Repeat Steps 2–5, N times.

Now, the approximate means of $S(\alpha_i, \beta_i)$ with respect to the posterior distribution are given, respectively, by

$$E\left(S \mid \mathbf{x}, \mathbf{y} \right) = \frac{1}{N - M} \sum_{i=M+1}^{N} S\left(\alpha_i, \beta_i \right),$$

$$E\left[\exp\left(-aS \right) \mid \mathbf{x}, \mathbf{y} \right] = \frac{1}{N - M} \sum_{i=M+1}^{N} \exp\left(-aS\left(\alpha_i, \beta_i \right) \right),$$

$$(32)$$

where M is the burn-in period. Therefore, the Bayes estimators of $S = S(\alpha_i, \beta_i)$ based on BSEL and BLINEX loss functions are given, respectively, by

$$\widehat{S}_{\text{BS}} = \omega \widehat{S}_{\text{ML}} + (1 - \omega) E\left(S\mathbf{x}, \mathbf{y} \right),$$

$$\widehat{S}_{\text{BL}} = -\frac{1}{a} \ln \left[\omega e^{-a \widehat{S}_{\text{ML}}} + (1 - \omega) E\left[\exp\left(-aS \right) \mid \mathbf{x}, \mathbf{y} \right] \right].$$

$$(33)$$

6. Simulation Study

In order to find the Bayes and likelihood estimates of the parameter S, a Monte Carlo study is performed following the algorithms as follows.

(1) For particular values of α_j and β_j, $j = 1, 2$, Lomax observations of various sizes are generated for different general progressive censored schemes.

TABLE 1: Censoring schemes used in the simulation study.

CS	n	m	r	R
(i)	50	35	5	$(6^*0, 5, 4^*0, 2, 3^*0, 3, 13^*0, 5)$
(ii)	30	20	2	$(6^*0, 5, 10^*0, 5)$
(iii)	25	20	2	$(8^*0, 5, 9^*0)$
(iv)	35	30	5	$(5^*0, 2, 3^*0, 3, 15^*0)$
(v)	40	35	3	$(12^*0, 5, 19^*0)$
(vi)	30	25	2	$(8^*0, 3, 1, 1, 12^*0)$
(vii)	80	75	5	$(30^*0, 1, 10^*0, 2, 5^*0, 2, 22^*0)$
(viii)	70	60	3	$(2^*0, 2, 15^*0, 3, 10^*0, 3, 21^*0, 2, 5^*0)$

TABLE 2: The simulation results and estimates of S when $\beta_1 = \beta_2 = \beta$.

CS (1)	CS (2)	ML		Bayes (MCMC)		
		S_{ML}	S_{BS}	S_{BL}		
				$a = -2$	$a = 1$	$a = 3$
(i)	(ii)	0.4871	0.4867	0.4871	0.4865	0.4860
		(0.0017)	(0.0016)	(0.0016)	(0.0016)	(0.0015)
(iii)	(iv)	0.5471	0.5389	0.5393	0.5387	0.5383
		(0.0092)	(0.0079)	(0.0079)	(0.0078)	(0.0078)
(v)	(vi)	0.5408	0.5386	0.5388	0.5384	0.5382
		(0.0080)	(0.0079)	(0.0079)	(0.0078)	(0.0078)
(vii)	(viii)	0.5548	0.5436	0.5440	0.5434	0.5430
		(0.0107)	(0.0088)	(0.0089)	(0.0088)	(0.0087)

(2) The ML estimates of α_j and β_j, $j = 1, 2$, are computed from the ML equations. The ML estimate of S is computed from (13) after replacing α_j and β_j, $j = 1, 2$, by their ML estimates.

(3) For $N = 20000$, $M = 2000$, the Bayes estimates of S are computed from (22), (23), (28), and (29) for BSEL and BLINEX loss functions based on MCMC algorithm.

(4) The squared deviations $(S^* - S)^2$ are calculated for different sample sizes and different schemes, where S^* is ML or Bayes estimates of S.

(5) The above steps are repeated 1000 times, and the estimated risk (ER) is computed by averaging the squared deviations over the 1000 repetitions.

The computational results are displayed in Tables 1, 2, and 3. Table 1 shows different censoring schemes used in the simulation study. In the case of $\beta_1 = \beta_2 = \beta$, we take $v_1 = 4$, $\delta_1 = 3$, $v_2 = 3$, $\delta_2 = 5$, $\omega = 0.5$, $\alpha_1 = 1.1339$, $\beta = 3$, $\alpha_2 = 0.7216$, and the true value of $S = 0.4516$. For this case, Table 2 presents simulation results and the MLE, Bayes estimate, and the corresponding mean squared error is reported within bracket. In the general case, we take $a_1 = 4$, $b_1 = 3$, $a_2 = 3$, $b_2 = 2$, $\omega = 0.5$, $\alpha_1 = 1.5788$, $\beta_1 = 1.6899$, $\alpha_2 = 1.5690$, $\beta_2 = 2.2002$, and the true value of $S = 0.6111$. The obtained simulation results for this case are shown in Table 3.

TABLE 3: The simulation results and estimates of S when $\beta_1 \neq \beta_2$.

CS (1)	CS (2)	ML		Bayes (MCMC)		
		S_{ML}	S_{BS}	S_{BL}		
				$a = -2$	$a = 1$	$a = 3$
(i)	(ii)	0.7217	0.6968	0.6977	0.6964	0.6954
		(0.0124)	(0.0077)	(0.0078)	(0.0076)	(0.0074)
(iii)	(iv)	0.6440	0.6085	0.6102	0.6077	0.6059
		(0.0013)	(0.0004)	(0.0004)	(0.0004)	(0.0005)
(v)	(vi)	0.7001	0.6731	0.6741	0.6726	0.6716
		(0.0080)	(0.0041)	(0.0042)	(0.0040)	(0.0039)
(vii)	(viii)	0.6771	0.6418	0.6434	0.6410	0.6395
		(0.0044)	(0.0010)	(0.0011)	(0.0010)	(0.0009)

7. Conclusions

In this paper, the estimation of the stress-strength parameter, S, for two Lomax distributions under general progressive type-II censoring has been considered. The maximum likelihood and Bayes estimators of the stress-strength parameter have been derived. The MCMC method is used for computing Bayes estimates. It is observed that Bayes estimators outperform the ML estimators in small samples, while the estimators are almost equally efficient in large samples. It may be noted, from Tables 2 and 3, that the Bayes estimates have the smallest mean squared errors as compared with their corresponding maximum likelihood estimates. Based on the obtained results in this study and because of the need to deal with small samples in life testing, we recommend to use Bayes estimators in place of ML estimators.

Acknowledgments

The authors are grateful to the editor for his valuable comments and suggestions which improved the presentation of the paper. This Project was funded by the Deanship of Scientific Research (DSR), King Abdulaziz University, Jeddah, under Grant no. 268/130/1432. The authors, therefore, acknowledge with thanks DSR support for Scientific Research.

References

[1] A. Balkema and L. de Haan, "Residual life time at great age," *Annals of Probability*, vol. 2, no. 5, pp. 792–804, 1974.

[2] M. C. Bryson, "Heavy-tailed distributions: properties and tests," *Technometrics*, vol. 16, no. 1, pp. 61–68, 1974.

[3] M. Ahsanullah, "Record values of Lomax distribution," *Statistica Nederlandica*, vol. 41, no. 1, pp. 21–29, 1991.

[4] N. Balakrishnan and M. Ahsanullah, "Relations for single and product moments of record values from Lomax distribution," *Sankhya B*, vol. 56, no. 2, pp. 140–146, 1994.

[5] A. Childs, N. Balakrishnan, and M. Moshref, "Order statistics from non-identical right-truncated Lomax random variables with applications," *Statistical Papers*, vol. 42, no. 2, pp. 187–206, 2001.

[6] B. C. Arnold, E. Castillo, and J. M. Sarabia, "Bayesian analysis for classical distributions using conditionally specified priors," *Sankhya B*, vol. 60, no. 2, pp. 228–245, 1998.

[7] M. M. El-Din, H. M. Okasha, and B. Al-Zahrani, "Empirical bayes estimators of reliability performances using progressive type-II censoring from Lomax model," *Journal of Advanced Research in Applied Mathematics*, vol. 5, no. 1, pp. 74–83, 2013.

[8] H. A. Howlader and A. M. Hossain, "Bayesian survival estimation of Pareto distribution of the second kind based on failure-censored data," *Computational Statistics and Data Analysis*, vol. 38, no. 3, pp. 301–314, 2002.

[9] M. E. Ghitany, F. A. Al-Awadhi, and L. A. Alkhalfan, "Marshall-Olkin extended lomax distribution and its application to censored data," *Communications in Statistics*, vol. 36, no. 10, pp. 1855–1866, 2007.

[10] E. Cramer and A. B. Schmiedt, "Progressively type-II censored competing risks data from Lomax distributions," *Computational Statistics and Data Analysis*, vol. 55, no. 3, pp. 1285–1303, 2011.

[11] N. Johnson, S. Kotz, and N. Balakrishnan, *Continuous Univariate Distribution*, vol. 1, John Wiley & Sons, New York, NY, USA, 2nd edition, 1994.

[12] B. C. Arnold, *Pareto Distributions*, International Cooperative, Silver Spring, Md, USA, 1983.

[13] S. Kotz, Y. Lumelskii, and M. Pensky, *The Stress-Strength Model and Its Generalizations: Theory and Applications*, World Scientific, New York, NY, USA, 2003.

[14] A. Baklizi, "Likelihood and Bayesian estimation of $Pr (X < Y)$ using lower record values from the generalized exponential distribution," *Computational Statistics and Data Analysis*, vol. 52, no. 7, pp. 3468–3473, 2008.

[15] M. Z. Raqab, M. T. Madi, and D. Kundu, "Estimation of $P(Y < X)$ for the three-parameter generalized exponential distribution," *Communications in Statistics*, vol. 37, no. 18, pp. 2854–2864, 2008.

[16] D. Kundu and M. Z. Raqab, "Estimation of $R = P(Y < X)$ for three-parameter Weibull distribution," *Statistics and Probability Letters*, vol. 79, no. 17, pp. 1839–1846, 2009.

[17] H. Panahi and S. Asadi, "Inference of stress-strength model for a lomax distribution," *Proceedings of World Academy of Science, Engineering and Technology*, vol. 7, no. 55, pp. 275–278, 2011.

[18] A. A. Soliman, A. Y. Al-Hossain, and M. M. Al-Harbi, "Predicting observables from Weibull model based on general progressive censored data with asymmetric loss," *Statistical Methodology*, vol. 8, no. 5, pp. 451–461, 2011.

[19] N. Balakrishnan and R. A. Sandhu, "A simple simulational algorithm for generating progressive type-II censored samples," *The American Statistician*, vol. 49, no. 2, pp. 229–230, 1995.

[20] B. Saraoglu, I. Kinaci, and D. Kundu, "On estimation of $R = P(Y < X)$ for exponential distribution under progressive type-II censoring," *Journal of Statistical Computation and Simulation*, vol. 82, no. 5, pp. 729–744, 2012.

[21] N. Balakrishnan and R. Aggarwala, *Progressive Censoring: Theory, Methods and Applications*, Birkhauser, Boston, Mass, USA, 2000.

[22] B. Nandi and A. B. Aich, "A note on confidence bounds for $P(Y > X)$ in bivariate normal samples," *Sankhya B*, vol. 56, no. 2, pp. 129–136, 1994.

[23] A. Barbiero, "Interval estimators for reliability: the bivariate normal case," *Journal of Applied Statistics*, vol. 39, no. 3, pp. 501–512, 2012.

[24] F. J. Rubio and M. F. Steel, "Bayesian inference for $P(X < Y)$ using asymmetric dependent distributions," *Bayesian Analysis*, vol. 7, no. 3, pp. 771–792, 2012.

[25] A. Zellner, "Bayesian and non-Bayesian estimation using balanced loss functions," in *Statistical Decision Theory and Methods*, J. O. Berger and S. S. Gupta, Eds., pp. 337–390, Springer, New York, NY, USA, 1994.

[26] M. J. Jozani, E. Marchand, and A. Parsian, "On estimation with weighted balanced-type loss function," *Statistics and Probability Letters*, vol. 76, no. 8, pp. 773–780, 2006.

[27] J. Ahmadi, M. J. Jozani, E. Marchand, and A. Parsian, "Bayes estimation based on k-record data from a general class of distributions under balanced type loss functions," *Journal of Statistical Planning and Inference*, vol. 139, no. 3, pp. 1180–1189, 2009.

[28] C. P. Robert and G. Casella, *Monte Carlo Statistical Methods*, Springer, New York, NY, USA, 2005.

[29] S. K. Upadhyay and A. Gupta, "A bayes analysis of modified Weibull distribution via Markov chain Monte Carlo simulation," *Journal of Statistical Computation and Simulation*, vol. 80, no. 3, pp. 241–254, 2010.

[30] Z. F. Jaheen and M. M. Al Harbi, "Bayesian estimation for the exponentiated Weibull model via Markov chain Monte Carlo simulation," *Communications in Statistics: Simulation and Computation*, vol. 40, no. 4, pp. 532–543, 2011.

11

The Relationship between Health and Household Economic Status Using Spatial Measures in Iraq, 2004

Faisal G. Khamis

Faculty of Economics and Administrative Sciences, AL-Zaytoonah University of Jordan, P.O. Box 130, Amman 11733, Jordan

Correspondence should be addressed to Faisal G. Khamis, faisal_ukm@yahoo.com

Academic Editor: Kai Yuan Cai

This study addresses spatial effects by applying spatial analysis in studying whether household economic status (HES) is related to health across governorates in Iraq. The aim is to assess variation in health and whether this variation is accounted for by variation in HES. A spatial univariate and bivariate autocorrelation measures were applied to cross-sectional data from census conducted in 2004. The hypothesis of spatial clustering for HES was confirmed by a positive global Moran's I of 0.28 with $P = 0.010$, while for health was not confirmed by a negative global Moran's I of -0.03. Based on local Moran's I_i, two and seven significant clusters in health and in HES were found respectively. Bivariate spatial correlation between health and HES wasn't found significant ($I_{xy} = -0.08$) with $P = 0.80$. In conclusion, geographical variation was found in each of health and HES. Based on visual inspection, the patterns formed by governorates with lowest health and those with lowest HES were partly identical. However, this study cannot support the hypothesis that variation in HES may spatially explain variation in health. Further research is needed to understand mechanisms underlying the influence of neighbourhood context.

1. Introduction

The economic status hypothesis proposes that HES in a community or population influences health because unfavorable comparisons lead to families with a lower position to experience negative emotions that cause stress and detrimentally impact health, and well-being, and individuals with different statuses are less likely to develop trust and cohesion with one another. These processes are important for individual and family health, and also because their results may detract from community level social resources. Research on neighborhoods and health is motivated by the idea that we live in places that represent more than physical locations. They are also the manifestation of the social, cultural, political, and geographic cleavages that shape a constellation of risks and resources. Research on neighborhood effects has reconnected public health with its earlier population foundations, showing that the social ecology and built environments are important "upstream" determinants of chronic and infectious disease.

The HES is most influenced and is more expressive of the deterioration of Iraq's economic conditions throughout the cities and rural areas and was heavily affected by the new developments during the year of survey and the years before. AL-Rubiay and AL-Rubaiy [1] studied the distribution of skin diseases in Basrah governorate, southern area in Iraq. They found that skin diseases are major problems in the community especially in those of low socioeconomic status. Their finding is consistent with what found in this paper, where Basrah was found as hot cluster in both health and HES.

Low socioeconomic status (SES), measured in low educational attainment and household income, are consistently related to greater disease severity, poorer lung function, and greater physical functional limitations in cross-sectional analysis [2]. The study of Arku et al. [3] investigated the relationship between housing and self-reported general and mental health in Ghana, where they found that housing conditions, demand and control residents have to where they live, emerged as significant predictors of self-reported general and mental health status. A $10 000 increase in income increased the odds of better self-rated health by 10% for those with two or more chronic conditions [4]. Allender et al. [5] concluded that rich wards surrounded by poor areas have higher coronary heart disease (CHD) mortality rates than rich wards surrounded by rich areas, and poor wards surrounded by rich areas have worse CHD mortality rates than poor wards surrounded by poor areas. Bassanesi [6] stated that within Brazilian cities, the disadvantaged social groups are spatially segregated, where this segregation is bad for the health of poor districts and good for the health of the rich districts. This process of segregation leading to divergent health outcomes depending on the socioeconomic profile of communities may intensify health inequalities. In Indian cities where the poor are more isolated in their neighbourhoods have higher mortality rates than cities where the poor are less isolated; whereas cities where the poor are clustered into fewer neighbourhoods have lower mortality rates than cities where the poor are more evenly spread out [7].

People of lower socioeconomic position face significant communication challenges which may negatively impact their health [8]. In South Korea, Jung-Choi et al. [9] found graded inverse association between income and mortality for most but not all, specific causes of death. The results of Fors et al. [10] showed that income inequalities were associated with mortality. As stated by Charloite et al. [11], chronic illnesses (CI) are one of the indicators that measure children's health, where their results in Denmark and Sweden were children in families with one or both parents without paid work had an increased prevalence of recurrent psychosomatic symptoms, and Bambra [12] showed that the relationship between work, worklessness, and health inequalities were influenced by the broader political and economic context in the form of welfare state regimes. High levels of household crowding and poor social, economic and environmental conditions in many Australian communities appear to place major constraints on the potential for building programs to impact on the occurrence of common childhood illness [13]. In all countries, rich and poor, there is an unequal distribution of health both within countries and within cities [14].

2. Materials and Methods

2.1. Data. The data were collected from Iraq household socioeconomic survey, based on census conducted in 2004. For each of ($N = 18$) governorate, health and HES data were applied. The percentage data of HES includes indicators of the financial position of the household, work, ownership, and the household's assessment of their overall economic status. The percentage data of health field includes indicators of child health, reproductive health, and chronic diseases as well as the time needed to reach a health center and the level of satisfaction with health services.

2.2. Analysis. Data analysis involved five steps. In step 1, the health and HES data are tested for normal distribution, where they were found to follow approximately normal distribution. In step 2, visual inspection based on the quantified gradients for each of health and HES data using quartiles were conducted. Step 3 included the calculation of global

Moran's I for each of health and HES variables to detect the global clustering and also the significance of I-statistic using permutation test for each variable was examined. Step 4 involved the calculation of local Moran's I_i for ith governorate and its P value using Monte Carlo simulation to detect the local clusters for health and HES. In step 5, using quartiles, visual inspection of local Moran's values for each variable was inspected based on choropleth mapping.

The health and HES data were categorized into four intervals. These intervals were used for all maps using lighter shades of gray to indicate increasing values of health and HES. Such approach enables qualitative evaluation of spatial pattern. In the neighbourhood researches, neighbours may be defined as governorates which border each other or within a certain distance of each other. In this paper, neighbouring structure was defined as governorates which share a boundary. The *second-order* method (queen pattern) which included both the first-order neighbours (rook pattern) and those diagonally linked (bishop pattern) was used. A neighbourhood system was given in Figure 1, where ID neighbour for each governorate was shown. Although maps allow us to visually assess spatial pattern, they have two important limitations: their interpretation varies from person to person, and there is the possibility that a perceived pattern is actually the result of chance factors, and thus not meaningful. For these reasons, it makes sense to compute a numerical measure of spatial pattern, which can be accomplished using spatial autocorrelation. Therefore, global spatial clustering and local spatial clusters were identified.

2.2.1. Identification of Global Spatial Clustering. The goal of a global index of spatial autocorrelation is to summarize the degree to which similar observations tend to occur near each other in geographic space. In this exploratory spatial analysis, the spatial autocorrelation using standard normal deviate (z-value) of Moran's I under normal assumption was tested. Moran's I is a coefficient used to measure the strength of spatial autocorrelation in regional data. The interpretation of the Moran's statistic is as follows. If $I > E(I)$, then a governorate tends to be connected to the governorates that have similar attribute values and vice versa. Global clustering test was used to determine whether clustering was existed throughout the study area, without determining statistical significance of local clusters. Moran's I is calculated as follows [15]:

$$I = \frac{N \sum_{i=1}^{N} \sum_{j=1}^{N} w_{ij}(x_i - \overline{x})\left(x_j - \overline{x}\right)}{S_0 \sum_{i=1}^{N} (x_i - \overline{x})^2},$$
$$S_0 = \sum_{i=1}^{N} \sum_{j=1}^{N} w_{ij},$$
$$i \neq j, \quad (1)$$

where $N = 18$ is the number of governorates, $w_{ij} = 1$ is a weight denoting the strength of the connection between two governorates i and j that shared a boundary; otherwise, w_{ij} = zero, x_i and x_j represent the health or HES in ith and jth governorate, respectively. The autocorrelation coefficient can be used to test the null hypothesis of no

FIGURE 1: Study area shows all governorates with their IDs and the neighbours of each governorate.

spatial autocorrelation or spatially independent versus the alternative of positive spatial autocorrelation:

H_0 : no clustering exists (no spatial autocorrelation);

H_1 : clustering exists (positive spatial autocorrelation).

$$(2)$$

A significant positive value of Moran's I indicates positive spatial autocorrelation, showing the overall pattern for the governorates having a high/low level of health or HES similar to their neighbouring governorates. To test the significance of global Moran's I, z-statistic which follows a standard normal distribution was applied. It was calculated as follows [16]:

$$z = \frac{I - E(I)}{\sqrt{\mathrm{var}(I)}}.$$

$$(3)$$

Permutation test was used. What a permutation test tells us is that a certain pattern in data was or was not likely to have arisen by chance. The observations of each health and HES were randomly reallocated 1000 times with 1000 spatial autocorrelations calculated in each time to test the null hypothesis of randomness. That is to say, the hypothesis under investigation suggests that there will be a tendency for

a certain type of spatial pattern to appear in data, whereas the null hypothesis says that if this pattern was present, then this was a purely chance effect of observations in a random order. The analysis suggests an evidence of clustering if the result of the global test is significant, but it does identify the locations of any particular clusters. Besides, the existence and location of localized spatial clusters in the study population are of interest in geographic sociology. Accordingly, local spatial statistic was advocated for identifying and assessing potential clusters.

2.2.2. Identification of Local Spatial Clusters. A global index can suggest *clustering* but cannot identify individual *clusters* [17]. Anselin [18] proposed the local Moran's I_i statistic to test the local autocorrelation, where local spatial clusters, may be identified as those locations or sets of contiguous locations for which the local Moran's I_i was significant.

Clusters may be due to aggregations of high values, aggregations of low values, or aggregations of moderate values. Thereby, high value of I_i statistic suggests cluster of similar (but not necessarily large) values across several governorates, and low value of I_i suggests an outlying cluster in a single governorate i (being different from most or all of its neighbours). A positive local Moran's value indicates local stability, such as governorate that has high/low health surrounded by governorate that has high/low health. A negative local Moran's value indicates local instability, such as governorate has low health surrounded by governorate has high health or vice versa. However, each governorate's I_i value can be mapped to provide insight into the location of governorates with comparatively high or low local association with its neighbouring values.

Anselin stated that the indication of local patterns of spatial association may be in line with a global indication although this is not necessarily the case. It is quite possible that the local pattern is an aberration that the global indicator would not pick up, or it may be that a few local patterns run in the opposite direction. Thereby, this is what found for health variable, where global Moran's I of health was found not significant, but some local clusters of high health were found significant. However, Moran's I_i for ith governorate may be defined as [17]:

$$I_i = \frac{(x_i - \bar{x})}{S} \sum_{j=1}^{N} \left[\left(\frac{w_{ij}}{\sum_{j=1}^{N_i} w_{ij}} \right) \frac{(x_j - \bar{x})}{S} \right], \quad i = 1, 2, \ldots, 18,$$

$$(4)$$

where, analogous to the global Moran's I, the x_i and x_j represents the health or HES in ith and jth governorate, respectively, N_i is the number of neighbours for ith governorate, and S is the standard deviation. It was noteworthy that the number of neighbours for ith governorate were taken into account in I_i statistic by the amount: $(w_{ij}/\sum_{j=1}^{N_i} w_{ij})$, where w_{ij} was measured in the same manner as in Moran's I statistic. Local Moran's statistic was used to test the null hypothesis of *no clusters*. However, local Moran's statistic is a decomposition of global Moran's I into the contributions of small areas. In the statistical analysis, all programs performed in $S + 8$ Software.

2.2.3. Bivariate Spatial Association. So far, only univariate spatial association is presented that quantifies the spatial structure of one variable at a time. There is much discussion about what is an appropriate measure for bivariate spatial association. However, spatial dependence or spatial clustering causes losing in the information that each observation carries. When N observations are made on a variable that is spatially dependent (and that dependence is positive so that nearby values tend to be similar), the amount of information carried by the sample is less than the amount of information that would be carried if the N observations are independent because a certain amount of the information carried by each observation is duplicated by other observations in the cluster. A general consequence of this is that the sampling variance of statistics is underestimated. As the level of spatial dependence increases the underestimation increases. The problem is that when spatial autocorrelation is present, the variance of the sampling distribution of, for example, Pearson's correlation coefficient, which is a function of the number of pairs of observations, is underestimated. Spatial autocorrelation

coefficient can be modified to estimate the bivariate spatial correlation between two variables [19]:

$$I_{xy} = \frac{1}{S_0} \frac{\sum_{i=1}^{N} \sum_{j=1}^{N} w_{ij}(x_i - \bar{x})\left(y_j - \bar{y}\right)}{\left[\sqrt{\sum_{i=1}^{N} (x_i - \bar{x})^2 / N} \right] \left[\sqrt{\sum_{j=1}^{N} \left(y_j - \bar{y}\right)^2 / N} \right]}, \quad (5)$$

where x and y are the health and HES variables, respectively. Although the mathematics is quite straightforward, very few software packages offer the option of computing I_{xy}. Thus, programming was used to find I_{xy}. To test the significance of I_{xy}, z-statistic was applied: $z = I_{xy}\sqrt{N-1}$, which follows approximately standard normal distribution.

3. Results

Descriptive analyses were performed to assess the demographic characteristics of data set. The mean and standard deviations for health were found 22.30 and 6.91, respectively; skewness and kurtosis were found −0.40 and −0.46, respectively. The five-number summary of health data consists of the minimum, maximum, and quartiles written in increasing order: Min = 7.60, Q_1 = 16.95, Q_2 = 22.80, Q_3 = 27.73, and Max = 32.70. From the five-number summary, the variations of the four quarters were found 9.35, 5.85, 4.93, and 4.97, respectively, where the first quarter has the greatest variation of all. Also, descriptive statistics were calculated for HES, where the mean and standard deviations were found 56.19 and 9.17, respectively; skewness and kurtosis were found 0.86 and 1.51 respectively. The five-number summary of HES data set was found: Min = 42.80, Q_1 = 48.85, Q_2 = 56.05, Q_3 = 62.63, and Max = 80.40. From the five-number summary, the variations of the four quarters were found 6.05, 7.2, 6.58, and 17.77, respectively, where the fourth quarter has the greatest variation of all.

Basrah governorate accounted for the lowest health (7.60%). It was followed by the governorates Missan and Erbil, which accounted for (12.4%) and (15.7%), respectively. The highest rate was in the governorate of Suleimaniya with (32.7%). This can be explained by the persistent growth of economic activity in most fields, which provide more job opportunities, and this will lead to better levels of health. Kirkuk governorate accounted for the lowest HES (42.8%). It was followed by Erbil and Diala governorates which accounted for (43.9%) and (47.1%), respectively. The highest HES was found in the governorate AL-Muthanna with (80.4%). Figure 1 shows the study area explaining all governorates with their identification numbers (IDs).

Maps can display geographical inequality across governorates of Iraq. Since local Moran's I_i vary by location, it is easier to interpret it visually by color coding. Figures 2(a), 2(b), 2(c), and 2(d) show visual insight for health, and its local Moran's values, HES and its local Moran's values respectively, with darkest shade corresponding to lowest quartile. Based on visual inspection, an overall worsening pattern (lower scores) for health was found in northern, central, and eastern-southern governorates, such as 4, 5, 6, and 18. The suggestion of global clustering for health was not confirmed by a negative global Moran's I of −0.03 with

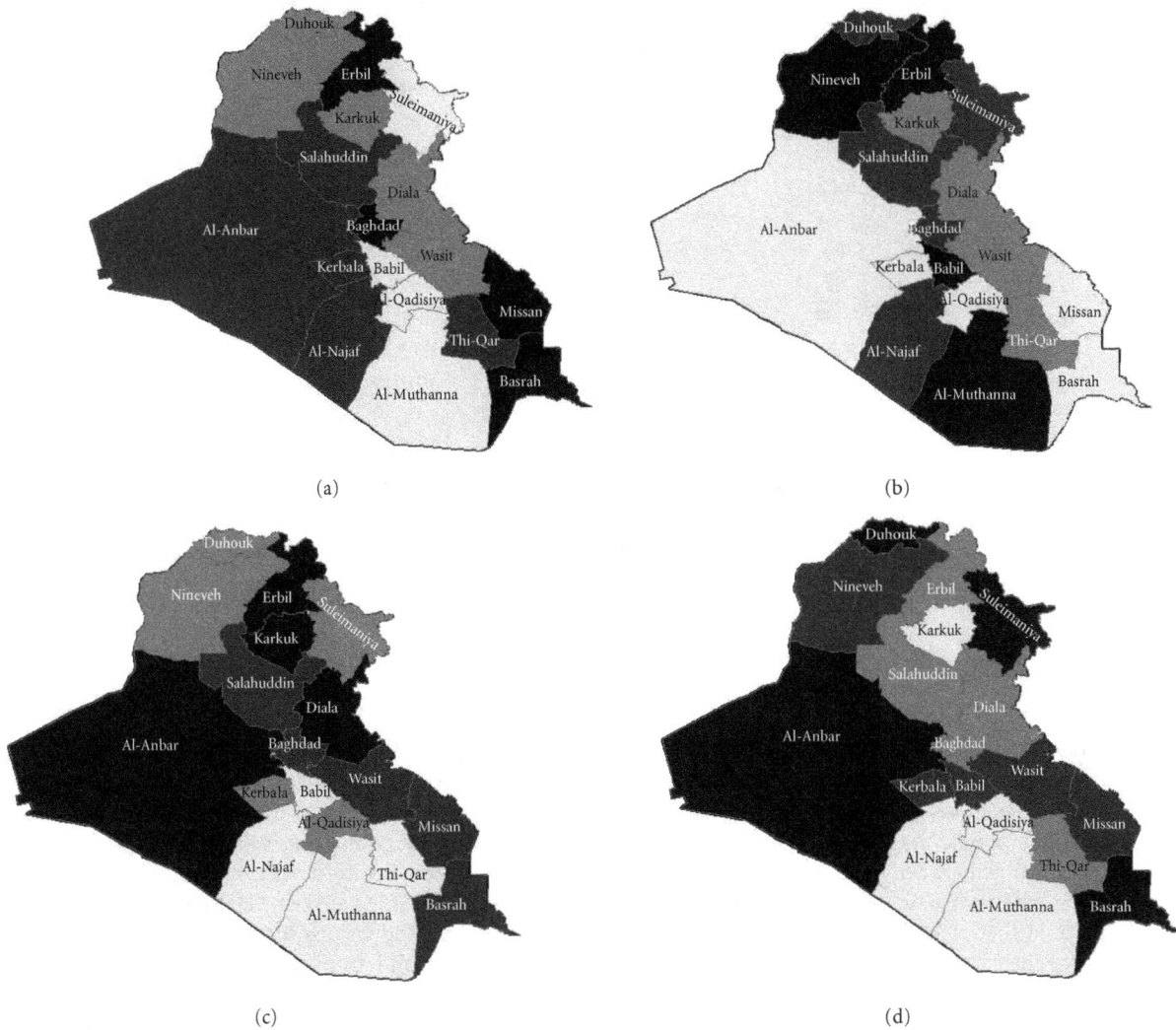

FIGURE 2: Choropleth maps show (a) health variable, (b) local Moran's values of health variable, (c) HES variable, and (d) local Moran's values of HES variable.

an associated $z = 0.20$ and $P = 0.840$; but confirmed with some local significant clusters of high health. Also, based on visual inspection, an overall worsening pattern (lower scores) for HES was found in the northern, western, and eastern-southern governorates, such as 1, 3, 7, and 18. The suggestion of global clustering for HES was confirmed by a positive global Moran's I of 0.28 with an associated $z = 2.57$ and $P = 0.010$.

To investigate global clustering, permutation test was done, where the permutation P value = 0.394 for health was found not significant; while permutation P value = 0.011 for HES was found significant. Thus, the null hypothesis of no spatial autocorrelation was not rejected for health but rejected for HES. The results of local Moran's I_i values for health and HES and their P values are reported in Table 1. Two significant clusters of high levels of health were found (14 and 17) as shown from their P values. The local significant clusters of high level of health in some governorates such as 14 and 17 could probably be contributed by the high level of HES in these governorates,

in some of their neighbours such as governorate 16, and/or by the HES inequality among their neighbours as shown in Figures 2(b) and 2(d). For HES, seven significant clusters were found (4, 5, 6, 12, 13, 14, and 15) as shown from their P-values.

Pearson's correlation coefficient between health and HES was found (0.38), which is not significant with ($P = 0.115$). Bivariate spatial correlation between health and HES was found ($I_{xy} = -0.08$) which is not significant with ($z = -0.33$ and $P = 0.80$). However, although both results were not significant, it is seen that Pearson coefficient is always over estimated when used in finding the spatial correlation. That is why, in investigating bivariate spatial correlation, it is recommended to use Wartenberg's [19] measure.

4. Discussion

The spatial association between spatial pattern of health and spatial pattern of HES was examined, allowing for the

TABLE 1: Shows both health (%) and HES (%), Local Moran's I_i values for health and HES, and their corresponding P values.

ID	Health	I_i for health	P value	HES	I_i for HES	P value
1	25.80	−.12	.625	58.90	−.18	.674
2	25.60	−.22	.761	57.20	−.07	.614
3	32.70	−.05	.578	57.90	−.21	.738
4	27.60	−.03	.548	42.80	.92	**.026**
5	15.70	−.49	.903	43.90	.43	**.062**
6	27.30	.04	.363	47.10	.38	**.098**
7	17.10	.07	.265	48.10	−.19	.767
8	16.50	−.09	.648	54.20	.11	.240
9	29.60	−.20	.766	63.80	.02	.377
10	19.40	.07	.308	58.10	.04	.352
11	23.80	.01	.401	49.10	−.11	.677
12	17.60	−.09	.654	49.50	.48	**.021**
13	21.10	−.07	.610	66.30	.77	**.013**
14	28.10	.40	**.066**	62.60	.63	**.024**
15	31.70	−.52	.896	80.40	1.50	**.001**
16	21.80	.02	.398	62.70	.31	.104
17	12.40	.95	**.024**	54.90	.01	.428
18	7.60	.10	.301	54.00	−.26	.753

effects of neighbouring governorates that share the boundary with a particular governorate. Findings allow policy makers to better identify what types of resources are needed and precisely where they should be employed. The rationale behind the relationship between health and HES is that people who live in or have a low level of HES usually suffer from financial strain that could lead to health problems for themselves and for their families.

After rejecting the null hypothesis, it becomes possible to conclude that there is some form of clustering, and it is of course of interest to know the exact nature of the clustering process. Is it only global type clustering or are there hot spot clusters? If the later, how many hot spots are there, and where are they located? In the analysis of the association between health and HES, exploratory tools are used such as descriptive tables and small-area choropleth mapping. Geographical distributions of health and HES were examined visually using maps.

The first wave of studies on neighbourhoods and health focused on showing the relevance of neighbourhoods and the effects beyond individual socioeconomic characteristics. These studies argued that neighbourhoods influence health by behavioral patterns such as collective socialization, peer-group influence, and institutional capacity. The second wave of the studies evaluated these mechanisms with latent measures of neighbourhood characteristics, such as level of segregation, collective social, and economic capacity [20]. However, health inequalities can only be reduced substantially if governorates have a democratic mandate to make the necessary policy changes, if demonstrably effective policies can be developed, and if these policies are implemented on the scale needed to reach the overall targets [21].

The HES may be associated with health reflecting the existing of individual income which provides good medical care, high quality of food, and acceptable household conditions. The usual correlation coefficients, such as Pearson coefficient, only test whether there is an association between two attributes by comparing values at the same location. Map comparison involves more than pairwise comparison between data recorded at the same locations as spatial units were arbitrary subdivisions of the study region and people could move around from one area to another and could be affected by HES levels in areas other than the area they live in, that is, the level of health in ith governorate was thought to be influenced by the levels of HES not just in ith governorate but also in neighbouring governorates. Neighbourhood residential turnover had been linked to poor child development, problem behavior, and health risks [22].

Permutation distribution can be used to test the significance of the global Moran's statistic. For this purpose, 1000 random permutations were applied. Simulated data is useful for validating the results of bivariate spatial analysis. However, using Monte Carlo simulation, 9999 random samples were simulated, 18 values for each sample, for each of health and HES. These samples (9999 matrices, each have two columns, one for health and the other for HES) were generated under bivariate standard normal distribution.

Whilst correlation obviously does not automatically imply any causation, there are two possibilities. First, low HES could cause low levels of health or second vice versa. Epidemiologic evidence suggests that the direction of causation from HES to health has a greater possibility than the converse (low health causes low HES). Although more research can be done to elucidate mechanisms and mediating factors, the present author found sufficient evidence to recommend that intervention research, to determine ways to reduce the adverse effect of HES on health. HES may exert detrimental effects on health through many mechanisms:

(1) by disrupting community and personal social relationships [23], (2) by leading to greater risk behavior, such as alcohol consumption and poor diet [24], (3) by causing stress [25], and (4) by precipitating reaction, like that caused by other losses [23]. It didnot assess the evidence for any particular mechanism or series of mechanisms since the main purpose in this study was to assess whether, not how, HES pattern is related to the pattern of health.

5. Conclusions

This study reports on the use of a particular form of spatial autocorrelation to group governorates according to how similar or dissimilar their health and HES are relative to surrounding areas. Exact causal mechanisms are not known but possibly include correlated health and HES. The findings support the common wisdom in the public health research domain that worsening pattern of health is more densely distributed in the areas where people have lower levels of HES. Findings, demonstrated that when health is associated with HES, thereby suggesting policies that improve HES growth may yield health returns. However, there is a possibility that a check of additional alternatives and a focus on other aggregate variables would have led to another conclusions.

Although, it cannot provide a causal relationship between health and HES, the results were conclusive in at least five aspects: First, based on mapping, low level of health was concentrated along the north-south axis, for instance in the governorates (5, 12, 7, 8, 17, and 18). Low HES was concentrated along the north-south axis, for instance, in the governorates (4, 5, 6, 7, 17, and 18). Based on *visual* inspection, the patterns formed by those governorates with lowest health and those with lowest HES were in general identical. Second, several governorates were not observed visually as hot spots for both health and HES, but after considering the information of their neighbours (i.e., calculating local Moran's I_i values), the pattern of hot spots, for example, governorates (9 and 15) for health, and governorates (1 and 9) for HES can obviously be seen. Third, based on global Moran's index, the clustering tendency showed that HES for each governorate can be spatially correlated with HES in neighbouring governorates, while the clustering tendency in health was not found significant. Fourth, the significance of bivariate spatial correlation didnot support the hypothesis that the spatial patterns of health and HES can be associated. Fifth, governorates which possess neighbours with high degree of inequality in health seem to show higher inequality in HES, for instance governorates (4 and 8). This was consistent with what Haining [26] stated, the levels of such variable in area i was thought to be influenced by the levels of another variable not just in area i but also in its neighbouring areas. This supports the hypothesis that the degree of variations in HES between these governorates and their neighbours could somewhat influence health. Global spatial pattern for health field was not found but some local clusters of high level of health were found in the southern part. Global spatial pattern for HES was found and several

local clusters of high level of HES were found in eastern-northern and western-southern parts.

References

[1] K. K. AL-Rubiay and L. K. AL-Rubaiy, "Dermatoepidemiology: a household survey among two urban areas in Basrah city, Iraq," *Internet Journal of Dermatology*, vol. 4, no. 2, pp. 1–10, 2006.

[2] M. D. Eisner, P. D. Blanc, T. A. Omachi et al., "Socioeconomic status, race and COPD health outcomes," *Journal of Epidemiology and Community Health*, vol. 65, no. 1, pp. 26–34, 2011.

[3] G. Arku, I. Luginaah, P. Mkandawire, P. Baiden, and A. B. Asiedu, "Housing and health in three contrasting neighbourhoods in Accra, Ghana," *Social Science and Medicine*, vol. 72, no. 11, pp. 1864–1872, 2011.

[4] F. I. Gunasekara, K. N. Carter, I. Liu, K. Richardson, and T. Blakely, "The relationship between income and health using longitudinal data from New Zealand," *Journal of Epidemiology and Community Health*. In press.

[5] S. Allender, P. Scarborough, T. Keegan, and M. Rayner, "Relative deprivation between neighbouring wards is predictive of coronary heart disease mortality after adjustment for absolute deprivation of wards," *Journal of Epidemiology and Community Health*. In press.

[6] S. L. Bassanesi, "Urbanization and spatial inequalities in health in Brazil," *Journal of Epidemiology and Community Health*. In press.

[7] T. Chandola, "Socioeconomic segregation in major Indian cities and mortality," *Journal of Epidemiology and Community Health*. In press.

[8] E. Z. Kontos, K. M. Emmons, E. Puleo, and K. Viswanath, "Determinants and beliefs of health information mavens among a lower-socioeconomic position and minority population," *Social Science and Medicine*, vol. 73, no. 1, pp. 22–32, 2011.

[9] K. Jung-Choi, Y. H. Khang, and H. J. Cho, "Socioeconomic differentials in cause-specific mortality among 1.4 million South Korean public servants and their dependents," *Journal of Epidemiology and Community Health*, vol. 65, no. 7, pp. 632–638, 2011.

[10] S. Fors, B. Modin, I. Koupil, and D. Vågerö, "Socioeconomic inequalities in circulatory and all-cause mortality after retirement: the impact of mid-life income and old-age pension. Evidence from the Uppsala Birth Cohort Study," *Journal of Epidemiology and Community Health*. In press.

[11] C. R. Pedersen, M. Madsen, and L. Köhler, "Does financial strain explain the association between children's morbidity and parental non-employment?" *Journal of Epidemiology and Community Health*, vol. 59, no. 4, pp. 316–321, 2005.

[12] C. Bambra, "Work, worklessness and the political economy of health inequalities," *Journal of Epidemiology and Community Health*, vol. 65, no. 9, pp. 746–750, 2011.

[13] R. S. Bailie, M. Stevens, and E. L. McDonald, "The impact of housing improvement and socio-environmental factors on common childhood illnesses: a cohort study in Indigenous Australian communities," *Journal of Epidemiology and Community Health*. In press.

[14] S. Friel, "The social and environmental determinants of urban health inequalities in low and middle income countries: findings from the Rockefeller foundation global research network on urban health equity," *Journal of Epidemiology and Community Health*. In press.

[15] A. D. Cliff and J. K. Ord, *Spatial Processes: Models & Applications*, Page Bros., London, UK, 1981.

[16] J. R. Weeks, *Population: An Introduction to Concepts and Issues*, Wadsworth Inc., 5th edition, 1992.

[17] L. A. Waller and C. A. Gotway, *Applied Spatial Statistics for Public Health Data*, John Wiley & Sons, Hoboken, NJ, USA, 2004.

[18] L. Anselin, "Local indicators of spatial association—LISA," *Geographical Analysis*, vol. 27, no. 2, pp. 93–115, 1995.

[19] D. Wartenberg, "Multivariate spatial correlation: a method for exploratory geographical analysis," *Geographical Analysis*, vol. 17, no. 4, pp. 263–283, 1985.

[20] R. J. Sampson, J. D. Morenoff, and T. Gannon-Rowley, "Assessing "neighborhood effects": social processes and new directions in research," *Annual Review of Sociology*, vol. 28, pp. 443–478, 2002.

[21] J. P. Mackenbach, "Can we reduce health inequalities? An analysis of the English strategy (1997–2010)," *Journal of Epidemiology and Community Health*, vol. 65, no. 7, pp. 568–575, 2011.

[22] T. Jelleyman and N. Spencer, "Residential mobility in childhood and health outcomes: a systematic review," *Journal of Epidemiology and Community Health*, vol. 62, no. 7, pp. 584–592, 2008.

[23] G. Jackson, "Alternative concepts and measures of unemployment," *Labour Force [Statistics Canada cat. no. 71-001]*, pp. 85–120, 1993.

[24] S. Morrell, R. Taylor, S. Quine, and C. Kerr, "Suicide and unemployment in Australia 1907–1990," *Social Science and Medicine*, vol. 36, no. 6, pp. 749–756, 1993.

[25] K. A. Moser, A. J. Fox, P. O. Goldblatt, and D. R. Jones, "Stress and heart disease: evidence of associations between unemployment and heart disease from the OPCS longitudinal study," *Postgraduate Medical Journal*, vol. 62, no. 730, pp. 797–799, 1986.

[26] R. Haining, *Spatial Data Analysis: Theory and Practice*, Cambridge University Press, Cambridge, UK, 2003.

12

A Robust Intelligent Framework for Multiple Response Statistical Optimization Problems Based on Artificial Neural Network and Taguchi Method

Ali Salmasnia,[1] Mahdi Bastan,[2] and Asghar Moeini[3]

[1] Department of Industrial Engineering, Faculty of Engineering, Tarbiat Modares University, Tehran, Iran
[2] Department of Industrial Engineering, Eyvanekey University, Semnan, Iran
[3] Department of Industrial Engineering, Faculty of Engineering, Shahed University, Tehran, Iran

Correspondence should be addressed to Ali Salmasnia, ali.salmasnia@modares.ac.ir

Academic Editor: Tadashi Dohi

An important problem encountered in product or process design is the setting of process variables to meet a required specification of quality characteristics (response variables), called a multiple response optimization (MRO) problem. Common optimization approaches often begin with estimating the relationship between the response variable with the process variables. Among these methods, response surface methodology (RSM), due to simplicity, has attracted most attention in recent years. However, in many manufacturing cases, on one hand, the relationship between the response variables with respect to the process variables is far too complex to be efficiently estimated; on the other hand, solving such an optimization problem with accurate techniques is associated with problem. Alternative approach presented in this paper is to use artificial neural network to estimate response functions and meet heuristic algorithms in process optimization. In addition, the proposed approach uses the Taguchi robust parameter design to overcome the common limitation of the existing multiple response approaches, which typically ignore the dispersion effect of the responses. The paper presents a case study to illustrate the effectiveness of the proposed intelligent framework for tackling multiple response optimization problems.

1. Introduction

Controllable input variables set to an industrial process to achieve proper operating conditions are one of the common problems in quality control. Taguchi method [1–3] is a widely accepted technique among industrial engineers and quality control practitioners for producing high quality products at low cost. In this regard, Ko et al. [4] employed Taguchi method and artificial neural network to perform design in multistage metal forming processes considering work ability limited by ductile fracture. Su et al. [5] proposed a new circuit design optimization method where genetic algorithm (GA) is combined with Taguchi method. Lo and Tsao [6] modified an analytical linkage-spring model based on neural network analysis and the Taguchi method to determine the design rules for reducing the loop height and the sagging altitude of gold wire-bonding process of

the integrated circuit (IC) package. In Taguchi's design method, the control variables (factors can be controlled by analyst) and noise variables (factors cannot be controlled by analyst) are considered influential on product quality. Therefore, the Taguchi method is to choose the levels of control variables and to reduce the effects of noise variables. That is, control variables setting should be determined with the intention that the quality characteristic (response variable) has minimum variation while its mean is close to the desired target. Nevertheless, so far, the Taguchi method can only be used for a single response problem; it cannot be used to optimize a multiple response optimization problem. But, in most industrial problems, we have dealt with more than one response variable and improving them simultaneously is very important. Common problem in the simultaneous optimization of response variables is to be different and sometimes contradictory to their optimality direction. Thus, optimizing the manufacturing process than one response variable led to nonoptimal amounts of other responses. So when dealing with multiresponse problems had better separately to optimize the response variables (Taguchi method) and finally, according to process engineer, is determined the optimum combination of design variables. Therefore, it is very important to design a method to optimize simultaneously responses. Another important point in the optimization process of the responses is to estimate the relationship between the response and control variables. In many cases, regression relationships do not have the ability to estimate properly the relationship between response and control variables and large amounts of mean square error (MSE) regression models can be seen that show the poor quality of these relationship descriptions [7]. In most cases, this problem occurs for two reasons: (i) reversal of the independence assumptions of input variables; (ii) being a complex relationship between response and control variables. In these cases, intelligent approaches (approach based on neural network and approach based on fuzzy) are an appropriate alternative to achieve a good estimation. In this regard, [8] proposed an approach based on neural networks to solve the quality optimization problem in Taguchi's dynamic experiment. However, this method is applicable only when there is a response variable.

Reference [9] proposed the neural network method and the data envelopment analysis (DEA) [10] to efficiently optimize the multiple response problem in the Taguchi method. With the neural network, the signal-to-noise (SN) ratios of responses are estimated by the known experimental data for each control variables combination, which also named decision making unit (DMU). Then, DEA is used to find each DMU's relative efficiency so that the optimal control variables combination can be found by relative efficiency value 100%. A three-step approach presented by [11] consists in (1) using neural networks to estimate mean square deviation (MSD) of responses for all possible combinations of control variable levels, (2) using DEA to compute the relative efficiency of all of those combinations, selecting those that are efficient, and (3) using DEA again to select among the efficient combinations the one which leads to a most robust quality loss penalization. A four step procedure to resolve the parameter design problem involving multiple responses is proposed by [12]. In this method, multiple signal-to-noise ratios are mapped into a single performance index called multiple response statistics (MRS) through neurofuzzy based model to identify the optimal level settings for each control variable. Analysis of variance is finally performed to identify control variables significant to the process. The above methods discuss only control variable values used in experimental trials; therefore, it cannot find the global optimal control variable settings considering all continual control variable values within the corresponding bounds.

Reference [13] presented the approach for solving problems with multiresponse surface using neural networks. In this approach, two neural networks are used, one for discovering optimal control factors vector and the other for estimating responses. Although parameter optimization can

be obtained, the effect of control variables on responses still cannot be achieved. A similar method based on artificial neural network (ANN) is presented by [14]. In this method, no matter whether the control variables are due to the level form or the real value, it can be employed. At the same time, the effect of the control variables multiple responses can be also obtained. Reference [15] proposed to use an artificial neural network to estimate the quantitative and qualitative response functions. In the optimization phase, a genetic algorithm (GA) in conjunction with a desirability function (DF) is used to determine the optimal control variable settings. Reference [16] presented a data mining approach to dynamic multiple response problem consisting of four stages which apply the methodologies of ANN, exponential desirability function (EDF), and simulated annealing (SA). First, an ANN is employed to construct the response model of a dynamic multiple response system by applying the experimental data to train the network. The response model is then employed to predict the corresponding quality responses by inputting specific control variable combinations. Second, each of the responses is evaluated by using EDF. Third, EDFs are integrated into an overall performance index (OPI) for evaluating a specific control variable combination. Finally, a SA is performed to obtain optimal control variable combination within experimental region. Another dynamic multiresponse approach is presented in [17]. In this method, similar to Chang's work [16], optimal phase is performed by GA, whereas optimal phase is performed by SA. Reference [18] focused on an optimization problem that involves multiple qualitative and quantitative responses in the thin quad flat pack (TQFP) modeling process. A fuzzy quality loss function is first employed to the qualitative responses. Neural network is then applied to estimate a nonlinear relationship between control and response variables. A GA together with EDF is applied to determine the optimal setting. Reference [19] presented the use of fuzzy-rule base reasoning and SN ratio for the optimization of multiple responses. The idea is to combine multiple SN ratios into a single performance index called multiple performance statistic (MPS) output, from which the optimum level settings of control variables can be obtained by maximizing MPS. A similar approach to [19] for optimizing the electrical discharge machining process with multiple performance characteristics has been reported by [20]. In this approach, several fuzzy rules are derived based on the performance requirement of the process. Next, the inference engine performs a fuzzy reasoning on fuzzy rules to generate a fuzzy value. Finally, the defuzzifier converts the fuzzy value into a single performance index and the optimal combination of the machining parameter levels can be determined based on maximizing performance index. Reference [21] formulated MRO problem as a multiobjective decision making problem and followed the basic idea of Zimmermann's [22] method. This approach first models the responses through multiple adaptive neurofuzzy inference system (MANFIS), then according to maximin approach, overall satisfaction is obtained by comprising via the use of membership functions among all the responses. Finally, a GA is applied to search the optimal solution on the response surfaces modeled by MANFIS.

A Robust Intelligent Framework for Multiple Response Statistical Optimization Problems Based on Artificial Neural
Network and Taguchi Method

99

With respect to the aforementioned approaches, it can be concluded that the major focus of these methods is on the location effect only, ignoring the dispersion effect of the responses. In other words, they assume that the variance for the responses is constant over the experimental space.

Reference [23] presented an integrated technique for experimental design of processes with multiple correlated responses, composed of three stages which (1) use expert system, designed for choosing an orthogonal array, to design an actual experiment, (2) use the Taguchi quality loss function to present relative significance of responses, principal component analysis (PCA) to uncorrelate responses, and gray relational analysis (GRA) to synthesize components into a single performance measure, (3) use neural networks to construct the response function model and genetic algorithms to optimize control variable design. An artificial intelligence technique that combines PCA, GRA, and GA with ANN and uses data collected from full factorial experimental design for optimization of Nd:YAG laser drilling of Ni-based superalloy sheets was proposed by [24]. We note that since principal components are linear combinations of original response variables, when PCA is conducted on quality loss values, their optimization directions might be lost. Regardless of this issue, aforementioned methods maximize the component values. In other words, they do not correctly consider the location effect of the responses. To overcome this problem, Salmasnia et al. [25] suggested a systematic procedure via PCA and desirability function that imposes specification limits on the responses to be achieved. Also, an AI tool, namely, ANFIS, is used to estimate the complicated relation between input (design variables) and outputs (responses), but this approach does not consider relative importance of responses in process optimization.

The purpose of this study is to develop a new intelligent approach that accommodates all of location and dispersion effects besides relative importance of responses in a single framework. It also does not depend on the type of relationship between response and control variables, hence making its application in cases where these relations are unknown. Another advantage of the proposed method which is in contrast to many other approaches considering discrete regions to search for optimal solution searches the experimental region continuously. We compare the characteristics of the different intelligent multiresponse approaches presented in literature to the proposed method in Table 1.

(i) Type of solution problem (TSP).

(ii) Aggregation approach (AA).

(iii) Location effect (LE).

(iv) Dispersion effect (DE).

(v) Relative importance of responses (RI).

(vi) Type of estimation (TE).

(vii) Type of search in the experimental region (TS).

The rest of the paper is organized in the following order. Section 2 describes the proposed general intelligent approach for the design of a multiple response process that uses the Taguchi signal-to-noise ratio function, ANN and GA. In Section 3, the application of the proposed model on a case study from literature is illustrated. Finally, conclusions are reported in Section 4.

2. The Proposed Method

This study proposes a robust intelligent optimization procedure for multiple response problems with complex relationship between response and process variables based on signal-to-noise ratio and artificial neural network. There are various methods to optimize multiple responses but most of them employ regression models to estimate relation function between response and process variables. Furthermore, they neglect dispersion effect of responses and assume that response variances are constant over the experimental space. This research proposes a new methodology which considers dispersion effect as well as location effect. In addition, the approach used to model building phase is artificial neural network (ANN), to resolve shortcomings of abovementioned regression models, to capture nonlinearity in relationship.

To develop the methodology, we first define the parameters and the variables used in the proposed approach. Then, the new methodology is described in detail.

2.1. The Parameters and Variables. The parameters and the variables used throughout this paper are defined as follows:

X: the design vector (a $p \times 1$ vector where p represents the number of controllable variables),

y_{ijk}: the observed value of the jth response under the ith experimental run in the kth replication,

$\overline{y}_{ij}$: the sample mean of the jth response under the ith experimental run,

S_{ij}: the sample standard deviation of the jth response under the ith experimental run,

SN_{ij}: the signal to noise (SN) ratio of the jth response under the ith experimental run,

NSN_{ij}: the normalized SN ration of the jth response under the ith experimental run,

$SN_{\min j}$: the minimum SN ratio for the jth response,

$SN_{\max j}$: the maximum SN ratio for the jth response,

w_j: the weight of the jth response,

Ω: the experimental region.

2.2. Model Development. The proposed method consists of three phases: (i) data gathering, (ii) response estimation, and (iii) optimization. In the first phase, by employing a proper experimental design, the significant factors are identified and then the required data are gathered. Next, in order to reduce the response variation and bring the response means close to the target values, signal-to-noise ratio and normalized values of them are calculated in each experimental run. The

TABLE 1: A characteristic comparison of the existing methods with the proposed approach.

Method	TSP	TS	LE	DE	RI	TE	AA
Su and Hsieh [8]	Single response	Continuous	✓	✓		Neural network	—
Ko et al. [4]	Single response	Continuous	✓	✓		Neural network	—
Lo and Tsao [6]	Single response	Discrete	✓	✓		Neural network	—
Hsieh and Tong [13]	Multiple response	Continuous	✓			Neural network	—
Hsieh [14]	Multiple response	Continuous	✓			Neural network	—
Liao [9]	Multiple response	Discrete	✓	✓		Neural network	DEA
Chiang and Su [18]	Multiple response	Continuous	✓			Neural network	EDF
Antony et al. [12]	Multiple response	Discrete	✓	✓		Neuro fuzzy	MRS
Cheng et al. [21]	Multiple response	Continuous	✓			MANFIS	—
Lin et al. [20]	Multiple response	Discrete	✓	✓		Fuzzy rule base	MPS
Tarng et al. [26]	Multiple response	Discrete	✓	✓		Fuzzy rule base	MPS
Lu and Antony [19]	Multiple response	Discrete	✓	✓		Fuzzy rule base	MPS
Noorossana et al. [15]	Multiple response	Continuous	✓		✓	Neural network	DF
Chang and Chen [17]	Multiple response	Continuous	✓			Neural network	EDF
Gutiérrez and Lozano [11]	Multiple response	Discrete	✓	✓		Neural network	DEA
Chatsirirungruang [27]	Multiple response	Continuous	✓			Linear regression	LF
Sibalija and Majstorovic [23]	Multiple response	Continuous		✓		Neural network	GRA
Salmasnia et al. [25]	Multiple response	Continuous	✓	✓		ANFIS	DF
The proposed method	Multiple response	Continuous	✓	✓	✓	Neural network	WSN

response estimation phase, an estimate of responses with respect to design variables, is calculated. To do this, artificial neural network is used as an estimator. Finally, the third phase consists of optimization of process using GA and finding the best solution. Figure 1 illustrates the conceptual framework of the proposed method.

Phase 1 (data gathering). This phase aims to gather the required data for training neural networks. This phase includes four steps that are described in the following.

Step 1 (identifying the significant control variables). The first step is to identify the process control variables that may influence the response(s) of interest which can be done by experts who are familiar to the area of system considered.

Step 2 (selecting a proper design of experiment). An experiment can be defined as a test or a set of tests in which purposeful changes are made on the control variables to identify the pattern of changes that may be observed in the response variables.

Step 3 (calculating the SN ratio for responses in each experimental run). Recently [1] introduced a family of performance measures called signal-to-noise (SN) ratios. The major aim of these criteria is to simultaneously reduce the response variation and bring the response means close to the target values. According to the Taguchi method, there are three types of responses. The responses with a fixed target are called the nominal of the best case (NTB). In addition, the cases in which the responses have a smaller-the-better target or larger-the-better target are called STB and LTB, respectively. For these cases, the SN ratios are defined as follows:

(i) nominal-the-best

$$SN_{ij} = 10 \log \left(\frac{\bar{y}_{ij}^2}{S_{ij}^2} \right),$$ (1)

(ii) larger-the-better

$$SN_{ij} = -10 \log \left(\frac{1}{m} \sum_{k=1}^{m} \frac{1}{y_{ijk}^2} \right),$$ (2)

(iii) smaller-the-better

$$SN_{ij} = -10 \log \left(\frac{1}{m} \sum_{k=1}^{m} y_{ijk}^2 \right).$$ (3)

Step 4 (normalizing the SN ratio for responses in each experimental run). The normalized SN ratio values can be computed using (4):

$$NSN_{ij} = \frac{SN_{ij} - SN_{\min j}}{SN_{\max j} - SN_{\min j}}.$$ (4)

The idea behind the normalization of SN ration values is to convert them into dimensionless numbers. This is simply because each response has different units of measurements. The NSN varies from a minimum of zero to a maximum of one (i.e., $0 \le NSN_{ij} \le 1$).

Phase 2 (response estimation). We suggest using BP neural networks to estimate the values of NSN of the different characteristics for all control variable combinations. To simplify training, we recommend using a separate BP neural network

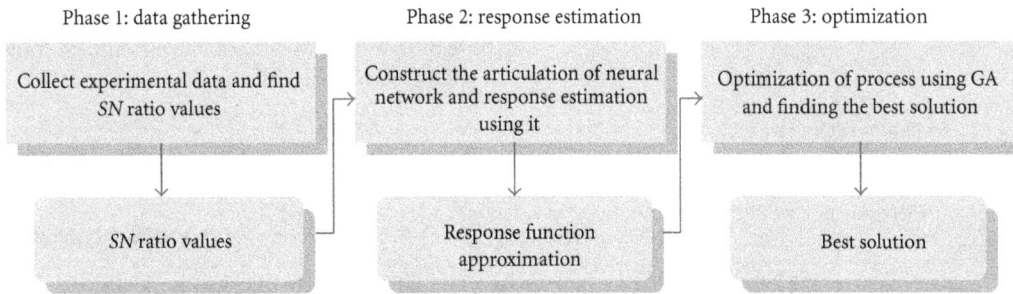

FIGURE 1: Conceptual framework of the proposed method.

for each quality characteristics. Each of these neural networks is trained with the data of the actual experiments. Each input pattern corresponds to a control variable combination, while the output is its associated SN ratio. The two main reasons for using neural networks for this task instead of other classical estimation (e.g., regression) are their non-parametric character and their generalization capability. Thus, on one hand, neural networks can approximate, without making any a prior assumption, any existing linear or nonlinear mapping between the control variables and SN ratios. On the other hand, well-trained neural networks are able to estimate, with acceptable error levels, the output values for any control variable combination, not just the ones experimentally tested. This phase consists of three steps as follow.

Step 1 (selection of the training and the testing data sets). It is usually that about one-fifth of the total data as the test data is randomly selected and the remaining data as the training data are considered [30].

Step 2 (determine the topology of neural network). Among the several conventional supervised learning neural networks are the perceptron, back propagation neural network (BPNN), learning vector quantization (LVQ), and counter propagation network (CPN). The BPNN model is employed due to its ability to achieve effective solutions for various industrial applications and neural networks power in modeling of a nonlinear and complex relationship between systems input and output in this study, to modeling the relationship between response and control variables.

At this step, a neural network would be trained for each response to estimate its relation with control variables. Thus, the number of input neurons equals the number of control variables; the output layer has one neuron corresponding to an NSN. The transfer function for all neurons in the hidden layer(s) is hyperbolic tangent activation function. According to definition of NSN, it can vary from zero to one; hence, the transfer function for the output neuron is tangent sigmoid function. The topology of the BP neural network with a single hidden layer-based process model used in the proposed approach is illustrated in Figure 2.

Step 3 (designing the most appropriate network's articulation to estimate each quality characteristic). As they are selected, the number of neurons of layers of input and output

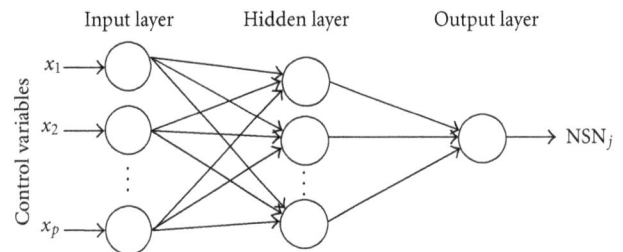

FIGURE 2: The topology of the BPNN with a hidden layer-based process model.

based on dimensions of the input and output vectors and appropriate number of hidden layer neurons often is set by using trial and error and based on indicators such as mean square error (MSE) or root mean square error (RMSE) laboratory, different back propagation networks will evaluate for discovering the appropriate network. Then, for each network is compared the network output for test data and training data with observations from experiments. Finally, a network with the lowest MSE is selected as optimal network.

Phase 3 (optimization). Once the BPNN has been properly trained and validated, they can be used to estimate the SN ratios for all possible control variable combinations. The next step is then to optimize process via GA. A GA is selected to perform the optimization for two important reasons. (1) Gradient-based optimization methods, like GRG, to calculate gradient and direction of improvement require response surface while in this method is used to estimate values instead of calculating the response surfaces from the neural network. (2) GA is known as a powerful heuristic search approach for optimization of complex and highly nonlinear functions. In the rest of this phase, first a robust parameter setting approach is suggested. Then, a brief introduction of GA and the implementation steps of it for finding optimal solution, shown in Figure 3, are given.

2.2.1. The Suggested Parameter Tuning Approach. Metaheuristics have a major drawback; they need some parameter tuning that is not easy to perform in a thorough manner. Those parameters are not only numerical values but may also involve the use of search components. Usually, metaheuristic designers tune one parameter at a time, and its optimal

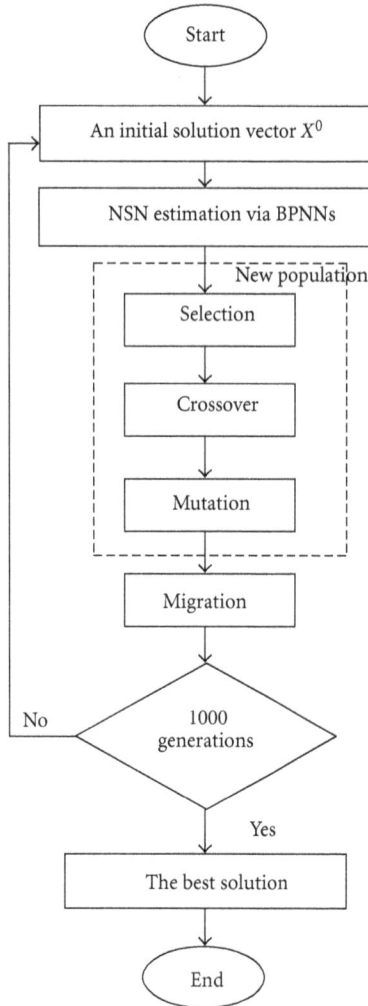

Figure 3: Optimization MRO via GA.

robustness in terms of the instances. These criteria usually have different scales. Hence, they should be transformed into a scale-free value as follows:

$$
d_{ijk} = \begin{cases} \dfrac{y_{ijk} - l_j}{u_j - l_j} & \text{for LTB type criteria,} \\[2ex] \dfrac{y_{ijk} - u_j}{l_j - u_j} & \text{for STB type criteria,} \end{cases} \tag{5}
$$

where u_j and l_j are desired upper and lower levels for the jth criteria, and d_{ijk} is dimensionless value corresponding to the observed value of the jth criterion under the ith experimental run in the kth replication that is called desirability value. It assigns values from 0 to 1 to the possible value of each objective function, in which a number closer to 1 is more desirable.

To aggregate several individual desirability values, the overall desirability (D) can be defined by taking the geometric mean of the individual desirability values. Therefore, overall desirability function yields a value less than or equal to the lowest individual desirability value. If this value is 0, one or more criterion is unacceptable. The most important feature of this approach is that an obtained optimal solution does not include any objective that lies outside the acceptable limits.

Step 6 (selection optimal setting). In order to reduce simultaneously the quality variation and bring the mean criteria close to the corresponding target values, signal-to-noise ratio should be conducted on overall desirability value. Next, the main effects on signal-to-noise ratios are determined. Thus, the corresponding diagram plots the factor effect on SN. The optimal factor/level combination produces the maximum SN value.

value is determined empirically. In this case, no interaction between parameters is studied. This sequential optimization strategy (i.e., one-by-one parameter) does not guarantee to find the optimal setting even if an exact optimization setting is performed.

To overcome this problem, a robust parameter tuning approach based on design of experiment, desirability function, and signal-to-noise ratio Taguchi is suggested. The proposed method consists of three steps: (1) design of experiment, (2) aggregation of objective functions, and (3) selection of optimal setting.

Step 4 (design of experiment). In this step, effective parameters such as mutation and crossover probabilities, search operators such as the type of selection strategy in evolutionary algorithms, the type, and so on are recognized. Next, a proper experimental design according to the number of effective parameters is selected.

Step 5 (aggregation of objective functions). Performance analysis of metaheuristics may be with respect to different criteria such as search time, quality of solutions, and

2.2.2. Genetic Algorithm for Solution Searching. The optimal solution is a set of control variables that maximizes weighted NSN (WNSN). After estimation of NSNs over control variables, we should apply a method to deal with the optimization segment. To this end, we implement genetic algorithm. Genetic algorithm was firstly introduced based on the Darwinian theory by [31]. It is one of the powerful stochastic search approaches and is widely employed for solving complex problems. In this method, a random initial population is created and probabilistic operations are used for evolving the subsequent generations. Through crossover and mutation operations, the algorithm directs the population towards the optimal solution. The quality of each individual is assessed with a fitness function which deals with the objective function of problem at hand. Each chromosome with better level of fitness has higher WNSN to generate the offspring. Through the evolution procedure the quality of offspring will be enhanced until a predefined stopping criterion is met. The major components of a genetic algorithm are as follows:

(1) initialization including parameters calibrations,

(2) determining a way to encode the solutions,

A Robust Intelligent Framework for Multiple Response Statistical Optimization Problems Based on Artificial Neural
Network and Taguchi Method

103

(3) generation of initial population,

(4) defining the operations that should be applied to the parents to generate next populations,

(5) a way to determine the fitness function that returns the quality of founded solutions.

Now here, we perform the genetic algorithm to solve MRO problem.

Solution Encoding. To represent each solution, a string of real numbers in the interval $[-1, 1]$ with size of control variable numbers in which each gene indicates the amount allocated to the corresponding control variable.

Initial Solution. Initial population provides the main algorithm with a starting point that can be created by some tailored heuristics. Here, we select to carry out the randomly generated initial solution.

Evaluation. In this section, the control variable combinations founded by GA should be evaluated. To do this, estimate the NSN values by inputting the control variable combination to trained BPNNs. Then, apply (6) to synthesize the obtained NSNs into a synthetic performance measure which is referred to as a fitness function. This value is returned to the main algorithm:

$$\text{WNSN} = \sum_{j=1}^{n} w_j \text{NSN}_j. \tag{6}$$

Selection. Selection operator chooses two individuals from population as parents to produce the offspring by crossover and mutation operators. The mechanism used in this section is based on the value of chromosome fitness. Chromosome with better fitness will have a greater WNSN to be chosen as a parent.

Crossover and Mutation Operators. Crossover is a GA operator that establishes new chromosomes by exchanging some parts of parents to create children which have characteristics from both parents. In this study, we carry out the uniform crossover in which a probability vector is produced. The size of vector is equal to number of control variables. If the value of probability is less than 0.5, the element from first parent is moved into child, otherwise the corresponding gene from second one is selected.

In order to increase the diversification of population, mutation operator is used to make random variations in chromosomes. In this study, random mutation is adopted where a random gene is selected from chromosome and a random number between 0 and 1 is replaced with its current element. See example for random mutation as follows.

Chromosome : $[0.54, 0.31, -0.84, 0.98, -0.21, -0.05]$

Random position : 4

Random factor level : -0.61

Offspring : $[0.54, 0.31, -0.84, -0.61, -0.21, -0.05]$

Termination Criterion. After predefined number of iterations, the algorithm terminates.

3. Numerical Illustration

In order to demonstrate the application of the proposed approach, in this section, a simulation study is carried out on the example given in [28]. In this example, there are two response variables (y_1, y_2) and five control variables $(x_1, x_2, x_3, x_4, x_5)$. It is assumed that y_1 and y_2 have the same relative importance and are smaller-the-better and larger-the-better, respectively. Five control variables, each with three levels, are allocated sequentially to an L_{18} orthogonal array. The experiments are conducted randomly.

The experimental data was analyzed by following the proposed method strictly. Table 2 shows the experimental observations. Table 3 displays SN ratios and NSN ratios for each response resulting from formula of data gathering phase.

According to the proposed method, next step is estimation of the NSNs of the different characteristics for control variables using neural networks. Since the neural networks with one or two hidden layers have ability to describe any nonlinear relationship between inputs and outputs and, on the other hand, increasing the number of layers leads to the rapid growth of the number of network parameters as a result in the process of identifying suitable neural network [32], we limit our studies with two layers networks. In order to discover the appropriate neural networks, response variables of the feed forward back propagation networks were tested with different parameters. Appropriate networks with the lowest MSE values are presented in Table 6. In both networks, the middle layers use from activation function of tangent hyperbolic and output layers use from activation function of sigmoid. Training algorithm in both networks is Levenberg-Marquardt, and ratio of the test data to the whole data is for both networks 22.22%.

The regression model considered for simulating the process and generating the data is illustrated in Tables 4 and 5 that are fitted using MINITAB 15 software.

The MSE of the two regression models is computed and presented in Table 7. As can be seen, the computed MSE from the regression models is high and this represents a poor fitness of the models. However, the two neural networks produce absolutely lower MSE. Therefore, neural networks can estimate the process function more accurately.

A GA optimization algorithm is performed on the Matlab platform at the final stage. The GA program is usually time consuming and needs many iterations to obtain convergence. However, for the present experiment, a relatively good solution was almost always obtained within 1000 iterations or in approximately 6 minutes. GA program is executed over 20 runs to set optimal control variable and the best solution obtained is $(x_1, x_2, x_3, x_4, x_5) = (0.9, 1, -1, -0.9, 0.88)$.

As mentioned before, RI is a main issue in MRO but it is considered less in intelligent approaches in literature. In order to illustrate the effect of it in process optimization, we resolve the problem with different weight vectors that are

TABLE 2: Experimental data.

Experimental number	Control variable					Response variable	
	x_1	x_2	x_3	x_4	x_5	y_1	y_2
1	−1	−1	−1	−1	−1	14.3	4
2	−1	0	0	0	0	15.7	4.3
3	−1	1	1	1	1	23.2	5.6
4	0	−1	0	0	0	12.1	3.7
5	0	0	1	1	1	8.7	4.9
6	0	1	−1	−1	−1	6.5	6.1
7	1	−1	0	−1	1	8.99	4.2
8	1	0	1	0	−1	11.8	4.3
9	1	1	−1	1	0	12.4	5.3
10	−1	−1	1	1	0	16.2	4.6
11	−1	0	−1	−1	1	26.9	4.1
12	−1	1	0	0	−1	10.5	5.3
13	0	−1	0	1	−1	16.9	3.9
14	0	0	1	−1	0	5.06	4.7
15	0	1	−1	0	1	7.08	5.4
16	1	−1	1	0	1	8.76	5.2
17	1	0	−1	1	−1	15.1	4.6
18	1	1	0	−1	0	5	5.8

TABLE 3: Signal-to-noise ratios and normalized values of them.

Experimental number	Control variable					SN ratio		NSN ratio	
	x_1	x_2	x_3	x_4	x_5	SN_1	SN_2	NSN_1	NSN_2
1	−1	−1	−1	−1	−1	−23.11	12.04	0.375513	0.156322
2	−1	0	0	0	0	−23.92	12.67	0.320109	0.301149
3	−1	1	1	1	1	−27.31	14.96	0.088235	0.827586
4	0	−1	0	0	0	−21.66	11.36	0.474692	0
5	0	0	1	1	1	−18.79	13.8	0.670999	0.56092
6	0	1	−1	−1	−1	−16.26	15.71	0.844049	1
7	1	−1	0	−1	1	−19.08	12.46	0.651163	0.252874
8	1	0	1	0	−1	−21.44	12.67	0.48974	0.301149
9	1	1	−1	1	0	−21.87	14.49	0.460328	0.71954
10	−1	−1	1	1	0	−24.19	13.26	0.301642	0.436782
11	−1	0	−1	−1	1	−28.6	12.26	0	0.206897
12	−1	1	0	0	−1	−20.42	14.49	0.559508	0.71954
13	0	−1	0	1	−1	−24.56	11.82	0.276334	0.105747
14	0	0	1	−1	0	−14.08	13.44	0.99316	0.478161
15	0	1	−1	0	1	−17	14.65	0.793434	0.756322
16	1	−1	1	0	1	−18.85	14.32	0.666895	0.68046
17	1	0	−1	1	−1	−23.58	13.26	0.343365	0.436782
18	1	1	0	−1	0	−13.98	15.27	1	0.898851

depicted in Table 8. As it was expected, the optimal value of factors and WNSN vary with respect to the weight vector.

Now, a comparative study between the proposed method and some major studies is represented. It shows the effectiveness of the proposed method against popular approaches in the literature. The comparison is conducted on WNSN that the higher the value, the more desirable the result. The results are summarized in the Table 9.

As mentioned before, Noorossana et al. [15] and Chang and Chen [17] are two approaches that emphasize only on the location effect of responses. Consequently, these approaches have poor performance in reducing variances of responses and also WNSN value.

Although Lin et al. [20], Tong et al. [28], and Tong et al. [29] have approved their results by considering variance in their method, they only consider discrete level combination

A Robust Intelligent Framework for Multiple Response Statistical Optimization Problems Based on Artificial Neural
Network and Taguchi Method

105

TABLE 4: Estimated effects and coefficients for NSN_1.

Trem	Effect	Coef	SE Coef	T
Constant		0.4931	0.05385	9.16
x_1	0.4588	0.2294	0.09007	2.55
x_2	0.3008	0.1504	0.06411	2.35
x_3	0.4991	0.2495	0.11179	2.23
x_4	−0.6258	−0.3129	0.12656	−2.47
x_5	0.0059	0.003	0.07864	0.04
$x_1 \times x_2$	−0.126	−0.063	0.10564	−0.6
$x_1 \times x_3$	−0.5102	−0.2551	0.16573	−1.54
$x_1 \times x_4$	0.4623	0.2312	0.16803	1.38
$x_1 \times x_5$	0.5489	0.2745	0.09386	2.92
$x_2 \times x_3$	−0.1459	−0.0729	0.12936	−0.56
$x_2 \times x_4$	−0.3425	−0.1713	0.10893	−1.57
$x_2 \times x_5$	0.3832	0.1916	0.12432	1.54
$x_3 \times x_4$	0.1434	0.0717	0.11759	0.61

TABLE 5: Estimated effects and coefficients for NSN_2.

Trem	Effect	Coef	SE Coef	T
Constant		0.4309	0.05067	8.5
x_1	0.3466	0.1733	0.08475	2.04
x_2	0.6185	0.3092	0.06032	5.13
x_3	0.3748	0.1874	0.10519	1.78
x_4	−0.2953	−0.1477	0.11908	−1.24
x_5	−0.0101	−0.0051	0.07399	−0.07
$x_1 \times x_2$	−0.2119	−0.106	0.0994	−1.07
$x_1 \times x_3$	−0.5436	−0.2718	0.15594	−1.07
$x_1 \times x_4$	0.4182	0.2091	0.1581	1.32
$x_1 \times x_5$	0.3681	0.1841	0.08831	2.08
$x_2 \times x_3$	−0.2392	−0.1196	0.12172	−0.98
$x_2 \times x_4$	−0.153	−0.0765	0.10249	−0.75
$x_2 \times x_5$	0.2316	0.1158	0.11698	0.99
$x_3 \times x_4$	0.4833	0.2416	0.11064	2.18

TABLE 6: Properties of the final neural networks.

Network	NSN ratio	Number of neurons in the hidden layers	MSE	
			Test	Train
1	NSN_1	4,6	0.034	1.41×10^{-13}
2	NSN_2	4,6	0.027	5.63×10^{-13}

TABLE 7: Computed MSE from the regression models.

NSN ratio		NSN_1	NSN_1
MSE	Test	0.1823	0.098786
	Train	7.56×10^{-9}	1.137×10^{-8}

of design variables used in experimental trials. Observed results indicate that their performance is dominated to the proposed method in WNSN.

Results of the numerical example support the claim that the proposed method, in contrast to other methods, considers mean and variance of responses and search experimental region continuously.

4. Conclusion

To overcome weakness of polynomial regression models in estimating the appropriate relationships between control and response variables in complex processes and difficulties of accurate optimization methods in the solution of such problems was presented a new approach based on neural network and genetic algorithm. The approach presented in addition to covering the weaknesses mentioned provides four other merits: (1) reduction of uncertainty in the process, (2) estimation of the relationship between control and response variables using traditional statistical approaches requires some statistical assumptions while the proposed approach without any assumptions is able to estimate such

TABLE 8: Optimal solution with different weights.

Method	w_1	w_2	x_1	x_2	x_3	x_4	x_5	WNSN
	0.1	0.9	1	1	0.48	0.3	0.92	1.118
Proposed method	0.5	0.5	0.9	1	−0.9	−0.7	0.88	1.431
	0.9	0.1	1	0.68	1	−1	0.73	1.103

TABLE 9: Comparison results for different methods in the field of MRO.

Method	x_1	x_2	x_3	x_4	x_5	WNSN
Lin et al. [20]	1	1	0	−1	−1	0.57
Noorossana et al. [15]	0.46	0.72	−0.2	0.5	0.32	0.628
Tong et al. [28]	1	1	1	−1	1	1.038
Tong et al. [29]	1	1	0	0	−1	0.426
Chang and Chen [17]	0.64	0.68	−0.72	0.63	0.78	0.734
Proposed method	0.9	1	−0.9	−0.7	0.88	1.431

a relationship, (3) to solve the optimization problems with multisurface responses using this general method considering relative importance of the response variables unlike more existing approaches, (4) there is no any undesirable mathematical complexities in the proposed approach.

As a future research, the qualitative variables can be considered as well as quantitative ones. Furthermore, it could be interesting to incorporate correlation among responses and also variance of the predicted responses into the proposed approach.

References

[1] G. Taguchi, *Introduction to Quality Engineering*, Asian Productivity Organization (Distributed by American Supplier Institute Inc.), Dearborn, Mich, USA, 1986.

[2] G. S. Peace, *Taguchi Methods: A Hands-On Approach*, Addison-Wesley, Boston, Mass, USA, 1993.

[3] M. S. Phadke, *Quality Engineering Using Robust Design*, Prentice-Hall, New York, NY, USA, 1989.

[4] D. C. Ko, D. H. Kim, and B. M. Kim, "Application of artificial neural network and Taguchi method to preform design in metal forming considering workability," *International Journal of Machine Tools and Manufacture*, vol. 39, no. 5, pp. 771–785, 1999.

[5] Y. Su, Z. Bao, F. Wang, and T. Watanabe, "Efficient GA approach combined with Taguchi method for mixed constrained circuit design," in *International Conference on Computational Science and Its Applications (ICCSA '11)*, pp. 290–293, 2011.

[6] Y. L. Lo and C. C. Tsao, "Integrated Taguchi method and neural network analysis of physical profiling in the wirebonding process," *IEEE Transactions on Components and Packaging Technologies*, vol. 25, no. 2, pp. 270–277, 2002.

[7] K. J. Kim, J. H. Byun, D. Min, and I. J. Jeong, *Multiresponse Surface Optimization: Concept, Methods, and Future Directions*, Tutorial, Korea Society for Quality Management, 2001.

[8] C. T. Su and K. L. Hsieh, "Applying neural networks to achieve robust design for dynamic quality characteristics," *International Journal of Quality and Reliability Management*, vol. 15, pp. 509–519, 1998.

[9] H. C. Liao, "A data envelopment analysis method for optimizing multi-response problem with censored data in the Taguchi method," *Computers and Industrial Engineering*, vol. 46, no. 4, pp. 817–835, 2004.

[10] A. Charnes, W. W. Cooper, and E. Rhodes, "Measuring the efficiency of decision making units," *European Journal of Operational Research*, vol. 2, no. 6, pp. 429–444, 1978.

[11] E. Gutiérrez and S. Lozano, "Data envelopment analysis of multiple response experiments," *Applied Mathematical Modelling*, vol. 34, no. 5, pp. 1139–1148, 2010.

[12] J. Antony, R. B. Anand, M. Kumar, and M. K. Tiwari, "Multiple response optimization using Taguchi methodology and neuro-fuzzy based model," *Journal of Manufacturing Technology Management*, vol. 17, no. 7, pp. 908–925, 2006.

[13] K. L. Hsieh and L. I. Tong, "Optimization of multiple quality responses involving qualitative and quantitative characteristics in IC manufacturing using neural networks," *Computers in Industry*, vol. 46, no. 1, pp. 1–12, 2001.

[14] K. L. Hsieh, "Parameter optimization of a multi-response process for lead frame manufacturing by employing artificial neural networks," *International Journal of Advanced Manufacturing Technology*, vol. 28, no. 5-6, pp. 584–591, 2006.

[15] R. Noorossana, S. Davanloo Tajbakhsh, and A. Saghaei, "An artificial neural network approach to multiple-response optimization," *International Journal of Advanced Manufacturing Technology*, vol. 40, no. 11-12, pp. 1227–1238, 2009.

[16] H. H. Chang, "A data mining approach to dynamic multiple responses in Taguchi experimental design," *Expert Systems with Applications*, vol. 35, no. 3, pp. 1095–1103, 2008.

[17] H. H. Chang and Y. K. Chen, "Neuro-genetic approach to optimize parameter design of dynamic multiresponse experiments," *Applied Soft Computing Journal*, vol. 11, no. 1, pp. 436–442, 2011.

[18] T. L. Chiang and C. T. Su, "Optimization of TQFP molding process using neuro-fuzzy-GA approach," *European Journal of Operational Research*, vol. 147, no. 1, pp. 156–164, 2003.

[19] D. Lu and J. Antony, "Optimization of multiple responses using a fuzzy-rule based inference system," *International Journal of Production Research*, vol. 40, no. 7, pp. 1613–1625, 2002.

[20] J. L. Lin, K. S. Wang, B. H. Yan, and Y. S. Tarng, "Optimization of the electrical discharge machining process based on the Taguchi method with fuzzy logics," *Journal of Materials Processing Technology*, vol. 102, no. 1, pp. 48–55, 2000.

[21] C. B. Cheng, C. J. Cheng, and E. S. Lee, "Neuro-fuzzy and genetic algorithm in multiple response optimization,"

A Robust Intelligent Framework for Multiple Response Statistical Optimization Problems Based on Artificial Neural Network and Taguchi Method

107

Computers and Mathematics with Applications, vol. 44, no. 12, pp. 1503–1514, 2002.

[22] H. J. Zimmermann, "Fuzzy programming and linear programming with several objective functions," *Fuzzy Sets and Systems*, vol. 1, no. 1, pp. 45–55, 1978.

[23] T. V. Sibalija and V. D. Majstorovic, "An integrated approach to optimize parameter design of multi-response processes based on Taguchi method and artificial intelligence ," *Journal Intelligent Manufacture*. In press.

[24] T. V. Sibalija, S. Z. Petronic, V. D. Majstorovic, R. Prokic-Cvetkovic, and A. Milosavljevic, "Multi-response design of Nd:YAG laser drilling of Ni-based superalloy sheets using Taguchi's quality loss function, multivariate statistical methods and artificial intelligence," *International Journal of Advanced Manufacturing Technology*, vol. 54, no. 5–8, pp. 537–552, 2011.

[25] A. Salmasnia, R. B. Kazemzadeh, and M. M. Tabrizi, "A novel approach for optimization of correlated multiple responses based on desirability function and fuzzy logics," *Neurocomputing*, vol. 91, pp. 56–66, 2012.

[26] Y. S. Tarng, W. H. Yang, and S. C. Juang, "Use of fuzzy logic in the Taguchi method for the optimization of the submerged arc welding process," *International Journal of Advanced Manufacturing Technology*, vol. 16, no. 9, pp. 688–694, 2000.

[27] P. Chatsirirungruang, "Application of genetic algorithm and Taguchi method in dynamic robust parameter design for unknown problems," *International Journal of Advanced Manufacturing Technology*, vol. 47, no. 9–12, pp. 993–1002, 2009.

[28] L. I. Tong, C. H. Wang, and H. C. Chen, "Optimization of multiple responses using principal component analysis and technique for order preference by similarity to ideal solution," *International Journal of Advanced Manufacturing Technology*, vol. 27, no. 3-4, pp. 407–414, 2005.

[29] L. I. Tong, C. C. Chen, and C. H. Wang, "Optimization of multi-response processes using the VIKOR method," *International Journal of Advanced Manufacturing Technology*, vol. 31, no. 11-12, pp. 1049–1057, 2007.

[30] Neural Ware, *Neural Works Professional II/Plus and Neural Works Explorer*, Neural Ware, Carnegie, Pa, USA; Penn Centre West, Beverly Hills, Calif, USA, 1990.

[31] J. H. Holland, *Adaptation in Natural and Artificial Systems*, University of Michigan Press, Ann Arbor, Mich, USA, 1975.

[32] M. T. Hagan, H. B. Demuth, and M. H. Beale, *Neural Network Design*, PWS Publishing, Boston, Mass, USA, 1996.

13

Distributions of Patterns of Pair of Successes Separated by Failure Runs of Length at Least k_1 and at Most k_2 Involving Markov Dependent Trials: GERT Approach

Kanwar Sen,[1] Pooja Mohan,[2] and Manju Lata Agarwal[3]

[1] *Department of Statistics, University of Delhi, Delhi 7, India*
[2] *RMS India, A-7, Sector 16, Noida 201 301, India*
[3] *The Institute for Innovation and Inventions with Mathematics and IT (IIIMIT),*
   Shiv Nadar University, Greater Noida 203207, India

Correspondence should be addressed to Kanwar Sen; kanwarsen2005@yahoo.com

Academic Editor: Tadashi Dohi

We use the Graphical Evaluation and Review Technique (GERT) to obtain probability generating functions of the waiting time distributions of 1st, and mth nonoverlapping and overlapping occurrences of the pattern $\Lambda_f^{k_1,k_2} = S\underset{k_1 \le k_f \le k_2}{\underline{FF \cdots F}}S$ ($k_1 > 0$), involving homogenous Markov dependent trials. GERT besides providing visual picture of the system helps to analyze the system in a less inductive manner. Mean and variance of the waiting times of the occurrence of the patterns have also been obtained. Some earlier results existing in literature have been shown to be particular cases of these results.

1. Introduction

Probability generating functions of waiting time distributions of runs and patterns have been studied and utilized in various areas of statistics and applied probability, with applications to statistical quality control, ecology, epidemiology, quality management in health care sector, and biological science to name a few. A considerable amount of literature treating waiting time distributions have been generated, see Fu and Koutras [1], Aki et al. [2], Koutras [3], Antzoulakos [4], Aki and Hirano [5], Han and Hirano [6], Fu and Lou [7], and so forth. The books by Godbole and Papastavridis [8], Balakrishnan and Koutras [9], Fu and Lou [10] provide excellent information on past and current developments in this area.

The probability generating function is very important for studying the properties of waiting time distributions of runs and patterns. Once a potentially problem-specific statistic involving runs and patterns has been defined, the task of deriving its distribution can be very complex and nontrivial. Traditionally, combinatorial methods were used to find the

exact distributions for the numbers of runs and patterns. By using the theory of recurrent events, Feller [11] obtained the probability generating function for waiting time of a success run $A = \underset{k}{\underline{SS \cdots S}}$ of size k in a sequence of Bernoulli trials. Fu and Chang [12] developed general method based on the finite Markov chain imbedding technique for finding the mean and probability generating functions of waiting time distributions of compound patterns in a sequence of i.i.d. or Markov dependent multistate trials. Ge and Wang [13] studied the consecutive-k-out-of-n: F system involving Markov Dependence.

Graphical Evaluation and Review Technique (GERT) has been a well-established technique applied in several areas. However, application of GERT in reliability studies has not been reported much. It is only recently that Cheng [14] analyzed reliability of fuzzy consecutive-k-out-of-n: F system using GERT. Agarwal et al. [15, 16], Agarwal & Mohan [17] and Mohan et al. [18] have also studied reliability of m-consecutive-k-out-of-n: F system and its various generalizations using GERT, and illustrated the efficiency of GERT in

Distributions of Patterns of Pair of Successes Separated by Failure Runs of Length at Least k₁ and at Most k₂
Involving Markov Dependent Trials: GERT Approach

109

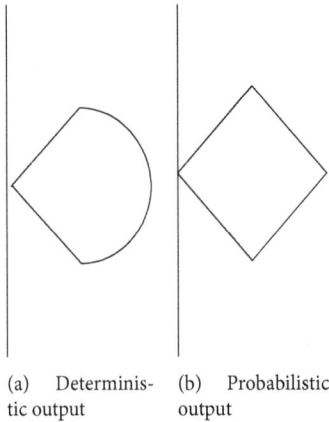

(a) Determinis-
tic output

(b) Probabilistic
output

FIGURE 1: Type of GERT nodes.

reliability analysis. Mohan et al. [19] studied waiting time distributions of 1st, and mth nonoverlapping and overlapping occurrences of the pattern $\Lambda^k = S\underset{k}{\underline{FF\cdots F}}S$, involving Markov dependent trials, using GERT. In this paper, probability generating functions of the waiting time distributions of 1st, and mth nonoverlapping and overlapping occurrences of a pattern involving pair of successes separated by a run of failures of length at least k_1 and at most k_2, $\Lambda_f^{k_1,k_2} = S\underset{k_1 \leq k_f \leq k_2}{\underline{FF\cdots F}}S$ $(k_1 > 0)$, involving Homogenous Markov Dependence, that is, probability that component i fails depends only upon the state of component $(i-1)$ and not upon the state of the other components, (Ge and Wang [13]) have been studied using GERT. Mean and variance of the time of their occurrences can then be obtained easily. Some earlier results existing in literature have been shown to be particular cases.

2. Notations and Assumptions

$\Lambda_f^{k_1,k_2} = S\underset{k_1 \leq k_f \leq k_2}{\underline{FF\cdots F}}S$: pair of successes separated by a run of failures of length at least k_1 and at most k_2, $k_1 > 0$.

X_i: indicator random variable for state of trial i, $X_i = 0$ or 1 according as trial i is success or failure.

p_0, q_0: $\Pr\{X_1 = 0\}$; $q_0 = 1 - p_0$

p_1, q_1: $\Pr\{X_i = 0 \mid X_{i-1} = 0\}$, probability that trial i is a success given that preceding trial $(i-1)$ is also a success, for $i = 2, 3, \ldots$; $q_1 = 1 - p_1$.

p_2, q_2: $\Pr\{X_i = 0 \mid X_{i-1} = 1\}$, probability that trial i is a success given that preceding trial $(i-1)$ is a failure, for $i = 2, 3, \ldots$; $q_2 = 1 - p_2$.

3. Brief Description of GERT and Definitions

GERT is a procedure for the analysis of stochastic networks having logical nodes (or events) and directed branches (or activities). It combines the disciplines of flow graph theory,

MGF (Moment Generating Function), and PERT (Project Evaluation and Review Technique) to obtain a solution to stochastic networks having logical nodes and directed branches. The nodes can be interpreted as the states of the system and directed branches represent transitions from one state to another. A branch has the probability that the activity associated with it will be performed. Other parameters describe the activities represented by the branches.

A GERT network in general contains one of the following two types of *logical nodes* (Figure 1):

(a) nodes with *Exclusive-Or input* function and *Deterministic output* function and

(b) nodes with *Exclusive-Or input* function and *Probabilistic output* function.

Exclusive-Or Input. The node is realized when any arc leading into it is realized. However, one and only one of the arcs can be realized at a given time.

Deterministic Output. All arcs emanating from the node are taken if the node is realized.

Probabilistic Output. Exactly one arc emanating from the node is taken if the node is realized.

In this paper type (b) nodes are used.

The transmittance of an arc in a GERT network, that is, the generating function of the waiting time for the occurrence of required system state is the corresponding W-function. It is used to obtain the information of a relationship, which exists between the nodes.

If we define $W(s \mid r)$, as the conditional W function associated with a network when the branches tagged with a z are taken r times, then the equivalent W generating function can be written as follows:

$$W(s, z) = \sum_{r=0}^{\infty} W(s \mid r) z^r. \qquad (1)$$

The function $W(0, z)$ is the generating function of the waiting time for the network realization.

Mason's Rule (Whitehouse [20], pp. 168–172). In an open flow graph, write down the product of transmittances along each path from the independent to the dependent variable. Multiply its transmittance by the sum of the nontouching loops to that path. Sum these modified path transmittances and divide by the sum of all the loops in the open flow graph yielding transmittance T as follows:

$$T = \frac{\left[\sum (\text{path} * \sum \text{nontouching loops})\right]}{\sum \text{loops}}, \qquad (2)$$

where

$$\sum \text{loops} = 1 - \left(\sum \text{first order loops}\right)$$

$$+ \left(\sum \text{second order loops}\right) - \cdots$$

$$\sum \text{nontouching loops}$$

$$= 1 - \left(\sum \text{first order nontouching loops}\right)$$

$$+ \left(\sum \text{second order nontouching loops}\right)$$

$$- \left(\sum \text{third order nontouching loops}\right) + \cdots \quad (3)$$

For more necessary details about GERT, one can see Whitehouse [20], Cheng [14] and Agarwal et al. [15].

4. Waiting Time Distribution of the Pattern $\Lambda_f^{k_1,k_2}$

Theorem 1. $W_{\underset{k_1 \leq k_f \leq k_2}{SFF\cdots FS}}(0,z)$, *the probability generating function for the waiting time distribution of the 1st occurrence of the pattern $\Lambda_f^{k_1,k_2}$ involving Homogenous Markov Dependence is given by*

$$W_{\underset{k_1 \leq k_f \leq k_2}{SFF\cdots FS}}(0,z)$$

$$= \left(q_0 p_2 z^2 + p_0 z (1 - q_2 z)\right)$$

$$\times \left(q_1 p_2 q_2^{k_1-1} z^{k_1+1}\left(1 - (q_2 z)^{k_2-k_1+1}\right)\right) \quad (4)$$

$$\times \left((1 - q_2 z)\left(1 - q_2 z - p_1 z - q_1 p_2 z^2 + q_2 p_1 z^2\right.\right.$$

$$\left.\left.- q_1 p_2 q_2^{k_2} z^{k_2+2} + q_1 p_2 q_2^{k_1-1} z^{k_1+1}\right)\right)^{-1}.$$

Then,

$$E\left[W\left(\Lambda_f^{k_1,k_2}\right)\right]$$

$$= E \left[\text{minimum number of trials required}\right.$$

$$\left.\text{to obtain the pattern } \Lambda_f^{k_1,k_2}\right] \quad (5)$$

$$= \frac{q_2(q_1 + p_2) + q_1 q_2^{k_1}(q_0 + p_2)\left(1 - q_2^{k_2-k_1+1}\right)}{q_1 p_2 q_2^{k_1}\left(1 - q_2^{k_2-k_1+1}\right)},$$

$$\text{Var}\left[W\left(\Lambda_f^{k_1,k_2}\right)\right] = \left.\frac{d^2 W_{\underset{k_1 \leq k_f \leq k_2}{SFF\cdots FS}}(0,z)}{dz^2}\right|_{z=1}$$

$$+ E\left[W\left(\Lambda_f^{k_1,k_2}\right)\right] - \left(E\left[W\left(\Lambda_f^{k_1,k_2}\right)\right]\right)^2. \quad (6)$$

The GERT network for this pattern is represented by Figure 2 where each node represents a specific state as described below:

S: initial node

1^A: a trial resulting in failure

0_i: a trial resulting in success corresponding to beginning of the ith occurrence of the pattern $\Lambda_f^{k_1,k_2}$, $i = 1,2,\ldots m$

1: a component is in failed state preceded by working component

j: jth contiguous failed trial preceded by a success trial, $j = 2,\ldots,k_1 - 1, k_1, k_1 + 1,\ldots, k_2, k_2 + 1$

0^i: ith occurrence of the required pattern, $i = 1,2,\ldots,m$.

There are in all $k_2 + 5$ nodes designated as $S, 1^A, 0_1, 1, 2,\ldots,k_1 - 1, k_1, k_1 + 1,\ldots, k_2 - 1, k_2, k_2 + 1$ and 0^1 representing specific states of the system, that is, a sequence of homogenous Markov dependent trials. GERT network can be summarized as follows. If the first trial results in failure then state 1^A is reached from S with conditional probability q_0 otherwise state 0_1 with conditional probability p_0. Further, if the preceding trial is failure (state 1^A) followed by a contiguous success trial then state 0_1 is reached from state 1^A with conditional probability p_2 otherwise it continues to move in state 1^A with conditional probability q_2. State 0_1 represents first occurrence of a success trial (may or may not be preceded by failed trials) and if next contiguous trial(s) is (are) also success then it continues to move in state 0_1 with conditional probability p_1 otherwise it moves in state 1 with conditional probability q_1. Again if the next contiguous trial results in failure then system state 2 occurs with conditional probability q_2 otherwise state 0_1 with conditional probability p_2. Similar procedure is followed till k_1 contiguous failed trials occur preceded by a success trial. Now, if the next contiguous trial is failure then state $k_1 + 1$ is reached with conditional probability q_2 otherwise state 0^1, first occurrence of required pattern. However, if the system is in state $k_1 + 1$ and next contiguous trial is failure system moves to state $k_1 + 2$ with conditional probability q_2. Again a similar procedure is followed till node k_2. However, if there occur $k_2 - k_1 + 1$ contiguous failed trials after state k_1 then the system moves to node $k_2 + 1$ and continues to move in that state until a success trial occurs at which it moves to state 0_1 and again a similar procedure is followed for the remaining trials till state 0^1, first occurrence of the required pattern. Details on the derivation of the theorem are given in the appendix.

Distributions of Patterns of Pair of Successes Separated by Failure Runs of Length at Least k_1 and at Most k_2
Involving Markov Dependent Trials: GERT Approach

111

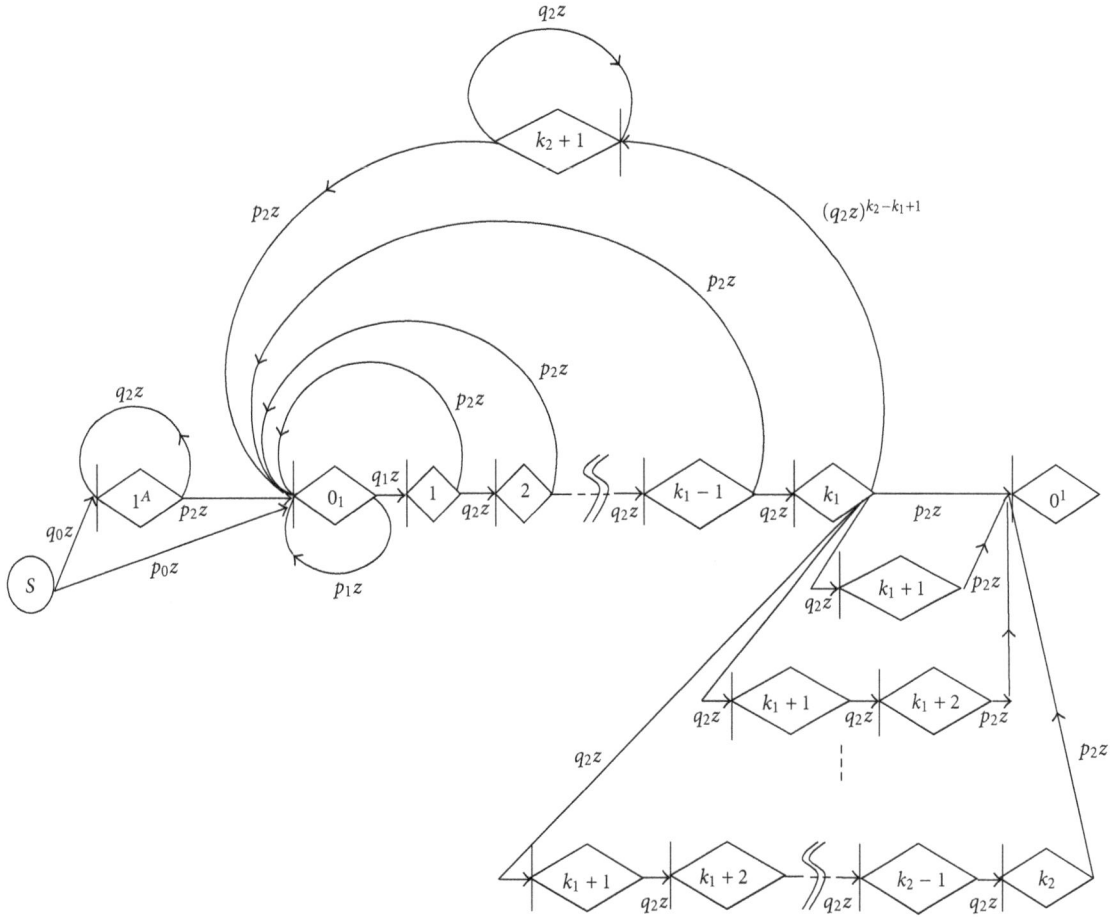

FIGURE 2: GERT network representing 1st occurrence of the pattern $\Lambda_f^{k_1,k_2} = \underset{k_1 \le k_f \le k_2}{S\underline{FF}\cdots\underline{F}S}$.

Theorem 2. $W_{\underset{k_1 \le k_f \le k_2}{S\underline{FF}\cdots\underline{F}S}}^{no}(0,z)_m$, the probability generating function for the waiting time distribution of the mth nonoverlapping occurrence of the pattern $\Lambda_f^{k_1,k_2}$ involving Homogenous Markov Dependence is given by

$$W_{\underset{k_1 \le k_f \le k_2}{S\underline{FF}\cdots\underline{F}S}}^{no}(0,z)_m$$

$$= \left(q_0 p_2 z^2 + p_0 z \left(1 - q_2 z \right) \right)$$

$$\times \left(p_2 q_1 q_2^{k_1-1} z^{k_1+1} \left\{ 1 - \left(q_2 z \right)^{k_2-k_1+1} \right\} \right)^m$$

$$\times \left(p_1 z \left(1 - q_2 z \right) + q_1 p_2 z^2 \right)^{m-1} \left(1 - q_2 z \right)^{-m}$$

$$\times \left(1 - q_2 z - p_1 z + q_2 p_1 z^2 - q_1 p_2 z^2 \right.$$

$$\left. + q_1 p_2 q_2^{k_1-1} z^{k_1+1} - q_1 p_2 q_2^{k_2} z^{k_2+2} \right)^{-m}.$$

(7)

Then,

$$E\left[W^{no} \left(\Lambda_f^{k_1,k_2} \right) \right]$$

$$= E\left[\text{minimum number of trials required to obtain} \right.$$

$$\left. \text{the } m\text{th nonoverlapping occurrence of pattern } \Lambda_f^{k_1,k_2} \right]$$

$$= q_1 q_2^{k_1} \left(1 - q_2^{k_2-k_1+1} \right)$$

$$\times \left(p_1 - p_0 + m \left(q_1 + p_2 \right) \right) + m \, q_2 \left(q_1 + p_2 \right)$$

$$\times \left(q_1 p_2 q_2^{k_1} \left(1 - q_2^{k_2-k_1+1} \right) \right)^{-1},$$

(8)

and $\mathrm{Var}[W^{no}(\Lambda_f^{k_1,k_2})]$ can be obtained by applying (6).

The GERT network for $m = 2$ (say) nonoverlapping occurrence of the pattern $\Lambda_f^{k_1,k_2}$ is represented by Figure 3. Each node represents a specific state as described earlier in Figure 2.

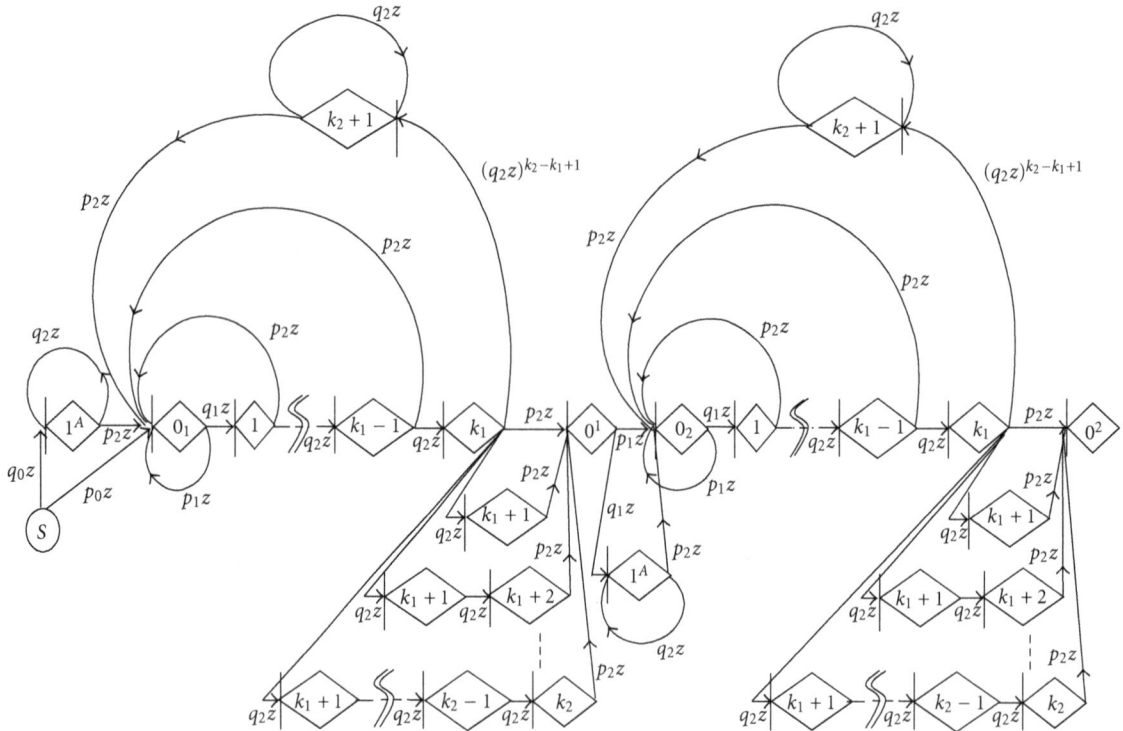

FIGURE 3: GERT network representing the mth ($m = 2$, here) nonoverlapping occurrence of the pattern $\Lambda_f^{k_1,k_2} = S\underset{k_1 \leq k_f \leq k_2}{\underline{FF\cdots FS}}$.

Theorem 3. $W^o_{\underset{k_1 \leq k_f \leq k_2}{\underline{SFF\cdots FS}}}(0,z)_m$, *the probability generating function for the waiting time distribution of the mth overlapping occurrence of the pattern $\Lambda_f^{k_1,k_2}$ involving Homogenous Markov Dependence is given by*

$$W^o_{\underset{k_1 \leq k_f \leq k_2}{\underline{SFF\cdots FS}}}(0,z)_m$$

$$= \left(q_0 p_2 z^2 + p_0 z \left(1 - q_2 z\right)\right)$$

$$\times \left(q_1 p_2 q_2^{k_1-1} z^{k_1+1} \left\{1 - \left(q_2 z\right)^{k_2-k_1+1}\right\}\right)^m$$

$$\times \left(\left(1 - q_2 z\right)\left(1 - q_2 z - p_1 z - q_1 p_2 z^2 + q_2 p_1 z^2\right.\right.$$

$$\left.\left. - q_1 p_2 q_2^{k_2} z^{k_2+2} + q_1 p_2 q_2^{k_1-1} z^{k_1+1}\right)\right)^{-m}.$$

$$(9)$$

Then,

$$E\left[W^o\left(\Lambda_f^{k_1,k_2}\right)\right]$$

$$= E\left[\text{minimum number of trials required to obtain}\right.$$

$$\left.\text{the mth overlapping occurrence of pattern } \Lambda_f^{k_1,k_2}\right]$$

$$= \frac{q_1 q_2^{k_1}\left(1 - q_2^{k_2-k_1+1}\right)(q_0 + p_2) + m q_2 (q_1 + p_2)}{q_1 p_2 q_2^{k_1}\left(1 - q_2^{k_2-k_1+1}\right)}.$$

$$(10)$$

and Var $[W^o(\Lambda_f^{k_1,k_2})]$ *can be obtained by applying* (6).

The GERT network for mth ($m = 2$, say), overlapping occurrence of the pattern $\Lambda_f^{k_1,k_2}$ is represented by Figure 4. Each node represents a specific state as described earlier in Figure 2.

Particular Cases. (i) For $k_1 = k_f = k_2 = k - 2$, $p_0 = p_1 = p_2 = p$, and $q_0 = q_1 = q_2 = q$, in the pattern $\Lambda_f^{k_1,k_2}$, that is, for a run of at most $k - 2$ failures bounded by successes, the probability generating function becomes

$$W_{\underset{k-2}{\underline{SFF\cdots FS}}}(0,z)$$

$$= \frac{(pz)^2\left(1 - (qz)^{k-1}\right)}{(1 - qz)\left(1 - z + \left(pz\left(1 - (qz)^{k-1}\right)\right)\right)},$$

$$(11)$$

verifying the results of Koutras [3, Theorem 3.2].

(ii) If $k_1 = 1$, $k_f = k_2 = k$, that is, pair of successes are separated by a run of failures of length at least one and at most k, that is, $\underset{1 \leq k}{\underline{SFF\cdots FS}}$ then (4), (7), and (9), respectively, become

$$W_{\underset{1 \leq k}{\underline{SFF\cdots FS}}}(0,z)$$

$$= \frac{\left(q_0 p_2 z^2 + p_0 z \left(1 - q_2 z\right)\right)\left(p_2 q_1 z^2 \left(1 - (q_2 z)^k\right)\right)}{(1 - q_2 z)\left(1 - q_2 z - p_1 z + q_2 p_1 z^2 - q_1 p_2 q_2^k z^{k+2}\right)}.$$

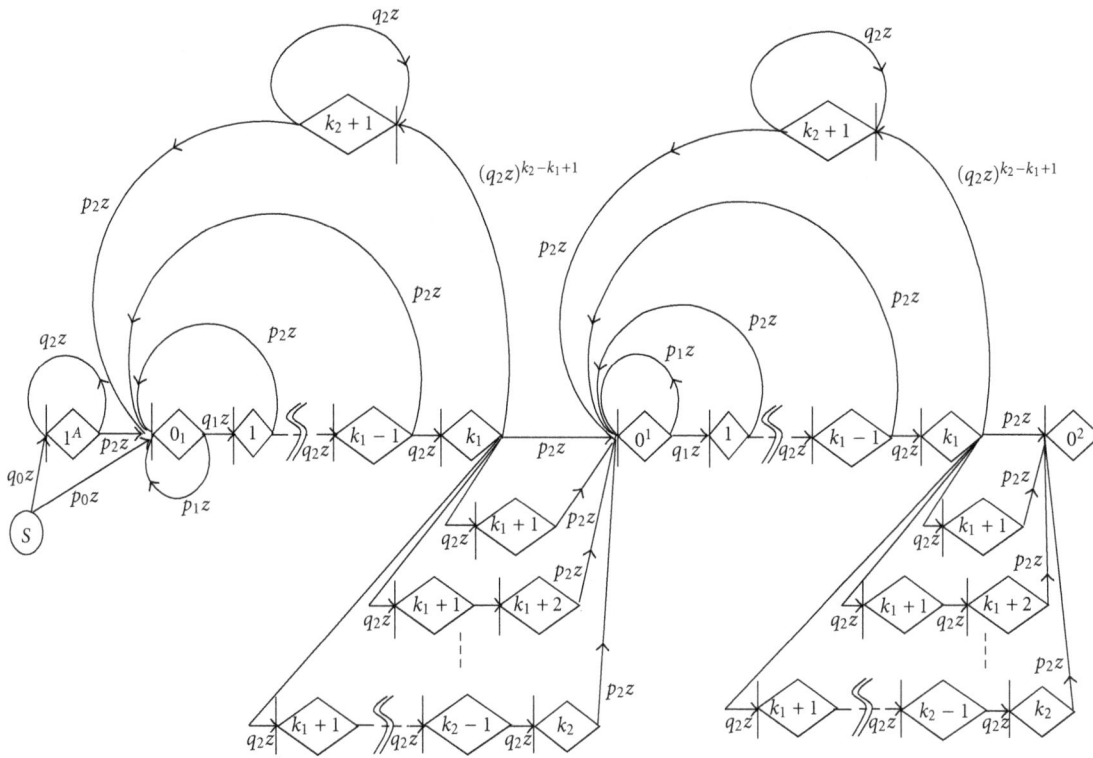

FIGURE 4: GERT network representing the mth ($m = 2$, here) overlapping occurrence of the pattern $\Lambda_f^{k_1, k_2} = \underset{k_1 \le k_f \le k_2}{S\underline{FF\cdots F}S}$.

$W_{\underset{1 \le k}{S\underline{FF\cdots F}S}}^{no}$

$= \left(q_0 p_2 z^2 + p_0 z \left(1 - q_2 z \right) \right) \left(q_1 p_2 z^2 + p_1 z \left(1 - q_2 z \right) \right)^{m-1}$

$\times \left(q_1 p_2 z^2 \left(1 - (q_2 z)^k \right) \right)^m \left(1 - q_2 z \right)^{-m}$

$\times \left(1 - p_1 z - q_2 z + p_1 q_2 z^2 - p_2 q_1 q_2^k z^{k+2} \right)^{-m}.$

$W_{\underset{1 \le k}{S\underline{FF\cdots F}S}}^{o} (0, z)$

$= \dfrac{\left(q_0 p_2 z^2 + p_0 z \left(1 - q_2 z \right) \right) \left(p_2 q_1 z^2 \left(1 - (q_2 z)^k \right) \right)^m}{\left(1 - q_2 z \right) \left(1 - q_2 z - p_1 z + q_2 p_1 z^2 - p_2 q_1 q_2^k z^{k+2} \right)^m}.$

$\hfill (12)$

(iii) If $k_1 = k$ and $k_2 \to \infty$, then it reduces to the pattern $\Lambda^{\ge k} = \underset{\ge k}{S\underline{FF\cdots F}S}$, that is,

$W_{\underset{\ge k}{S\underline{FF\cdots F}S}} (0, z)$

$pt = \left(q_0 p_2 z^2 + p_0 z \left(1 - q_2 z \right) \right) \left(p_2 q_1 q_2^{k-1} z^{k+1} \right)$

$\times \left(\left(1 - q_2 z \right) \left(1 - q_2 z - p_1 z + q_2 p_1 z^2 \right. \right.$

$\left. \left. - q_1 p_2 z^2 \left(1 - (q_2 z)^{k-1} \right) \right) \right)^{-1}.$

$\hfill (13)$

(iv) If $k_1 = k_f = k_2 = k$,
then (4), (7), (9) reduce to the results of Mohan et al. [19], that is,

$W_{\underset{k}{S\underline{FF\cdots F}S}} (0, z)$

$= \left(q_0 p_2 z^2 + p_0 z \left(1 - q_2 z \right) \right) \left(q_1 p_2 q_2^{k-1} z^{k+1} \right)$

$\times \left(1 - q_2 z - p_1 z - p_2 q_1 z^2 + p_1 q_2 z^2 - q_1 p_2 q_2^k z^{k+2} \right.$

$\left. + q_1 p_2 q_2^{k-1} z^{k+1} \right)^{-1},$

$W_{\underset{k}{S\underline{FF\cdots F}S}}^{no} (0, z)_m$

$= \left(q_0 p_2 z^2 + p_0 z \left(1 - q_2 z \right) \right) \left(p_2 q_1 q_2^{k-1} z^{k+1} \right)^m$

$\times \left[p_1 z \left(1 - q_2 z \right) + p_2 q_1 z^2 \right]^{m-1}$

$\times \left[1 - q_2 z - p_1 z - q_1 p_2 z^2 + q_2 p_1 z^2 - q_1 p_2 q_2^k z^{k+2} \right.$

$$+ q_1 p_2 q_2^{k-1} z^{k+1} \big]^{-m},$$

$$W^o_{\underset{k}{SFF\cdots FS}}(0,z)_m$$

$$= \left(q_0 p_2 z^2 + p_0 z \left(1 - q_2 z\right)\right) \left(p_2 q_1 q_2^{k-1} z^{k+1}\right)^m \left(1 - q_2 z\right)^{m-1}$$

$$\times \left(1 - q_2 z - p_1 z - q_1 p_2 z^2 + q_2 p_1 z^2 - q_1 p_2 q_2^k z^{k+2}\right.$$

$$\left. + q_1 p_2 q_2^{k-1} z^{k+1}\right)^{-m}.$$

(14)

5. Conclusion

In this paper we proposed Graphical Evaluation and Review Technique (GERT) to study probability generating functions of the waiting time distributions of 1st, and mth nonoverlapping and overlapping occurrences of the pattern $\Lambda_f^{k_1,k_2} = \underset{k_1 \le k_f \le k_2}{SFF\cdots FS}(k_1 > 0)$, involving homogenous Markov dependent trials. We have also demonstrated the flexibility and usefulness of our approach by validating some earlier results existing in literature as particular cases of these results.

Appendix

Proof of Theorems

From the GERT network (Figure 2), it can be observed that there are $2(k_2 - k_1 + 1)$ paths to reach state 0^1 from the starting node S, see Table 1.

Now, to apply Mason's rule, we must also locate all the loops. However only first and second order loops exist. First order loops are given in Table 2.

However, paths at Nos. $k_2 - k_1 + 2, k_2 - k_1 + 3, \cdots, 2(k_2 - k_1 + 1)$ also contain first order nontouching loop (1^A to 1^A), whose value is given by $q_2 z$. Further, first order loop No. $k_1 + 3$ forms first order nontouching loop to both paths, whose value is given by $q_2 z$.

Also second order loops corresponding to first order loops from No. 2 to No. $k_1 + 3$ are given (since first order loop no. 1, that is, 1^A to 1^A forms nontouching loop with each of the other mentioned first order loops and can be taken separately, Whitehouse [20] pp. 257) in Table 3.

Thus, by applying Mason's rule we obtain the following generating function $W_{\underset{k_1 \le k_f \le k_2}{SFF\cdots FS}}(0,z)$ of the waiting time for the occurrence of the pattern $\Lambda_f^{k_1,k_2}$ as follows:

$$W_{\underset{k_1 \le k_f \le k_2}{SFF\cdots FS}}(0,z)$$

$$= \left(q_0 p_2 z^2 + p_0 z \left(1 - q_2 z\right)\right)\left(1 - q_2 z\right)$$

$$\times \left[q_1 p_2 q_2^{k_1-1} z^{k_1+1} \left\{1 + (q_2 z) + (q_2 z)^2 + \cdots + (q_2 z)^{k_2-k_1}\right\}\right]$$

$$\times \left(\left(1 - q_2 z\right)\left(1 - q_2 z - p_1 z + q_2 p_1 z^2 - q_1 p_2 z^2\right.\right.$$

$$\left.\left. + q_1 p_2 q_2^{k_1-1} z^{k_1+1} - q_1 p_2 q_2^{k_2} z^{k_2+2}\right)\right)^{-1},$$

(A.1)

which yields (4).

Similarly, for the mth nonoverlapping occurrence of the pattern (Theorem 2) once when state 0^1 is reached, that is, the first occurrence of required pattern, if the next contiguous trial results in success then state 0_2 is directly reached with conditional probability p_1 otherwise state 1^A with conditional probability q_1 and continues to move in that state with conditional probability q_2 until a success occurs at which it moves to state 0_2 with conditional probability p_2. Thus, there are two paths to reach node 0_2 from node 0^1. Now, on reaching node 0_2 again the same procedure is followed for the second occurrence of the required pattern, represented by node 0^2. Thus, for the second occurrence of the pattern node 0^1 acts as a starting node following a similar procedure as for the first occurrence.

It can be observed from the GERT network that there are $4 \cdot (k_2 - k_1 + 1)^2$ paths (in general for any m there are $2 \cdot 2^{m-1}(k_2 - k_1 + 1)^m$ paths) for reaching state 0^2 (0^m) from the starting node S. Thus, proceeding as in Theorem 1 and applying Mason's rule we obtain the following generating function $W^{no}_{\underset{k_1 \le k_f \le k_2}{SFF\cdots FS}}(0,z)_2$ of the waiting time for the 2nd nonoverlapping occurrence of the required pattern $\Lambda_f^{k_1,k_2}$ as follows:

$$W^{no}_{\underset{k_1 \le k_f \le k_2}{SFF\cdots FS}}(0,z)_2$$

$$= \left(q_0 p_2 z^2 + p_0 z \left(1 - q_2 z\right)\right)\left(p_1 z \left(1 - q_2 z\right) + q_1 p_2 z^2\right)$$

$$\times \left(p_2 q_1 q_2^{k_1-1} z^{k_1+1} \left\{1 - (q_2 z)^{k_2-k_1+1}\right\}\right)^2$$

$$\times \left(1 - q_2 z\right)^{-2} \left(1 - q_2 z - p_1 z + q_2 p_1 z^2 - q_1 p_2 z^2\right.$$

$$\left. + q_1 p_2 q_2^{k_1-1} z^{k_1+1} - q_1 p_2 q_2^{k_2} z^{k_2+2}\right)^{-2}.$$

(A.2)

Proceeding similarly as above, we can obtain generating function $W^{no}_{\underset{k_1 \le k_f \le k_2}{SFF\cdots FS}}(0,z)_m$ as given by (7).

For mth overlapping occurrence of the pattern (Theorem 3), proceeding as in Theorem 1, once when state 0^1 is reached, that is, the first occurrence of the required pattern, if the next contiguous trial results in failure then state 1 is directly reached with conditional probability q_1 otherwise it continues to move in state 0^1 with conditional probability p_1 until a failed trial occurs resulting in occurrence of state 1 with conditional probability q_1. Now, on reaching node 1 if the next contiguous trial results in failure then state 2 occurs with conditional probability q_2 otherwise state 0^1 with conditional probability p_2. Again similar procedure is followed for the second occurrence of the required pattern. Thus, for the second occurrence, node 0^1 acts as a starting

TABLE 1

No.	Paths	Value
1	S to 1^A to 0_1 to 1 to 2$\cdots$to k_1 to 0^1	$(q_0z)(q_1z)(p_2z)^2(q_2z)^{k_1-1}$
2	S to 1^A to 0_1 to 1 to 2$\cdots$to k_1 to k_1+1 to 0^1	$(q_0z)(q_1z)(p_2z)^2(q_2z)^{k_1}$
3	S to 1^A to 0_1 to 1 to 2$\cdots$to k_1 to k_1+1 to k_1+2 to 0^1	$(q_0z)(q_1z)(p_2z)^2(q_2z)^{k_1+1}$
$\vdots$	$\vdots$	$\vdots$
k_2-k_1	S to 1^A to 0_1 to 1 to 2$\cdots$to k_1 to$\ldots$to k_2-1 to 0^1	$(q_0z)(q_1z)(p_2z)^2(q_2z)^{k_2-2}$
k_2-k_1+1	S to 1^A to 0_1 to 1 to 2$\cdots$to k_1 to$\ldots$to k_2-1 to k_2 to 0^1	$(q_0z)(q_1z)(p_2z)^2(q_2z)^{k_2-1}$
k_2-k_1+2	S to 0_1 to 1 to 2$\cdots$to k_1 to 0^1	$(p_0z)(q_1z)(p_2z)(q_2z)^{k_1-1}$
k_2-k_1+3	S to 0_1 to 1 to 2$\cdots$to k_1 to k_1+1 to 0^1	$(p_0z)(q_1z)(p_2z)(q_2z)^{k_1}$
k_2-k_1+4	S to 0_1 to 1 to 2$\cdots$to k_1 to k_1+1 to k_1+2 to 0^1	$(p_0z)(q_1z)(p_2z)(q_2z)^{k_1+1}$
$\vdots$	$\vdots$	$\vdots$
$2k_2-2k_1+1$	S to 0_1 to 1 to 2$\cdots$to k_1 to$\ldots$to k_2-1 to 0^1	$(p_0z)(q_1z)(p_2z)(q_2z)^{k_2-2}$
$2(k_2-k_1+1)$	S to 0_1 to 1 to 2$\cdots$to k_1 to$\ldots$to k_2-1 to k_2 to 0^1	$(p_0z)(q_1z)(p_2z)(q_2z)^{k_2-1}$

TABLE 2

No.	First order loops	Value
1	1^A to 1^A	(q_2z)
2	0_1 to 0_1	(p_1z)
3	0_1 to 1 to 0_1	$(q_1z)(p_2z)$
4	0_1 to 1 to 2 to 0_1	$(q_1z)(p_2z)(q_2z)$
$\vdots$	$\vdots$	$\vdots$
k_1+1	0_1 to 1 to 2 to$\ldots$to k_1-1 to 0_1	$(q_1z)(p_2z)(q_2z)^{k_1-2}$
k_1+2	0_1 to 1 to 2 to$\ldots$to k_1-1 to k_1 to k_2+1 to 0_1	$(q_1z)(p_2z)(q_2z)^{k_2}$
k_1+3	k_2+1 to k_2+1	(q_2z)

TABLE 3

Second order loops	Value
No. 2 and No. k_1+3	$(p_1z)(q_2z)$
No. 3 and No. k_1+3	$(q_1z)(p_2z)(q_2z)$
No. 4 and No. k_1+3	$(q_1z)(p_2z)(q_2z)^2$
$\vdots$	$\vdots$
No. k_1+1 and No. k_1+3	$(q_1z)(p_2z)(q_2z)^{k_1-1}$

node following a similar procedure as for the first occurrence (except that there is only one path to reach node 1 from node 0^1).

It can be observed that in the GERT network for 2nd overlapping occurrence of the required pattern (Figure 4), there are $2\cdot(k_2-k_1+1)^2$ (in general for any m, there are $2\cdot(k_2-k_1+1)^m$) paths for reaching state 0^2 (0^m) from the starting node S.

Now pairs $(0_1,0^1)$, $(0^1,0^2)$ are identical. Thus, proceeding as in the Theorem 1 and by applying Mason's rule the generating function $W^o_{\underset{k_1\le k_f\le k_2}{SFF\ldots FS}}(0,z)_2$ of the waiting time for the 2nd overlapping occurrence of the required pattern is given by

$$W^o_{\underset{k_1\le k_f\le k_2}{SFF\ldots FS}}(0,z)_2 = \frac{\left(q_0p_2z^2+p_0z(1-q_2z)\right)\left(q_1p_2q_2^{k_1-1}z^{k_1+1}\left\{1-\left(q_2z\right)^{k_2-k_1+1}\right\}\right)^2}{\left(1-q_2z\right)\left(1-q_2z-p_1z+q_2p_1z^2-q_1p_2z^2-q_1p_2q_2^{k_2}z^{k_2+2}+q_1p_2q_2^{k_1-1}z^{k_1+1}\right)^2}.\tag{A.3}$$

Similarly, proceeding as above we can obtain generating function $W^o_{\underset{k_1 \leq k_f \leq k_2}{SFF \cdots FS}}(0, z)_m$ as given by (9).

Acknowledgments

The authors would like to express their sincere thanks to the editor and referees for their valuable suggestions and comments which were quite helpful to improve the paper.

References

[1] J. C. Fu and M. V. Koutras, "Distribution theory of runs: a Markov chain approach," *Journal of the American Statistical Association*, vol. 89, no. 427, pp. 1050–1058, 1994.

[2] S. Aki, N. Balakrlshnan, and S. G. Mohanty, "Sooner and later waiting time problems for success and failure runs in higher order Markov dependent trials," *Annals of the Institute of Statistical Mathematics*, vol. 48, no. 4, pp. 773–787, 1996.

[3] M. V. Koutras, "Waiting time distributions associated with runs of fixed length in two-state Markov chains," *Annals of the Institute of Statistical Mathematics*, vol. 49, no. 1, pp. 123–139, 1997.

[4] D. L. Antzoulakos, "On waiting time problems associated with runs in Markov dependent trials," *Annals of the Institute of Statistical Mathematics*, vol. 51, no. 2, pp. 323–330, 1999.

[5] S. Aki and K. Hirano, "Sooner and later waiting time problems for runs in markov dependent bivariate trials," *Annals of the Institute of Statistical Mathematics*, vol. 51, no. 1, pp. 17–29, 1999.

[6] Q. Han and K. Hirano, "Sooner and later waiting time problems for patterns in Markov dependent trials," *Journal of Applied Probability*, vol. 40, no. 1, pp. 73–86, 2003.

[7] J. C. Fu and W. Y. W. Lou, "Waiting time distributions of simple and compound patterns in a sequence of rth order markov dependent multi-state trials," *Annals of the Institute of Statistical Mathematics*, vol. 58, no. 2, pp. 291–310, 2006.

[8] A. P. Godbole and S. G. Papastavridis, *Runs and Patterns in Probability: Selected Papers*, Kluwer Academic Publishers, Amsterdam, The Netherlands, 1994.

[9] N. Balakrishnan and M. V. Koutras, *Runs and Scans With Applications*, John Wiley and Sons, New York, NY, USA, 2002.

[10] J. C. Fu and W. Y. Lou, *Distribution Theory of Runs and Patterns and its Applications: A Finite Markov Chain Imbedding Approach*, World Scientific Publishing Co, River Edge, NJ, USA, 2003.

[11] W. Feller, *An Introduction to Probability Theory and its Applications*, vol. 1, John Wiley and Sons, New York, NY, USA, 3rd edition, 1968.

[12] J. C. Fu and Y. M. Chang, "On probability generating functions for waiting time distributions of compound patterns in a sequence of multistate trials," *Journal of Applied Probability*, vol. 39, no. 1, pp. 70–80, 2002.

[13] G. Ge and L. Wang, "Exact reliability formula for consecutive-k-out-of-n:F systems with homogeneous Markov dependence," *IEEE Transactions on Reliability*, vol. 39, no. 5, pp. 600–602, 1990.

[14] C. H. Cheng, "Fuzzy consecutive-k-out-of-n:F system reliability," *Microelectronics Reliability*, vol. 34, no. 12, pp. 1909–1922, 1994.

[15] M. Agarwal, K. Sen, and P. Mohan, "GERT analysis of m-consecutive-k-out-of-n systems," *IEEE Transactions on Reliability*, vol. 56, no. 1, pp. 26–34, 2007.

[16] M. Agarwal, P. Mohan, and K. Sen, "GERT analysis of m-consecutive-k-out-of-n: F System with dependence," *International Journal for Quality and Reliability (EQC)*, vol. 22, pp. 141–157, 2007.

[17] M. Agarwal and P. Mohan, "GERT analysis of m-consecutive-k-out-of-n:F system with overlapping runs and $(k-1)$-step Markov dependence," *International Journal of Operational Research*, vol. 3, no. 1-2, pp. 36–51, 2008.

[18] P. Mohan, M. Agarwal, and K. Sen, "Combined m-consecutive-k-out-of-n: F & consecutive k_c-out-of-n: F systems," *IEEE Transactions on Reliability*, vol. 58, no. 2, pp. 328–337, 2009.

[19] P. Mohan, K. Sen, and M. Agarwal, "Waiting time distributions of patterns involving homogenous Markov dependent trials: GERT approach," in *Proceedings of the 7th International Conference on Mathematical Methods in Reliability: Theory, Methods, Applications (MMR '11)*, L. Cui and X. Zhao, Eds., pp. 935–941, Beijing, China, June 2011.

[20] G. E. Whitehouse, *Systems Analysis and Design Using Network Techniques*, Prentice-Hall, Englewood Cliffs, NJ, USA, 1973.

14

Inference on Reliability of Stress-Strength Models for Poisson Data

Alessandro Barbiero

Department of Economics, Management and Quantitative Methods, University of Milan, 20122 Milan, Italy

Correspondence should be addressed to Alessandro Barbiero; alessandro.barbiero@unimi.it

Academic Editor: Shey-Huei Sheu

Researchers in reliability engineering regularly encounter variables that are discrete in nature, such as the number of events (e.g., failures) occurring in a certain spatial or temporal interval. The methods for analyzing and interpreting such data are often based on asymptotic theory, so that when the sample size is not large, their accuracy is suspect. This paper discusses statistical inference for the reliability of stress-strength models when stress and strength are independent Poisson random variables. The maximum likelihood estimator and the uniformly minimum variance unbiased estimator are here presented and empirically compared in terms of their mean square error; recalling the delta method, confidence intervals based on these point estimators are proposed, and their reliance is investigated through a simulation study, which assesses their performance in terms of coverage rate and average length under several scenarios and for various sample sizes. The study indicates that the two estimators possess similar properties, and the accuracy of these estimators is still satisfactory even when the sample size is small. An application to an engineering experiment is also provided to elucidate the use of the proposed methods.

1. Introduction

A stress-strength model, in the simplest terms, considers a unit/system that is subjected to an external stress, modeled by r.v. X, against which the unit sets its own strength, modeled by r.v. Y, in order to properly operate. The probability that the unit withstands the stress is then given by $R = P(X < Y)$, which is usually called reliability.

A great deal of work has been done about this topic: most of it deals with the computation of reliability, if the distributions of stress and strength are known, or its estimation under various parametric assumptions on X and Y, when samples from X and Y are available. A complete review is available in [1]. Many applications of the stress-strength model, for its own nature, are related to engineering or military problems, where it is also referred to as a load-strength model [2]. However, there are also natural applications in medicine or psychology, which involve the comparison of two r.v., representing, for example, the effect of a specific drug or treatment administered to two groups (control and test); here, reliability assumes a wider meaning.

Almost all of these papers consider continuous distributions for X and Y, since many practical applications of the stress-strength model in engineering fields presuppose continuous quantitative data. A relatively small amount of work is devoted to discrete or categorical data. Data may be discrete by nature, for example, the number of events occurring in a certain spatial or temporal interval; sometimes discrete data are derived from continuous ones by grouping or discretization or censoring, and then, instead of numerical measurements on X and Y, they are presented in a form of ordered categories.

Among the r.v. modeling discrete data, the Poisson can be of interest in several practical applications. The Poisson r.v. is often used to model rare events such as the number of claims in automobile insurance, the number of times a website is accessed, the number of calls to a phone operator, the number of words mistyped per page in a book, and so forth [3, 4]. The distribution of the difference between two independent r.v. each having a Poisson distribution has already attracted some attention [5]. Strackee and van der Gon [6] stated that "in a steady state the number of light quanta, emitted or absorbed in a definite time, is distributed according to a Poisson distribution. In view thereof, the physical limit of perceptible contrast in vision can be studied in terms of the difference between two independent variates each following a Poisson distribution". Irwin [7] studied the case when the two variables X and Y each have the same expected value; Skellam [8] was the first to discuss the problem when $E(X) = \lambda_1 \neq E(Y) = \lambda_2$. Strackee and van der Gon [6] gave tables of the approximate values of the cumulative probability $P(Y - X < d)$ for several combinations of the values of the parameters λ_1 and λ_2. More recently, Karlis and Ntzoufras [9] used the Poisson difference distribution to model the difference in the decayed, missing, and filled teeth index before and after treatment; Karlis and Ntzoufras [10] applied it to model the difference in the number of goals in football games.

In this paper, we examine point and interval estimation for the reliability of the stress-strength model with independent Poisson stress and strength. Although the maximum likelihood (ML) and uniformly minimum variance unbiased (UMVU) estimators of reliability have a known analytical expression, their statistical properties cannot be easily derived and thus need to be assessed through a Monte Carlo simulation study. Confidence intervals for reliability based on approximate expression for variance are also presented, and their performances in terms of coverage rate and average width are empirically investigated.

The paper is laid out as follows: in Section 2 reliability for Poisson stress-strength model and its ML and UMVU estimators are presented and discussed. Section 3 introduces approximate variance estimators and confidence intervals for reliability. Section 4 is devoted to a Monte Carlo (MC) study, which empirically investigates the performance of ML and UMVU estimators, and the corresponding confidence intervals for different combinations of distributional parameters and sample sizes. Section 5 describes an application, and Section 6 gives final remarks.

2. Point Estimators

Let X and Y be independent r.v. modeling stress and strength, respectively, with $X \sim \text{Poisson}(\lambda_1)$ and $Y \sim \text{Poisson}(\lambda_2)$. Then, the reliability $R = P(X < Y)$ of the stress-strength model is given by (see [1])

$$R = P(X < Y) = \sum_{x=0}^{+\infty} \frac{\lambda_1^x e^{-\lambda_1}}{x!} \left[1 - \sum_{y=0}^{x} \frac{\lambda_2^y e^{-\lambda_2}}{y!} \right]$$

$$= \lim_{k \to \infty} \sum_{x=0}^{k} \frac{\lambda_1^x e^{-\lambda_1}}{x!} \left[1 - \sum_{y=0}^{x} \frac{\lambda_2^y e^{-\lambda_2}}{y!} \right]. \tag{1}$$

The terms of the external sum rapidly converge to zero: reliability can be actually computed taking into account only its first terms. As an example, we compute the reliability R when $\lambda_1 = 1$ and $\lambda_2 = 4$; the partial sums are reported in Table 1: the value of R is already stable at the 7th decimal digit when $k = 10$.

TABLE 1: Partial sums for the computation of R for a Poisson stress-strength model ($\lambda_1 = 1$, $\lambda_2 = 4$).

$k = 1$	$k = 2$	$k = 3$	$k = 4$	$k = 5$
0.6953312	0.8354743	0.87021	0.8758994	0.876558
$k = 6$	$k = 7$	$k = 8$	$k = 9$	$k = 10$
0.8766146	0.8766183	0.8766185	0.8766186	0.8766186

TABLE 2: Values of the UMVU estimator of R for a Poisson stress-strength model ($\overline{x} = 1$, $\overline{y} = 2; 3; 4$) with varying sample sizes.

n_1	n_2	$\overline{x} = 1, \overline{y} = 2$	$\overline{x} = 1, \overline{y} = 3$	$\overline{x} = 1, \overline{y} = 4$
10	10	0.612649	0.786485	0.887889
10	20	0.607790	0.781051	0.883313
10	30	0.606199	0.779271	0.881805
20	10	0.613908	0.785907	0.886621
20	20	0.609078	0.780562	0.882117
20	30	0.607496	0.778811	0.880632
30	10	0.614334	0.785732	0.886212
30	20	0.609512	0.780416	0.881731
30	30	0.607932	0.778674	0.880254
50	50	0.607031	0.777192	0.878786
100	100	0.606364	0.776097	0.877697
	$\widetilde{R}$	0.605703	0.775015	0.876619

If two simple random samples $\mathbf{x}$ of size n_1 and $\mathbf{y}$ of size n_2 from X and Y, respectively, are available, reliability can be estimated with the ML estimator, obtained by substituting in (1) the maximum likelihood estimators of the unknown parameters λ_1 and λ_2:

$$\widetilde{R} = \sum_{x=0}^{+\infty} \frac{\overline{x}^x e^{-\overline{x}}}{x!} \left[1 - \sum_{y=0}^{x} \frac{\overline{y}^y e^{-\overline{y}}}{y!} \right]. \tag{2}$$

Otherwise, one can use the UMVU estimator [1]:

$$\widehat{R} = \sum_{x=0}^{U} \binom{n_1 \overline{x}}{x} \frac{(n_1-1)^{n_1 \overline{x} - x}}{n_1^{n_1 \overline{x}}} \left[1 - \sum_{y=0}^{x} \binom{n_2 \overline{y}}{y} \frac{(n_2-1)^{n_2 \overline{y} - y}}{n_2^{n_2 \overline{y}}} \right], \tag{3}$$

where $U = \min(n_1 \overline{x}, n_2 \overline{y} - 1)$. Note that formula (3) is represented via a finite sum, whereas formula (2) contains a rapidly converging series. The number of calculations that formula (3) performs depends on the sample means and the sample sizes, which jointly define the number of terms of the external sum; in formula (2) the terms of the external sum rapidly converge to zero, so that it may practically need fewer calculations than (3).

In Table 2, the values for the UMVU estimator are reported when $\overline{x} = 1$ and $\overline{y} = 2; 3; 4$, for different combinations of sample sizes n_1 and n_2. Note that the values of $\widehat{R}$ are very close to the value of $\widetilde{R}$ even for small sample sizes and get closer as the sample sizes increase. These results are pictorially displayed in Figure 1 for $\overline{x} = 1$ and $\overline{y} = 4$.

Due to the complex expressions involved, the bias of $\widetilde{R}$ and the variance of either estimators $\widetilde{R}$ and $\widehat{R}$ cannot be

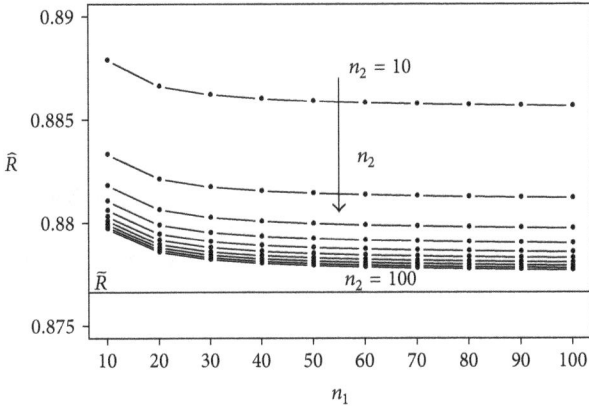

FIGURE 1: Values of the UMVU estimator of R for a Poisson stress-strength model ($\overline{x} = 1, \overline{y} = 4$) with varying sample sizes. The ML estimator is represented by the horizontal line.

analytically derived; a comparison of their performance (in terms of mean square error) can be carried out through MC simulations.

3. Variance Estimators and Confidence Intervals

Whereas the exact value of the variance or the mean square error of either estimator introduced in Section 2 is almost impracticable to derive, an approximate value can be easily supplied recalling the delta method [11]. For the ML estimator $\widetilde{R}(\overline{x}, \overline{y})$, since $\overline{X}$ and $\overline{Y}$ are independent estimators of λ_1 and λ_2, the variance of $\widetilde{R}$ can be approximated as

$$V\left(\widetilde{R}\right) \approx \left(\frac{\partial \widetilde{R}}{\partial \overline{x}}\right)^2\bigg|_{\overline{x}=\lambda_1, \overline{y}=\lambda_2} V\left(\overline{x}\right)$$
$$+ \left(\frac{\partial \widetilde{R}}{\partial \overline{y}}\right)^2\bigg|_{\overline{x}=\lambda_1, \overline{y}=\lambda_2} V\left(\overline{y}\right), \tag{4}$$

with $V(\overline{x}) = \lambda_1/n_1$, $V(\overline{y}) = \lambda_2/n_2$ and

$$\frac{\partial \widetilde{R}}{\partial \overline{x}} = \sum_{x=0}^{\infty} \frac{e^{-\overline{x}} \overline{x}^{x-1}}{x!} (x - \overline{x}) \left[1 - \sum_{y=0}^{x} \frac{\overline{y}^y e^{-\overline{y}}}{y!}\right],$$
$$\frac{\partial \widetilde{R}}{\partial \overline{y}} = -\sum_{x=0}^{\infty} \frac{\overline{x}^x e^{-\overline{x}}}{x!} \sum_{y=0}^{x} \frac{e^{-\overline{y}} \overline{y}^{y-1}}{y!} (y - \overline{y}). \tag{5}$$

An analogous approximation can be carried out for the variance of the UMVU estimator; remembering that $x! = \Gamma(x+1)$, $\widehat{R}$ can be rewritten as

$$\widehat{R} = \sum_{x=0}^{U} \frac{\Gamma(n_1\overline{x}+1)}{\Gamma(x+1)\Gamma(n_1\overline{x}-x+1)} \frac{(n_1-1)^{n_1\overline{x}-x}}{n_1^{n_1\overline{x}}}$$
$$\times \left[1 - \sum_{y=0}^{x} \frac{\Gamma(n_2\overline{y}+1)}{\Gamma(y+1)\Gamma(n_2\overline{y}-y+1)} \frac{(n_2-1)^{n_2\overline{y}-y}}{n_2^{n_2\overline{y}}}\right], \tag{6}$$

and then the two first-order partial derivatives are given by

$$\frac{\partial \widehat{R}}{\partial \overline{x}} = \sum_{x=0}^{U} \left[1 - \sum_{y=0}^{x} \binom{n_2\overline{y}}{y} \frac{(n_2-1)^{n_2\overline{y}-y}}{n_2^{n_2\overline{y}}}\right]$$
$$\times \frac{1}{(n_1-1)^x} n_1 \binom{n_1\overline{x}}{x} \left(\frac{n_1-1}{n_1}\right)^{n_1\overline{x}}$$
$$\times \left[\frac{\Gamma'(n_1\overline{x}+1)}{\Gamma(n_1\overline{x}+1)} - \frac{\Gamma'(n_1\overline{x}-x+1)}{\Gamma(n_1\overline{x}-x+1)} + \log\left(\frac{n_1-1}{n_1}\right)\right],$$
$$\frac{\partial \widehat{R}}{\partial \overline{y}} = -\sum_{x=0}^{U} \sum_{y=0}^{x} \binom{n_1\overline{x}}{x} \frac{(n_1-1)^{n_1\overline{x}-x}}{n_1^{n_1\overline{x}}}$$
$$\times \frac{1}{(n_2-1)^y} n_2 \binom{n_2\overline{y}}{y} \left(\frac{n_2-1}{n_2}\right)^{n_2\overline{y}}$$
$$\times \left[\frac{\Gamma'(n_2\overline{y}+1)}{\Gamma(n_2\overline{y}+1)} - \frac{\Gamma'(n_2\overline{y}-y+1)}{\Gamma(n_2\overline{y}-y+1)} + \log\left(\frac{n_2-1}{n_2}\right)\right]. \tag{7}$$

The approximate variances of $\widetilde{R}$ and $\widehat{R}$ derived through the delta method can be estimated substituting in (4) the sample means to the unknown parameters and thus getting

$$v\left(\widetilde{R}\right) = \left(\frac{\partial \widetilde{R}}{\partial \overline{x}}\right)^2 \frac{\overline{x}}{n_1} + \left(\frac{\partial \widetilde{R}}{\partial \overline{y}}\right)^2 \frac{\overline{y}}{n_2} \tag{8}$$

and an analogous result for $v(\widehat{R})$.

The Gamma function $\Gamma(x)$ and its first derivative, $\Gamma'(x) = \int_0^{+\infty} t^{x-1} e^{-t} \log t\, dt$, involved in the partial derivatives of $\widehat{R}$, have to be numerically computed. In the R software environment [12] this task is easily performed through the gamma and digamma functions, the latter providing the ratio $\Gamma'(x)/\Gamma(x)$.

Once one has computed $v(\widetilde{R})$, an approximate $(1-\alpha)\cdot 100\%$ confidence interval for R can be built, recalling the asymptotic normality of $\widetilde{R}$:

$$\left(\widetilde{R} + z_{\alpha/2}\sqrt{v\left(\widetilde{R}\right)}, \widetilde{R} + z_{1-\alpha/2}\sqrt{v\left(\widetilde{R}\right)}\right), \tag{9}$$

and in an analogous way for $\widehat{R}$. Since R is bounded in $[0, 1]$, special care has to be given when $\widehat{R}$ is close to one (close to zero) and/or sample sizes are small: the upper bound may exceed one (the lower bound may fall below zero), and then the CI in (9) will be modified as follows:

$$\left(\max\left(0, \widetilde{R} + z_{\alpha/2}\sqrt{v\left(\widetilde{R}\right)}\right),\right.$$
$$\left.\min\left(1, \widetilde{R} + z_{1-\alpha/2}\sqrt{v\left(\widetilde{R}\right)}\right)\right). \tag{10}$$

More sophisticated asymptotic confidence intervals for R can be built recalling some normalizing transformations, such as logit and arcsine [13].

TABLE 3: Parameter values for the X and Y distributions and corresponding reliability explored in the simulation study of a Poisson stress-strength model.

λ_1	λ_2	λ_1	λ_2	λ_1	λ_2	λ_1	λ_2	R
1	1.547	2	2.522	5	5.509	10	10.504	0.5
1	1.973	2	3.089	5	6.361	10	11.683	0.6
1	2.497	2	3.764	5	7.342	10	13.015	0.7
1	3.2	2	4.646	5	8.584	10	14.666	0.8
1	4.338	2	6.032	5	10.47	10	17.121	0.9

4. Simulation Study

The simulation study aims at empirically comparing the performance of the ML and UMVU estimators, in terms of bias and mean square error, and the confidence intervals based on them, in terms of the coverage rate and average length. Since the approximation of the variance derived through the delta method (4) holds for large samples, we will investigate to what extent it still holds for small and moderate sample sizes, and how it affects inferential results. In this MC study, the value of the parameter λ_1 of the Poisson distribution for stress X is first set equal to a "reference" value, 1, and the parameter λ_2 of the Poisson distribution modeling strength is allowed to vary in order to obtain four different levels of reliability R, namely, 0.5, 0.6, 0.7, 0.8, and 0.9. Note that a value of $\lambda_2 = 1.547$ is needed in order to get $R = P(X < Y) = 0.5$ while $\lambda_2 = \lambda_1 = 1$ lead only to $P(X < Y) + 0.5 \cdot P(X = Y) = 0.5$. Then, λ_1 is set equal to greater values (namely 2, 5, and 10), and λ_2 is allowed to vary in order to ensure the five values of reliability R above. The corresponding values of λ_2 for each combination of R and λ_1 values are reported in Table 3.

For each couple (λ_1, λ_2), a huge number ($S = 2,000$) of samples **x** of size n_1 and **y** of size n_2 are drawn from $X \sim$ Poisson(λ_1) and $Y \sim$ Poisson(λ_2) independently. Different and unequal sample sizes are here considered (all the nine possible combinations between the values $n_1 = 10, 20, 50$, and $n_2 = 10, 20, 50$). The ML and UMVU estimators are computed on each sample, their approximate variances are calculated, and the corresponding 95% confidence intervals for R are built. Some measures of performance for these estimators are supplied. In more detail, the MC root mean square error and the percentage relative bias of the ML estimator are provided:

$$\text{RMSE}_{\text{MC}}\left(\widetilde{R}\right) = \sqrt{\frac{1}{S}\sum_{s=1}^{S}\left(\widetilde{R}(s) - R\right)^2}$$

$$\text{RB}_{\text{MC}}\left(\widetilde{R}\right) = \frac{\left((1/S)\sum_{s=1}^{S}\widetilde{R}(s) - R\right)}{R} \cdot 100\%,$$

(11)

where $\widetilde{R}(s)$ denotes the value of $\widetilde{R}$ for the sth sample. Analogous indexes are derived for $\widehat{R}$, whose bias is null, and for which we then expect the MC relative bias to be close to zero.

Regarding estimating the variance, the true variance $V(\widetilde{R})$ is approximated by its MC mean:

$$V\left(\widetilde{R}\right) \approx V_{\text{MC}}\left(\widetilde{R}\right) = E_{\text{MC}}\left[\left(\widetilde{R} - \overline{\widetilde{R}}\right)^2\right] \quad (12)$$

with $\overline{\widetilde{R}} = \sum_{s=1}^{S}\widetilde{R}(s)/S$, and then the MC relative bias and RMSE of $v(\widetilde{R})$ are calculated the same way as for $\widetilde{R}$.

The MC coverage rate of the CIs is simply defined as follows:

$$\frac{1}{S}\sum_{s=1}^{S}I\left[\widetilde{R}(s) + z_{\alpha/2}\sqrt{v\left(\widetilde{R}(s)\right)} \leq R \leq \widetilde{R}(s) + z_{1-\alpha/2}\sqrt{v\left(\widetilde{R}(s)\right)}\right],$$

(13)

where $I(E)$ is the indicator function, taking value 1 if E is true, 0 otherwise. The length of the confidence interval is then equal to $2z_{1-\alpha/2}\sqrt{v(\widetilde{R}(s))}$. The same performance indexes are derived for $\widehat{R}$.

The simulation results for $\lambda_1 = 1$ are reported in Table 4 (RB and RMSE for ML and UMVU point estimators), Table 5 (RB and RMSE for variance estimators), and Table 6 (coverage rate and average length of confidence intervals).

The simulation results show that the ML estimator $\widetilde{R}$ always presents a very small bias even for small samples: in absolute value, the MC percentage relative bias is always smaller than 1.841% for all the scenarios considered (whereas the maximum absolute MC percentage relative bias for $\widehat{R}$, which is theoretically unbiased, is 0.552%). In 42 scenarios out of 45, $\widetilde{R}$ underestimates R. Regarding the RMSE, the ML estimator performs better than UMVU in 27 cases out of 45, worse in 7 cases, and in 11 cases the RMSE is equal at the third decimal digit. However, under each scenario, even for smaller sample sizes, the values of RMSE for the ML and UMVUE estimators are very close. The ML outperforms the UMVU estimator as the value of R gets close to 0.5; their performances tend to be alike as n_1 and n_2 increase. For both estimators, for fixed sample sizes, the RMSE increases as R decreases; for a fixed R, the RMSE increases, as the sample sizes decrease (as expected). Figure 2 displays the MC distribution of the ML and UMVU estimators in the case $R = 0.7$, for three values of sample size; it highlights their very similar behaviour.

Regarding the approximate variance estimators, surprisingly their performance is good even for the moderate sample sizes considered in this study; the percentage relative bias, in absolute value, is never greater than 8%: the worst performance occurs for $n_1 = n_2 = 50$ and $R = 0.9$. Indeed, when both sample sizes equal 50, the RB is greater than for small sample sizes, whereas one would expect that the RB decreases in absolute value when sample sizes increase. The results of further simulations not reported here show that the RB actually decreases to zero for $n_1 = n_2 = 100$. For both estimators, the rate of underestimates is almost equal to the rate of overestimates. Under each scenario, and especially when $R = 0.5$, the value of RB of the variance estimator $v(\widehat{R})$ is quite close to the corresponding value of the RB of $v(\widetilde{R})$, whereas the RMSE of $v(\widetilde{R})$ is smaller than the RMSE

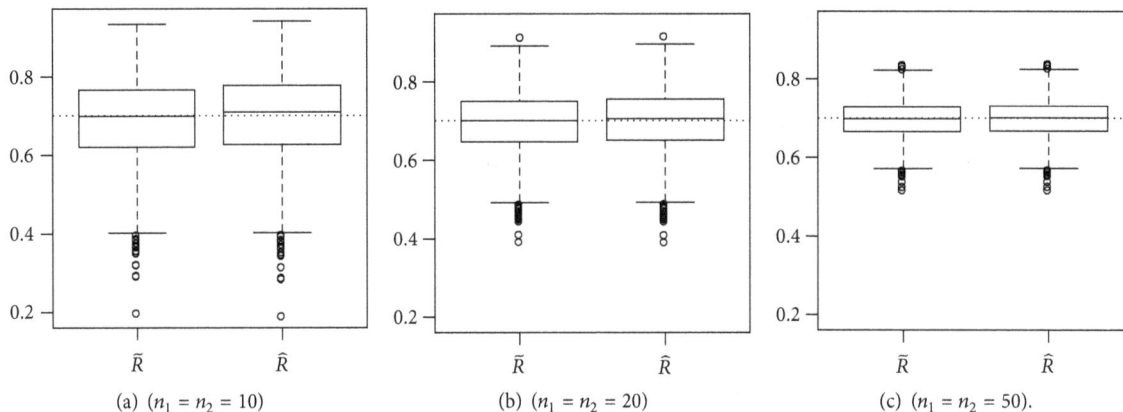

FIGURE 2: MC distribution of $\tilde{R}$ (MLE) and $\hat{R}$ (UMVUE) when $\lambda_1 = 1$, $\lambda_2 = 2.497$ ($R = 0.7$).

of $v(\hat{R})$ in each of the 45 cases considered. The RB of the two approximate variance estimators does not present a clear trend in terms of R; while their RMSEs, for each of the couples (n_1, n_2) here explored, seem to present a maximum near $R = 0.8$ and a minimum for $R = 0.5$.

The confidence intervals built upon the point estimators and these variance estimators present coverage that is always greater than 87% for UMVU and 90.5% for ML: the lowest value is obtained for $n_1 = n_2 = 10$ and $R = 0.9$. They attain the nominal level (95%) for $n_1 = n_2 = 50$; in 31 and 21 cases out of 45, respectively, the coverage rate of the ML and UMVU interval estimators is greater than or equal to 92.5%. Overall, the CIs present better coverage when R is close to 0.5. In fact, in this case, the distributions of $\hat{R}$ and $\tilde{R}$ tend to be symmetrical and are more finely approximated by the normal distribution; then, the confidence intervals (9), which assume an underlying normal distribution, show a better performance. The CIs based on the ML estimator almost always show a coverage rate greater than those based on the UMVU estimator; moreover, the latter are always a bit wider, unless when $R = 0.9$. This feature tends to be negligible when the sample sizes are increased. As one would expect, the average length decreases as sample sizes increase, for fixed R, and as R increases, for fixed sample sizes.

The results for $\lambda_1 > 1$, which are not reported here for the sake of brevity, confirm the previous findings. Even if the study is obviously not exhaustive, since only several scenarios have been covered, nevertheless these general features can be outlined.

5. An Example of Application

In this section, we apply the inferential techniques presented in Section 3 to a real dataset. The application is based on the data from an engineering experiment discussed in [3], carried out in an electric company, under several experimental conditions (called "runs"), corresponding to different combinations of 8 factors. The blackening experiment was conducted in a three-layer oven; when each run was completed, 30 masks from each layer in the oven were collected to examine the number of defects in each mask. The total number of defects in the 30 masks from the upper layer for each experimental run is observed (see Table 7). We focus on runs 1 and 2, where 317 and 184 defects, respectively, are counted.

Since the number of defects in a mask is either zero or a positive integer, the appropriate distribution is the Poisson. Denoting with Y and X the variables modeling this number for run 1 and run 2, respectively, we are interested in determining a point estimate and an interval estimator for the probability that the number of defects in run 1 is smaller than in run 2, that is, $P(X < Y)$. Since the sample size is 30 for both variables, then $\bar{x} = 6.13$ and $\bar{y} = 10.57$; then, supposing that X and Y follow a Poisson distribution, the ML and UMVU estimators and their corresponding approximate variances can be computed according to (2), (3), and (8); the associated confidence intervals can be estimated recalling (10).

The results are presented in Table 8 and show the closeness between the two approaches. All the confidence intervals for $P(X < Y)$ always exclude 0.5, thus meaning that the difference in sample means testifies to the statistical dominance of Y on X: the number of defects under run 1 is stochastically larger than the number of defects under run 2.

6. Conclusions

In this paper, point and interval estimators for the reliability of a Poisson stress-strength model are presented, discussed, and empirically compared through a Monte Carlo simulation study. The results show that the maximum likelihood and uniformly minimum variance unbiased estimators possess similar sampling properties, and the first is slightly preferable to the second in terms of dispersion around the true value of reliability. Moreover, although the variance estimators proposed here are approximate (i.e., biased), being based on the delta method for asymptotically normal r.v., the empirical results emphasize that the estimators' bias is small even for moderate sample sizes, and these estimators can be usefully employed to build approximate confidence intervals, whose coverage is shown to be overall close to the fixed nominal

TABLE 4: Simulation results: bias and root mean square error of ML and UMVU estimators.

	R = 0.9	R = 0.8	R = 0.7	R = 0.6	R = 0.5
	$(n_1, n_2) = (10, 10)$				
RB($\tilde{R}$)	−1.214	−1.539	−1.539	−1.216	−0.932
RB($\hat{R}$)	−0.111	−0.219	−0.275	−0.189	−0.310
RMSE($\tilde{R}$)	0.061	0.091	0.109	0.119	0.124
RMSE($\hat{R}$)	0.059	0.091	0.111	0.123	0.129
	$(n_1, n_2) = (10, 20)$				
RB($\tilde{R}$)	−0.753	−0.906	−0.796	−0.492	−0.073
RB($\hat{R}$)	−0.079	−0.182	−0.221	−0.207	−0.204
RMSE($\tilde{R}$)	0.049	0.075	0.091	0.100	0.103
RMSE($\hat{R}$)	0.048	0.075	0.093	0.102	0.106
	$(n_1, n_2) = (10, 50)$				
RB($\tilde{R}$)	−0.386	−0.349	−0.138	0.197	0.673
RB($\hat{R}$)	0.029	0.017	0.023	0.040	0.093
RMSE($\tilde{R}$)	0.041	0.063	0.077	0.084	0.086
RMSE($\hat{R}$)	0.040	0.064	0.078	0.086	0.088
	$(n_1, n_2) = (20, 10)$				
RB($\tilde{R}$)	−1.151	−1.576	−1.676	−1.665	−1.598
RB($\hat{R}$)	−0.154	−0.297	−0.335	−0.409	−0.552
RMSE($\tilde{R}$)	0.055	0.082	0.098	0.109	0.113
RMSE($\hat{R}$)	0.052	0.082	0.100	0.111	0.116
	$(n_1, n_2) = (20, 20)$				
RB($\tilde{R}$)	−0.557	−0.725	−0.716	−0.564	−0.259
RB($\hat{R}$)	0.001	−0.050	−0.068	−0.040	0.055
RMSE($\tilde{R}$)	0.040	0.062	0.077	0.084	0.088
RMSE($\hat{R}$)	0.039	0.062	0.078	0.085	0.089
	$(n_1, n_2) = (20, 50)$				
RB($\tilde{R}$)	−0.280	−0.313	−0.235	−0.060	0.121
RB($\hat{R}$)	0.017	−0.004	−0.010	0.013	−0.020
RMSE($\tilde{R}$)	0.031	0.049	0.060	0.066	0.068
RMSE($\hat{R}$)	0.031	0.049	0.061	0.067	0.069
	$(n_1, n_2) = (50, 10)$				
RB($\tilde{R}$)	−1.036	−1.438	−1.704	−1.622	−1.841
RB($\hat{R}$)	−0.106	−0.184	−0.316	−0.217	−0.528
RMSE($\tilde{R}$)	0.050	0.076	0.093	0.101	0.106
RMSE($\hat{R}$)	0.048	0.075	0.093	0.102	0.108
	$(n_1, n_2) = (50, 20)$				
RB($\tilde{R}$)	−0.403	−0.508	−0.450	−0.060	−0.330
RB($\hat{R}$)	0.084	0.134	0.242	0.013	0.254
RMSE($\tilde{R}$)	0.034	0.054	0.066	0.066	0.077
RMSE($\hat{R}$)	0.034	0.054	0.067	0.067	0.078
	$(n_1, n_2) = (50, 50)$				
RB($\tilde{R}$)	−0.284	−0.412	−0.480	−0.448	−0.427
RB($\hat{R}$)	−0.059	−0.138	−0.218	−0.238	−0.304
RMSE($\tilde{R}$)	0.024	0.038	0.047	0.053	0.054
RMSE($\hat{R}$)	0.024	0.038	0.048	0.053	0.055

TABLE 5: Simulation results: bias and root mean square error of variance estimators.

	R = 0.9	R = 0.8	R = 0.7	R = 0.6	R = 0.5
	$(n_1, n_2) = (10, 10)$				
RB($v(\tilde{R})$)	0.22	−2.64	−2.37	−3.28	−5.23
RB($v(\hat{R})$)	1.02	−2.22	−2.09	−3.17	−5.15
RMSE($v(\tilde{R})$)·10^4	25.9	36.1	34.0	25.3	19.8
RMSE($v(\hat{R})$)·10^4	27.5	40.4	39.1	29.5	22.9
	$(n_1, n_2) = (10, 20)$				
RB($v(\tilde{R})$)	3.99	−0.07	−3.34	−2.72	−4.11
RB($v(\hat{R})$)	4.67	0.29	−3.17	−2.63	−4.08
RMSE($v(\tilde{R})$)·10^4	15.5	22.7	21.9	15.8	10.7
RMSE($v(\hat{R})$)·10^4	16.1	24.7	24.1	17.6	11.9
	$(n_1, n_2) = (10, 50)$				
RB($v(\tilde{R})$)	6.22	2.33	0.70	0.88	−0.55
RB($v(\hat{R})$)	6.66	2.56	0.82	0.93	−0.54
RMSE($v(\tilde{R})$)·10^4	10.2	15.6	15.3	11.5	7.20
RMSE($v(\hat{R})$)·10^4	10.5	16.6	16.4	12.4	7.71
	$(n_1, n_2) = (20, 10)$				
RB($v(\tilde{R})$)	−2.80	−3.56	−2.55	−3.67	−5.45
RB($v(\hat{R})$)	−2.58	−3.43	−2.47	−3.57	−5.37
RMSE($v(\tilde{R})$)·10^4	18.4	26.3	24.8	18.3	14.6
RMSE($v(\hat{R})$)·10^4	19.1	28.5	27.6	20.7	16.1
	$(n_1, n_2) = (20, 20)$				
RB($v(\tilde{R})$)	4.76	2.51	−0.60	1.38	−0.72
RB($v(\hat{R})$)	5.09	2.67	−0.54	1.40	−0.71
RMSE($v(\tilde{R})$)·10^4	8.84	13.5	12.9	8.84	5.03
RMSE($v(\hat{R})$)·10^4	9.04	14.3	13.9	9.60	5.45
	$(n_1, n_2) = (20, 50)$				
RB($v(\tilde{R})$)	7.39	4.45	2.23	2.81	2.20
RB($v(\hat{R})$)	7.60	4.55	2.28	2.84	2.21
RMSE($v(\tilde{R})$)·10^4	4.67	7.41	7.24	5.13	2.68
RMSE($v(\hat{R})$)·10^4	4.74	7.69	7.58	5.39	2.81
	$(n_1, n_2) = (50, 10)$				
RB($v(\tilde{R})$)	−3.24	−4.60	−5.48	−3.24	−5.42
RB($v(\hat{R})$)	−3.08	−4.59	−5.52	−3.27	−5.45
RMSE($v(\tilde{R})$)·10^4	14.3	21.2	20.8	15.0	11.5
RMSE($v(\hat{R})$)·10^4	14.5	22.5	22.6	16.6	12.4
	$(n_1, n_2) = (50, 20)$				
RB($v(\tilde{R})$)	1.79	−0.23	1.15	2.81	0.47
RB($v(\hat{R})$)	1.93	−0.16	1.18	2.84	0.48
RMSE($v(\tilde{R})$)·10^4	5.64	9.00	8.78	5.13	3.21
RMSE($v(\hat{R})$)·10^4	5.68	9.34	9.25	5.39	3.40
	$(n_1, n_2) = (50, 50)$				
RB($v(\tilde{R})$)	8.02	7.02	6.14	5.35	5.68
RB($v(\hat{R})$)	8.08	7.05	6.15	5.36	5.69
RMSE($v(\tilde{R})$)·10^4	2.19	3.62	3.63	2.64	1.88
RMSE($v(\hat{R})$)·10^4	2.21	3.69	3.73	2.72	1.92

level, especially when the value of reliability is close to 0.5. However, when R is close to 1 (or, symmetrically, 0), the

intervals can show a poorer performance; then caution is needed when constructing a confidence interval based on a

TABLE 6: Simulation results: coverage rate and average length of confidence intervals based on ML and UMVU estimators.

	$R = 0.9$	$R = 0.8$	$R = 0.7$	$R = 0.6$	$R = 0.5$
	\multicolumn{5}{c}{$(n_1, n_2) = (10, 10)$}				
cov $(\tilde{R})$	0.905	0.916	0.917	0.918	0.916
cov $(\hat{R})$	0.870	0.893	0.913	0.916	0.914
length $(\tilde{R})$	0.220	0.338	0.413	0.456	0.471
length $(\hat{R})$	0.214	0.342	0.425	0.472	0.490
	\multicolumn{5}{c}{$(n_1, n_2) = (10, 20)$}				
cov $(\tilde{R})$	0.912	0.922	0.922	0.924	0.925
cov $(\hat{R})$	0.893	0.914	0.915	0.925	0.924
length $(\tilde{R})$	0.184	0.284	0.348	0.383	0.394
length $(\hat{R})$	0.181	0.287	0.355	0.394	0.406
	\multicolumn{5}{c}{$(n_1, n_2) = (10, 50)$}				
cov $(\tilde{R})$	0.911	0.919	0.924	0.926	0.927
cov $(\hat{R})$	0.899	0.918	0.921	0.926	0.926
length $(\tilde{R})$	0.158	0.246	0.300	0.330	0.337
length $(\hat{R})$	0.156	0.248	0.305	0.337	0.345
	\multicolumn{5}{c}{$(n_1, n_2) = (20, 10)$}				
cov $(\tilde{R})$	0.925	0.930	0.928	0.929	0.922
cov $(\hat{R})$	0.894	0.914	0.924	0.926	0.922
length $(\tilde{R})$	0.197	0.305	0.374	0.415	0.431
length $(\hat{R})$	0.191	0.306	0.381	0.426	0.443
	\multicolumn{5}{c}{$(n_1, n_2) = (20, 20)$}				
cov $(\tilde{R})$	0.927	0.936	0.935	0.938	0.933
cov $(\hat{R})$	0.913	0.927	0.927	0.937	0.933
length $(\tilde{R})$	0.154	0.241	0.297	0.330	0.342
length $(\hat{R})$	0.151	0.242	0.301	0.336	0.348
	\multicolumn{5}{c}{$(n_1, n_2) = (20, 50)$}				
cov $(\tilde{R})$	0.937	0.941	0.942	0.944	0.938
cov $(\hat{R})$	0.925	0.933	0.940	0.944	0.938
length $(\tilde{R})$	0.123	0.194	0.238	0.263	0.271
length $(\hat{R})$	0.122	0.195	0.241	0.267	0.275
	\multicolumn{5}{c}{$(n_1, n_2) = (50, 10)$}				
cov $(\tilde{R})$	0.926	0.928	0.927	0.927	0.921
cov $(\hat{R})$	0.892	0.916	0.917	0.928	0.924
length $(\tilde{R})$	0.180	0.281	0.347	0.387	0.403
length $(\hat{R})$	0.174	0.280	0.351	0.394	0.412
	\multicolumn{5}{c}{$(n_1, n_2) = (50, 20)$}				
cov $(\tilde{R})$	0.935	0.935	0.943	0.944	0.946
cov $(\hat{R})$	0.917	0.932	0.938	0.944	0.945
length $(\tilde{R})$	0.132	0.209	0.260	0.263	0.304
length $(\hat{R})$	0.130	0.209	0.262	0.267	0.308
	\multicolumn{5}{c}{$(n_1, n_2) = (50, 50)$}				
cov $(\tilde{R})$	0.946	0.953	0.948	0.954	0.948
cov $(\hat{R})$	0.941	0.949	0.947	0.955	0.953
length $(\tilde{R})$	0.097	0.153	0.190	0.212	0.219
length $(\hat{R})$	0.096	0.154	0.191	0.213	0.221

point estimate close to 1 (0). In this case, one can resort, for example, to some variance-stabilizing transformation of the estimate.

TABLE 7: Data for the blackening experiment in [3].

Run	\multicolumn{8}{c}{Factor}								Number of defects
	A	B	C	D	E	F	G	H	
1	1	1	1	1	1	1	1	1	317
2	1	1	2	2	2	2	2	2	184
3	1	1	3	3	3	3	3	3	528
4	1	2	1	2	1	2	3	3	163
5	1	2	2	3	2	3	1	1	96
6	1	2	3	1	3	1	2	2	300
7	1	3	1	3	2	1	2	3	177
8	1	3	2	1	3	2	3	1	182
9	1	3	3	2	1	3	1	2	75
10	2	1	1	3	3	2	1	2	146
11	2	1	2	1	1	3	2	3	135
12	2	1	3	2	2	1	3	1	232
13	2	2	1	2	3	3	2	1	543
14	2	2	2	3	1	1	3	2	101
15	2	2	3	1	2	2	1	3	282
16	2	3	1	1	2	3	3	2	90
17	2	3	2	2	3	1	1	3	288
18	2	3	3	3	1	2	2	1	554

TABLE 8: Results for the application: ML and UMVUE point estimates and asymptotic confidence intervals (L, U) for R.

	Point est.	\multicolumn{2}{c}{90% CI}		\multicolumn{2}{c}{95% CI}		\multicolumn{2}{c}{99% CI}	
		L	U	L	U	L	U
ML	0.834	0.759	0.909	0.745	0.923	0.717	0.951
UMVU	0.838	0.763	0.913	0.749	0.927	0.721	0.955

Acknowledgments

The author thanks the editor and two anonymous referees for their comments and suggestions on the original paper. Special thanks go to Riccardo Inchingolo for his moral support.

References

[1] S. Kotz, M. Lumelskii, and M. Pensky, *The Stress-Strength Model and Its Generalizations: Theory and Applications*, World Scientific, New York, NY, USA, 2003.

[2] B. V. Gnedenko and I. A. Ushakov, *Probabilistic Reliability Engineering*, John Wiley & Sons, New York, NY, USA, 1995.

[3] P. C. Wang, "Comparisons on the analysis of Poisson data," *Quality and Reliability Engineering International*, vol. 15, no. 5, pp. 379–383, 1999.

[4] F. A. Haight, *Handbook of the Poisson Distribution*, John Wiley & Sons, New York, NY, USA, 1967.

[5] N. L. Johnson, A. W. Kemp, and S. Kotz, *Univariate Discrete Distributions*, John Wiley & Sons, New York, NY, USA, 2005.

[6] J. Strackee and J. J. D. van der Gon, "The frequency distribution of the difference between two Poisson variates," *Statistica Neerlandica*, vol. 16, no. 1, pp. 17–23, 1962.

[7] J. O. Irwin, "The frequency distribution of the difference between two independent variates following the same Poisson

distribution," *Journal of the Royal Statistical Society*, vol. 100, no. 3, pp. 415–416, 1937.

[8] J. G. Skellam, "The frequency distribution of the difference between two Poisson variates belonging to different populations," *Journal of the Royal Statistical Society A*, vol. 109, no. 3, p. 296, 1946.

[9] D. Karlis and I. Ntzoufras, "Bayesian analysis of the differences of count data," *Statistics in Medicine*, vol. 25, no. 11, pp. 1885–1905, 2006.

[10] D. Karlis and I. Ntzoufras, "Bayesian modelling of football outcomes: using the Skellam's distribution for the goal difference," *IMA Journal Management Mathematics*, vol. 20, no. 2, pp. 133–145, 2009.

[11] L. R. Klein, *A Textbook of Econometrics*, Prentice Hall, New York, NY, USA, 1953.

[12] R Development Core Team, *R: A Language and Environment For Statistical Computing*, R Foundation for Statistical Computing, Vienna, Austria, 2011, http://www.R-project.org/.

[13] S. P. Mukherjee and S. S. Maiti, "Stress-strength reliability in the Weibull case," in *Frontiers in Reliability*, vol. 4, pp. 231–248, World Scientific, Singapore, 1998.

15

A Heuristic Methodology for Efficient Reduction of Large Multistate Event Trees

Eftychia C. Marcoulaki

System Reliability and Industrial Safety Laboratory, National Centre for Scientific Research "Demokritos", P.O. Box 60228, Agia Paraskevi, 15310 Athens, Greece

Correspondence should be addressed to Eftychia C. Marcoulaki; emarcoulaki@ipta.demokritos.gr

Academic Editor: Nikolaos E. Limnios

This work proposes a new methodology for the management of event tree information used in the quantitative risk assessment of complex systems. The size of event trees increases exponentially with the number of system components and the number of states that each component can be found in. Their reduction to a manageable set of events can facilitate risk quantification and safety optimization tasks. The proposed method launches a deductive exploitation of the event space, to generate reduced event trees for large multistate systems. The approach consists in the simultaneous treatment of large subsets of the tree, rather than focusing on the given single components of the system and getting trapped into guesses on their structural arrangement.

1. Introduction

For a given system, the scope of quantitative risk assessment is to investigate the circumstances giving rise to different modes of system operation and to quantify the risk for each operation mode. A system can be comprised of hardware, software, humans, or organizational components [1]. Each component can be found in various states of operation, leading to multiple modes of failure and normal operation for the overall system. Once this mapping of component states to system outcomes is known, it is theoretically possible to quantify the risks for different operation modes to occur, given the occurrence probabilities for all the component states [2, 3].

The computational effort and the memory requirements for risk evaluations increase exponentially as the system components and the number of component states increases. Exact calculations for binary systems are achieved faster by employing binary decision diagrams to effectively organize the evaluation procedure [4]. Since the logic behind multistate systems is not Boolean, multistate behavior cannot be represented by binary models without introducing additional variables and constraints [5]. Rocco and Muselli [6] developed a methodology based on machine-learning and hamming clustering to address multistate systems and any success criterion.

The required computational resources can be reduced using approximate risk estimations [7] or criticality analysis [8]. Event trees represent the combination of component states leading to each mode of system operation. Quantification of event trees enables faster exact risk evaluation [2] and is not limited to binary or two-terminal systems, it is, however, very computationally intensive. Clearly, implementation of event tree quantification to systems with many components in multiple states needs to be preceded by substantial reduction of the number of tree branches.

This work develops an algorithm that can efficiently exploit a large event space and generate a reduced event tree. It is assumed that every real system has an intrinsic logic behind the assignment of system outcomes to the system events. A simple and general methodology is suggested to robustly extract this knowledge, by exploiting the system outcome space information as this is stored in a table listing all the possible event combinations and their associated final system outcomes. The proposed algorithm is not biased by any prior information on the functionalities of the system in structural or algebraic form and seeks to acquire knowledge on the system logic and encapsulate it in the reduced tree. The algorithm can be generally applied to any given system and is not affected by the way that the supplied data may be organized or sorted. The procedure is deductive, starting from the entire event table and systematically organizing the system events in a set of clusters. The final set of clusters can be translated back to an event tree that is significantly reduced compared to original outcome space information supplied to the algorithm.

The paper is organized as follows: Section 2 presents some very basic system definitions used in the system representation and operations presented in Section 3. In Section 3 a suitable system representation is defined, using sets of events and based on the concepts of Cartesian products. Section 4 describes a set of clustering and declustering operations to be applied on the event sets. Section 5 discusses the implementation of the proposed developments into an algorithmic procedure. Section 6 presents an illustration example and a large case study to demonstrate the proposed methodology. Section 7 concludes the work.

2. Basic System Definitions

The system considered here is comprised of *blocks*. Each block can be found in various states, and the system response (or output) at every given instance of time, t, depends on the state that the blocks occupy at t. Each block relates to a component, a set of components or a part of a component of the system, regardless of physical conventions and according to the choices in the system modeling. The basic definitions are taken from Papazoglou [9, 10] where the blocks were interconnected to form functional block diagrams. In the present work, the blocks are stripped of their networking functionalities, and the definitions are simplified accordingly.

2.1. System Blocks and Their States. Consider a system of K-independent *blocks*. Each block b_k, $k \in \{1, 2, \ldots, K\}$, can be found in various *internal states* during the system operation period.

Let the *state set of block* b_k, denoted as $\mathbf{S}_k = \{s_k^1, s_k^2, \ldots, s_k^{K_{S_k}}\}$, be a partition over the possible instances of b_k, where $s_k^{i_k}$ denotes the i_kth state of block b_k, and K_{S_k} denotes the number of elements in $\mathbf{S}_k$, thus K_{S_k} is the cardinality of S_k, $|S_k|$.

It is assumed that, at every given time instance within the period of system operation, b_k is found in exactly one state (e.g., at 94% of maximum production level), and this state relates to exactly one member of $\mathbf{S}_k$ (e.g., s_k^2 = "at least 90% of maximum production level").

2.2. Event Definitions. A *basic event* is an instance of a single system block according to the state set partition of this block. A basic event for block b_k is denoted as $\{s_k^{i_k}\}$, $s_k^{i_k} \in \mathbf{S}_k$.

A *joint event* $\{s_{k_1}^{i_{k_1}}, s_{k_2}^{i_{k_2}}, \ldots, s_{k_m}^{i_{k_m}}\}$ is the combination of more than one basic events taking place in m different blocks of the system $b_{k_1}, b_{k_2}, \ldots, b_{k_m}$. Note that, the set $\{s_{k_1}^{i_{k_1}}, s_{k_2}^{i_{k_2}}, \ldots, s_{k_m}^{i_{k_m}}\}$ is the result of the Cartesian product $\{s_{k_1}^{i_{k_1}}\} \times \{s_{k_2}^{i_{k_2}}\} \times \cdots \times \{s_{k_m}^{i_{k_m}}\}$.

A *complete joint event* $\hat{e}$ is defined here as a joint event over all the system blocks $\{s_1^{i_1}, s_2^{i_2}, \ldots, s_K^{i_K}\}$. For simplicity, the term *event* is used here instead of complete joint event.

Note that, the above event definitions are simplified compared to Papazoglou [10] where blocks had functionalities not considered here.

2.3. Event Space, Subspaces, and Event Partitions. Let the *system event space* $\mathbf{E}$ be the set of all the possible complete joint events $\hat{e}$. Then $\mathbf{E} = \{\hat{e}_1, \hat{e}_2, \ldots, \hat{e}_{K_E}\}$, where K_E denotes the number of elements $\mathbf{E}$, thus $K_E = |E|$. Note that, $\mathbf{E}$ is the result of the Cartesian product $\mathbf{S}_1 \times \mathbf{S}_2 \times \cdots \times \mathbf{S}_K$, therefore $|\mathbf{E}| = \prod_{k=1}^{K} |\mathbf{S}_k|$.

Consider a nonempty subspace $\mathbf{A}$ of the system event space $\mathbf{A} \subseteq \mathbf{E}$: $\mathbf{A} \neq \emptyset$. Let $\mathbf{Q}(\mathbf{A})$ denote a *partition* applied over $\mathbf{A}$, composed of $K_{\mathbf{Q}(\mathbf{A})}$ disjoint subspaces of $\mathbf{A}$ denoted by $\mathbf{q}_i^Q(\mathbf{A})$, $i \in \{1, 2, \ldots, K_{\mathbf{Q}(\mathbf{A})}\}$.

2.4. Event Table. Let $\mathbf{R} = \{r_1, r_2, \ldots, r_{K_R}\}$ denote the set of the all possible *system outcomes* and K_R is their number. Therefore, K_R is the cardinality, $|\mathbf{R}|$, of $\mathbf{R}$.

Each event yields a unique system outcome, while different events may yield the same outcome. For instance, there might be more than one event that leads to system failure. In event trees, a complete joint event and its associated outcome are equivalent to a path [9, 10].

Let $T : \hat{e} \mapsto r = T(\hat{e})$, where $\hat{e} \in \mathbf{E}$ and $r \in \mathbf{R}$, denote the many-to-one mapping from the event space $\mathbf{E}$ to the set of system outcomes $\mathbf{R}$. The T mapping is recorded in the *system event table*, as a complete list of all the system events and their outcome.

The T defines a partition on $\mathbf{E}$, which is herein called the *outcome-based partition* and denoted by $\mathbf{Q}^T(\mathbf{E})$. The members of $\mathbf{Q}^T(\mathbf{E})$ are denoted by $\mathbf{q}_{r_i}^{Q^T}(\mathbf{E})$, $r_i \in \mathbf{R}$.

Table 1 gives the event table of an example taken from Papazoglou [10]. In this case, there are 32 events and 4 possible outcomes, and $\mathbf{Q}^T(\mathbf{E})$ is comprised of the 4 sets:

(1) $\mathbf{q}_{r_1}^{Q^T}(\mathbf{E}) = \{\hat{e}_1, \hat{e}_3, \hat{e}_4, \hat{e}_{11}, \hat{e}_{12}, \hat{e}_{17}, \hat{e}_{18}, \hat{e}_{19}, \hat{e}_{20}, \hat{e}_{25}, \hat{e}_{26}, \hat{e}_{27}, \hat{e}_{28}\}$,

(2) $\mathbf{q}_{r_2}^{Q^T}(\mathbf{E}) = \{\hat{e}_7, \hat{e}_{15}, \hat{e}_{16}, \hat{e}_{21}, \hat{e}_{22}, \hat{e}_{23}, \hat{e}_{24}, \hat{e}_{29}, \hat{e}_{30}, \hat{e}_{31}, \hat{e}_{32}\}$,

(3) $\mathbf{q}_{r_3}^{Q^T}(\mathbf{E}) = \{\hat{e}_5, \hat{e}_6, \hat{e}_8, \hat{e}_{13}, \hat{e}_{14}\}$,

(4) $\mathbf{q}_{r_4}^{Q^T}(\mathbf{E}) = \{\hat{e}_2, \hat{e}_9, \hat{e}_{10}\}$.

Note that the T mapping can be derived from the structural dependencies among the system blocks, as dictated by the rational and physical interconnections of components within the system and depicted in the form of a fault tree or a functional block diagram. However, such information may be unavailable or too difficult to attain or process. This work assumes that the only information available is the system event table.

3. System Representation

The system representation presented here aims to organize the information contained in the event table. The table data are partitioned and organized into vectors summarizing

TABLE 1: Event table for the motor-operated valve [10].

Complete joint events	Block states			System outcomes
	b_1	b_2	b_3	
$\widehat{e_1}$	s_1^1	s_2^1	s_3^1	r_1
$\widehat{e_2}$	s_1^1	s_2^1	s_3^2	r_4
$\widehat{e_3}$	s_1^1	s_2^1	s_3^3	r_1
$\widehat{e_4}$	s_1^1	s_2^1	s_3^4	r_1
$\widehat{e_5}$	s_1^1	s_2^2	s_3^1	r_3
$\widehat{e_6}$	s_1^1	s_2^2	s_3^2	r_3
$\widehat{e_7}$	s_1^1	s_2^2	s_3^3	r_2
$\widehat{e_8}$	s_1^1	s_2^2	s_3^4	r_3
$\widehat{e_9}$	s_1^1	s_2^3	s_3^1	r_4
$\widehat{e_{10}}$	s_1^1	s_2^3	s_3^2	r_4
$\widehat{e_{11}}$	s_1^1	s_2^3	s_3^3	r_1
$\widehat{e_{12}}$	s_1^1	s_2^3	s_3^4	r_1
$\widehat{e_{13}}$	s_1^1	s_2^4	s_3^1	r_3
$\widehat{e_{14}}$	s_1^1	s_2^4	s_3^2	r_3
$\widehat{e_{15}}$	s_1^1	s_2^4	s_3^3	r_2
$\widehat{e_{16}}$	s_1^1	s_2^4	s_3^4	r_2
$\widehat{e_{17}}$	s_1^2	s_2^1	s_3^1	r_1
$\widehat{e_{18}}$	s_1^2	s_2^1	s_3^2	r_1
$\widehat{e_{19}}$	s_1^2	s_2^1	s_3^3	r_1
$\widehat{e_{20}}$	s_1^2	s_2^1	s_3^4	r_1
$\widehat{e_{21}}$	s_1^2	s_2^2	s_3^1	r_2
$\widehat{e_{22}}$	s_1^2	s_2^2	s_3^2	r_2
$\widehat{e_{23}}$	s_1^2	s_2^2	s_3^3	r_2
$\widehat{e_{24}}$	s_1^2	s_2^2	s_3^4	r_2
$\widehat{e_{25}}$	s_1^2	s_2^3	s_3^1	r_1
$\widehat{e_{26}}$	s_1^2	s_2^3	s_3^2	r_1
$\widehat{e_{27}}$	s_1^2	s_2^3	s_3^3	r_1
$\widehat{e_{28}}$	s_1^2	s_2^3	s_3^4	r_1
$\widehat{e_{29}}$	s_1^2	s_2^4	s_3^1	r_2
$\widehat{e_{30}}$	s_1^2	s_2^4	s_3^2	r_2
$\widehat{e_{31}}$	s_1^2	s_2^4	s_3^3	r_2
$\widehat{e_{32}}$	s_1^2	s_2^4	s_3^4	r_2

the contribution of block states in each subspace of the partition. These vectors will provide the framework to apply the manipulations described in Section 4.

3.1. Cartesian Subspaces and Partitions.

Let $\mathbf{S}_k^{\mathbf{A}} \subseteq \mathbf{S}_k$, $\mathbf{A} \subseteq \mathbf{E} \wedge \mathbf{A} \neq \varnothing$ denote the set of b_k block states, $k \in \{1, 2, \dots, K\}$, in the events comprising $\mathbf{A}$. Note that, since $\mathbf{A} \neq \varnothing$, then $\mathbf{S}_k^{\mathbf{A}} \neq \varnothing$, for all k.

Let $\mathbf{C}(\mathbf{A})$, $\mathbf{A} \subseteq \mathbf{E} \wedge \mathbf{A} \neq \varnothing$, denote the Cartesian product $\mathbf{S}_1^{\mathbf{A}} \times \mathbf{S}_2^{\mathbf{A}} \times \cdots \times \mathbf{S}_K^{\mathbf{A}}$. In general, $\mathbf{C}(\mathbf{A})$ is a superset of $\mathbf{A}$, since it always contains all the elements of $\mathbf{A}$ and may contain events not included in $\mathbf{A}$.

Consider the system of Table 1. For the subspace $\mathbf{B} = \{\widehat{e_1}, \widehat{e_3}, \widehat{e_4}, \widehat{e_{17}}, \widehat{e_{18}}, \widehat{e_{19}}, \widehat{e_{20}}\}$ we get $\mathbf{S}_1^{\mathbf{B}} = \{s_1^1, s_1^2\}$, $\mathbf{S}_2^{\mathbf{B}} = \{s_2^1\}$,

$\mathbf{S}_3^{\mathbf{B}} = \{s_3^1, s_3^2, s_3^3, s_3^4\}$, and $\mathbf{C}(\mathbf{B}) = \{\widehat{e_1}, \widehat{e_2}, \widehat{e_3}, \widehat{e_4}, \widehat{e_{17}}, \widehat{e_{18}}, \widehat{e_{19}}, \widehat{e_{20}}\}$. As expected, $\mathbf{C}(\mathbf{B})$ is a superset of $\mathbf{B}$.

A *Cartesian subspace*, denoted as $\underline{\mathbf{A}}$, is herein defined as a nonempty subspace of $\mathbf{E}$, such that $\underline{\mathbf{A}} = \mathbf{C}(\underline{\mathbf{A}})$. Note that, all the singleton subspaces are Cartesian.

A *Cartesian partition* over $\mathbf{A}$, $\mathbf{A} \subseteq \mathbf{E} \wedge \mathbf{A} \neq \varnothing$, denoted as $\underline{\mathbf{Q}}(\mathbf{A})$, is herein defined as a partition comprised only of Cartesian subspaces of $\mathbf{A}$. The elements of $\underline{\mathbf{Q}}(\mathbf{A})$ are denoted as $\underline{\mathbf{q}}_i^{\mathbf{Q}}(\mathbf{A})$, $i \in \{1, 2, \dots, K_{\mathbf{Q}(\mathbf{A})}\}$. Note that, every subspace of $\mathbf{E}$ has at least one Cartesian partition comprised of singleton subspaces.

For instance, the subspace $\mathbf{B}$ defined above can be partitioned into the four subspaces $\underline{\mathbf{q}}_1^{\mathbf{Q}}(\mathbf{B}) = \{\widehat{e_1}\}$, $\underline{\mathbf{q}}_2^{\mathbf{Q}}(\mathbf{B}) = \{\widehat{e_3}\}$, $\underline{\mathbf{q}}_3^{\mathbf{Q}}(\mathbf{B}) = \{\widehat{e_4}\}$, and $\underline{\mathbf{q}}_4^{\mathbf{Q}}(\mathbf{B}) = \{\widehat{e_{17}}, \widehat{e_{18}}, \widehat{e_{19}}, \widehat{e_{20}}\}$. These are all Cartesian, since $\underline{\mathbf{q}}_1^{\mathbf{Q}}(\mathbf{B})$, $\underline{\mathbf{q}}_2^{\mathbf{Q}}(\mathbf{B})$, $\underline{\mathbf{q}}_3^{\mathbf{Q}}(\mathbf{B})$ are singleton, and $\underline{\mathbf{q}}_4^{\mathbf{Q}}(\mathbf{B}) = s_1^2 \times s_2^1 \times \mathbf{S}_3$.

3.2. Implicit Subspaces and Partitions.

The state set $\mathbf{S}_k^{\mathbf{A}}$ is called a *complete state set* when it contains all the possible states of the b_k block, that is, $\mathbf{S}_k^{\mathbf{A}} = \mathbf{S}_k$.

An *implicit subspace*, denoted as $\underset{\approx}{\mathbf{A}}$, is herein defined as a Cartesian subspace when all of its constituent block state subsets $\mathbf{S}_k^{\underset{\approx}{\mathbf{A}}}$ are either complete or singleton sets. Note that an implicit subspace corresponds to the implicant [11] containing only the block states in the singleton sets. The subspace $\underline{\mathbf{q}}_4^{\mathbf{Q}}(\mathbf{B})$ defined above is implicit and corresponds to the implicant $\{s_1^2, s_2^1\}$.

For example, the sets $\mathbf{B}_1 = s_1^2 \times s_2^2 \times \mathbf{S}_3$, $\mathbf{B}_2 = s_1^2 \times \{s_2^1, s_2^2\} \times \mathbf{S}_3$ and $\mathbf{B}_3 = s_1^2 \times \mathbf{S}_2 \times \mathbf{S}_3$ are all Cartesian (since they are Cartesian products) but $\mathbf{B}_2$ is not implicit.

An *implicit partition* over $\mathbf{A}$, denoted as $\underset{\approx}{\mathbf{Q}}(\mathbf{A})$, is herein defined as a partition comprised only of implicit subspaces. The elements of $\underset{\approx}{\mathbf{Q}}(\mathbf{A})$ are denoted as $\underset{\approx}{\mathbf{q}}_i^{\mathbf{Q}}(\mathbf{A})$. Every subspace $\mathbf{A}$ has at least one implicit partition, the $\underset{\approx}{\mathbf{Q}}(\mathbf{A}) = \mathbf{A}$, and the cardinalities of all possible partitions over $\mathbf{A}$ lie between one (when $\mathbf{A}$ is an implicit subspace) and $|\mathbf{A}|$.

For example, the partition over subspace $\mathbf{B}$ into $\underset{\approx}{\mathbf{q}}_i^{\mathbf{Q}}(\mathbf{B})$, $i \in \{1, 2, 3, 4\}$, is implicit. Also, the partition of $\mathbf{B}_2 = s_1^2 \times \{s_2^1, s_2^2\} \times \mathbf{S}_3$ into $s_1^2 \times s_2^1 \times \mathbf{S}_3$ and $s_1^2 \times s_2^2 \times \mathbf{S}_3$ is implicit.

An implicit partition conveys the same information as the set of its corresponding implicants. Therefore $\mathbf{B}_2$ can be derived from the prime implicants $\{s_1^2, s_2^1\}$ and $\{s_1^2, s_2^2\}$.

3.3. Contribution Vectors.

The *contribution vector* of subspace $\mathbf{A}$, denoted by $\overline{\mathbf{N}}(\mathbf{A})$, is defined here as a vector of nonnegative integer entries $n_k^{i_k}(\mathbf{A}) \in \mathbb{Z}_+$, reporting the sum of the contributions of each state $s_k^{i_k}$ of each block b_k in the events comprising $\mathbf{A}$, namely, the number of times that each block state contributes in the development of $\mathbf{A}$.

For the example of subspace $\mathbf{B}$, the contribution vector is $\overline{\mathbf{N}}(\mathbf{B}) = [3, 4; 7, 0, 0, 0; 2, 1, 2, 2]$, where the semicolons are used as separators between the three block compartments.

Properties of contribution vectors include the following.

(i) The vector length is $\lambda = \sum_{k=1}^{K} |\mathbf{S}_k|$, therefore $\overline{\mathbf{N}}(\mathbf{A}) \in \mathbb{N}_+^{\lambda}$, for all $\mathbf{A} \subseteq \mathbf{E}$.

(ii) For each block b_k, the sum of all the block entries in $\overline{\mathbf{N}}(\mathbf{A})$ equals $|\mathbf{A}|$.

In general, vector $\overline{\mathbf{N}}(\mathbf{A})$ is an abstract (inductive) representation of the information stored in $\mathbf{A}$. As the size of $\mathbf{A}$ increases, retrieving it from $\overline{\mathbf{N}}(\mathbf{A})$ is not trivial, it might even be impossible.

3.4. Bicontribution Vectors. Let the *bicontribution vector* $\overline{\mathbf{L}}(\mathbf{A})$ of subspace $\mathbf{A}$ be defined as a vector of Boolean entries $l_k^{i_k}(\mathbf{A}) \in \{O, I\}$, reporting the contribution or not of each state $s_k^{i_k}$ of each block b_k to the events of $\mathbf{A}$, namely, whether a certain block state contributes ("true" or I) or not ("false" or O) in the development of set $\mathbf{A}$.

Properties of bicontribution vectors include the following.

(i) The vector length is $\lambda = \sum_{k=1}^{K} |\mathbf{S}_k|$, so $\overline{\mathbf{L}}(\mathbf{A}) \in \{O, I\}^{\lambda}$, for all $\mathbf{A} \subseteq \mathbf{E}$.

(ii) The sum of all the entries corresponding to block b_k equals $|\mathbf{S}_k^{\mathbf{A}}|$, for all $k \in \{1, 2, \dots, K\}$.

For instance, the associated bicontribution vector of $\overline{\mathbf{N}}(\mathbf{B})$ is $\overline{\mathbf{L}}(\mathbf{B}) = [I, I; I, O, O, I; I, I, I, I]$. The sum of entries is 8 and equals the number of block state instances present in set $\mathbf{B}$.

Let $\Lambda[\overline{\mathbf{N}}(\mathbf{A})] : \{1, 2, \dots, |\mathbf{A}|\} \rightarrow \{I, O\}$ denote the operation applied on the contribution vector $\overline{\mathbf{N}}(\mathbf{A})$ to derive its associated bicontribution vector $\overline{\mathbf{L}}(\mathbf{A})$. Based on the definitions of contribution and bicontribution vectors:

$$\overline{\mathbf{L}}(\mathbf{A}) = \Lambda[\overline{\mathbf{N}}(\mathbf{A})] \iff l_k^{i_k}(\mathbf{A}) = \begin{cases} O, & n_k^{i_k}(\mathbf{A}) = 0, \\ I, & n_k^{i_k}(\mathbf{A}) > 0, \end{cases} \forall k, i_k.$$

(1)

The reverse operation $\Lambda^{-1}[\overline{\mathbf{N}}(\mathbf{A})]$ yields $\mathbf{C}(\mathbf{A})$ as follows:

$$\Lambda^{-1}[\overline{\mathbf{N}}(\mathbf{A})]$$

$$= \{\hat{e} = \{s_1^{i_1}, s_2^{i_2}, \dots, s_K^{i_K}\} \in \mathbf{E} : l_k^{i_k}(\mathbf{A}) = I, \forall k, i_k\}.$$

(2)

Therefore, from the definition of $\mathbf{C}(\mathbf{A})$, $\Lambda^{-1}[\overline{\mathbf{L}}(\mathbf{A})] = \mathbf{C}(\mathbf{A})$. Clearly, $\overline{\mathbf{L}}(\mathbf{A})$ always carries less information than its associated vector $\overline{\mathbf{N}}(\mathbf{A})$, unless $\mathbf{A}$ is Cartesian.

3.5. Cartesian Contribution Vectors. The *Cartesian contribution vector* of subspace $\mathbf{A}$, denoted as $\underline{\mathbf{N}}(\mathbf{A})$, is defined here as the contribution vector of $\mathbf{C}(\mathbf{A})$, thus $\underline{\mathbf{N}}(\mathbf{A}) \equiv \overline{\mathbf{N}}(\mathbf{C}(\mathbf{A}))$. Let $\underline{n}_k^{i_k}(\mathbf{A})$ denote the entries of $\underline{\mathbf{N}}(\mathbf{A})$.

For instance, starting from subspace $\mathbf{B}$ we get $\underline{\mathbf{N}}(\mathbf{B}) = [4, 4; 8, 0, 0, 0; 2, 2, 2, 2]$.

Let $\Xi[\overline{\mathbf{L}}(\mathbf{A})]$ denote the operation applied on the bicontribution vector $\overline{\mathbf{L}}(\mathbf{A})$ to derive the Cartesian contribution vector $\underline{\mathbf{N}}(\mathbf{A})$. Based on the above vector definitions:

$$\underline{\mathbf{N}}(\mathbf{A}) = \Xi[\overline{\mathbf{L}}(\mathbf{A})] \iff \underline{n}_k^{i_k}(\mathbf{A}) = \begin{cases} 0, & l_k^{i_k}(\mathbf{A}) = O, \\ \dfrac{|\mathbf{C}(\mathbf{A})|}{|\mathbf{S}_k^{\mathbf{A}}|}, & l_k^{i_k}(\mathbf{A}) = I, \end{cases}$$

$$\forall k, i_k.$$

(3)

For every subspace $\mathbf{A}$, each entry of $\mathbf{C}(\mathbf{A})$, $n_k^{i_k}(\mathbf{A})$, lays between zero and $\underline{n}_k^{i_k}(\mathbf{A})$. Nonzero entries for which $n_k^{i_k}(\mathbf{A}) = \underline{n}_k^{i_k}(\mathbf{A})$ are herein called *Cartesian entries*.

The vector $\overline{\mathbf{N}}(\mathbf{A})$ satisfies the *Cartesian property* if and only if is $\mathbf{A}$ a Cartesian subset. In this case $\overline{\mathbf{N}}(\mathbf{A}) = \underline{\mathbf{N}}(\mathbf{A})$.

Going back to set $\mathbf{B}$ and comparing $\overline{\mathbf{N}}(\mathbf{B})$ to $\underline{\mathbf{N}}(\mathbf{B})$:

(i) the entries corresponding to the block states s_1^4, s_4^1, s_4^3, and s_4^4 are Cartesian,

(ii) the $\overline{\mathbf{N}}(\mathbf{B})$ vector does not satisfy the Cartesian property since $n_k^{i_k}(\mathbf{B}) \neq \underline{n}_k^{i_k}(\mathbf{B})$ at $\{k, i_k\} = \{2, 1\}$ or $\{3, 2\}$.

The contribution vectors of implicit subspaces are herein called *implicit contribution vectors*. Since, implicit subspaces are always Cartesian, implicit contribution vectors always satisfy the Cartesian property.

3.6. Vector Partitions. Considering a partition $\mathbf{Q}(\mathbf{A})$, the *contribution vector partition* of $\mathbf{Q}(\mathbf{A})$, denoted as $\overline{\overline{\mathbf{N}}}(\mathbf{Q}(\mathbf{A}))$, is defined as the set of the contribution vectors $\overline{\mathbf{N}}(\mathbf{q}_i^{\mathbf{Q}}(\mathbf{A}))$.

Since $\sum_i n_k^{i_k}(\mathbf{q}_i^{\mathbf{Q}}(\mathbf{A})) = n_k^{i_k}(\mathbf{A})$, for all k, i_k, the vector $\overline{\mathbf{N}}(\mathbf{A})$ is specified as the *composite contribution vector* of $\overline{\overline{\mathbf{N}}}(\mathbf{Q}(\mathbf{A}))$. In general, composite contribution vectors carry less information than their vector partitions.

Likewise to subspace partitions, vector partitions can be classified as Cartesian (when all their associated vectors are Cartesian) or implicit (when all their associated vectors are implicit). Cartesian and implicit partitions are denoted as $\overline{\overline{\mathbf{N}}}(\bullet)$ and $\underline{\underline{\mathbf{N}}}(\bullet)$, respectively.

Considering subspace $\mathbf{B}$ and its partition $\mathbf{q}_1^{\mathbf{Q}}(\mathbf{B}) = \{\hat{e}_1\}$, $\mathbf{q}_2^{\mathbf{Q}}(\mathbf{B}) = \{\hat{e}_3\}$, $\mathbf{q}_3^{\mathbf{Q}}(\mathbf{B}) = \{\hat{e}_4\}$, and $\mathbf{q}_4^{\mathbf{Q}}(\mathbf{B}) = \{\hat{e}_{17}, \hat{e}_{18}, \hat{e}_{19}, \hat{e}_{20}\}$, the respective contribution vector partition is:

$$\overline{\overline{\mathbf{N}}}(\mathbf{Q}(\mathbf{B})) = \begin{cases} \overline{\mathbf{N}}(\mathbf{q}_1^{\mathbf{Q}}(\mathbf{B})) \\ \overline{\mathbf{N}}(\mathbf{q}_2^{\mathbf{Q}}(\mathbf{B})) \\ \overline{\mathbf{N}}(\mathbf{q}_3^{\mathbf{Q}}(\mathbf{B})) \\ \overline{\mathbf{N}}(\mathbf{q}_4^{\mathbf{Q}}(\mathbf{B})) \end{cases} = \begin{cases} [1, 0; 1, 0, 0, 0; 1, 0, 0, 0] \\ [1, 0; 1, 0, 0, 0; 0, 0, 1, 0] \\ [1, 0; 1, 0, 0, 0; 0, 0, 0, 1] \\ [0, 4; 4, 0, 0, 0; 1, 1, 1, 1] \end{cases}.$$

(4)

The $\mathbf{q}_i^{\overset{Q}{\approx}}(\mathbf{B})$ subsets are implicit, so $\overline{\overline{\mathbf{N}}}(\mathbf{Q}(\mathbf{B}))$ is an implicit contribution vector partition.

It should be highlighted that a Cartesian contribution vector partition is a disjoint-set data structure, as it contains all the information necessary to: (a) find which subspace includes a certain event and (b) reconstruct the event union set using the $\Lambda^{-1}[\overline{\mathbf{L}}(\bullet)]$ operations.

4. Decomposition and Recomposition Operations

Stating from a given outcome-based partition $\mathbf{Q}^T(\mathbf{E})$, the scope here is to derive an implicit partition featuring minimal cardinality for each of the $\mathbf{q}_{r_i}^{\mathbf{Q}^T}(\mathbf{E})$, $r_i \in \mathbf{R}$ subspaces. This would be equivalent to a set of prime implicants describing the event table. The scope is accomplished through the application of specific operations on the system vectors defined above. These operations extract knowledge from the information carried in the composition vectors and store this knowledge in the minimal possible schemes. The naming of these operations is after Shannon's decomposition [11].

4.1. Decomposition of Contribution Vectors. This section presents an operation for the systematic manipulation of a contribution vector to yield a contribution vector partition. The operation is called *decomposition*, since it is a case of multistate Shannon's decomposition applied on the contribution vectors. This operation decreases the abstraction of the information stored in the vector, since contribution vector partitions carry more information than their composite vectors. Starting from a contribution vector, subsequent decomposition actions generate a low cardinality implicit partition of this vector.

Consider a subspace $\mathbf{A} \neq \varnothing$: $\exists k$: $|\mathbf{S}_k^{\mathbf{A}}| \geq 2$ and a nonempty subset $\sigma_k \subset \mathbf{S}_k^{\mathbf{A}}$. The *decomposition operation* of $\mathbf{A}$ according to σ_k is a partition rule applied on $\mathbf{A}$ to generate the sets $\{\mathbf{A}^{\sigma_k}, \mathbf{A} - \mathbf{A}^{\sigma_k}\}$ such that:

$$\mathbf{A}^{\sigma_k} = \left\{ \hat{e} \in \mathbf{A} \cap \Delta_{\mathbf{A}}^{\sigma_k} \right\},$$

$$\mathbf{A} - \mathbf{A}^{\sigma_k} = \left\{ \hat{e} \in \mathbf{A} \cap \left(\mathbf{C}(\mathbf{A}) - \Delta_{\mathbf{A}}^{\sigma_k} \right) \right\}, \qquad (5)$$

where $\Delta_{\mathbf{A}}^{\sigma_k} = \left\{ \mathbf{S}_1^{\mathbf{A}} \times \mathbf{S}_2^{\mathbf{A}} \times \cdots \times \sigma_k \times \cdots \times \mathbf{S}_K^{\mathbf{A}} \right\}$.

Since $\sigma_k \neq \varnothing$ then $\mathbf{A}^{\sigma_k}, \mathbf{A} - \mathbf{A}^{\sigma_k} \neq \varnothing$.

The simplest way of applying the decomposition operation is to go through each one of the events in $\mathbf{A}$, as described in the operation definition. This procedure requires computational effort to decide whether a certain event belongs in $\mathbf{A}^{\sigma_k}$ or $\mathbf{A} - \mathbf{A}^{\sigma_k}$. If, on the other hand, $\overline{\mathbf{N}}(\mathbf{A})$ contains Cartesian entries, the operation can be applied directly on $\overline{\mathbf{N}}(\mathbf{A})$. Section 6 shows that decomposing $\overline{\mathbf{N}}(\mathbf{A})$ rather than $\mathbf{A}$ reduces the computational effort by several orders of magnitude.

Consider a contribution vector $\overline{\mathbf{N}}(\mathbf{A})$: $\exists k$: $|\mathbf{S}_k^{\mathbf{A}}| \geq 2 \wedge |\mathbf{S}_k^{\mathbf{A}}| \neq 0$ and a nonempty set $\sigma_k \subseteq \mathbf{S}_k^{\mathbf{A}}$. The *vector decomposition operation* according to σ_k is a partition rule

applied on $\overline{\mathbf{N}}(\mathbf{A})$ to yield the contribution vector partition $\{\overline{\mathbf{N}}(\mathbf{A}^{\sigma_k}), \overline{\mathbf{N}}(\mathbf{A}^{\overline{\sigma_k}})\}$ such that

$$\overline{\mathbf{N}}(\mathbf{A}^{\sigma_k}) = \overline{\mathbf{N}}\left(\mathbf{S}_1^{\mathbf{N}} \times \mathbf{S}_2^{\mathbf{N}} \times \cdots \times \sigma_k \times \cdots \times \mathbf{S}_K^{\mathbf{N}} \right),$$

$$\overline{\mathbf{N}}(\mathbf{A}^{\overline{\sigma_k}}) = \overline{\mathbf{N}}(\mathbf{A}) - \overline{\mathbf{N}}(\mathbf{A}^{\sigma_k}). \qquad (6)$$

The development of $\overline{\mathbf{N}}(\mathbf{A}^{\overline{\sigma_k}})$ ensures that this vector has the Cartesian property. Instead of the Cartesian product $\mathbf{S}_1^{\mathbf{N}} \times \mathbf{S}_2^{\mathbf{N}} \times \cdots \times \sigma_k \times \cdots \times \mathbf{S}_K^{\mathbf{N}}$ we can use the bicontribution vector:

$$\overline{\mathbf{L}}(\mathbf{A}^{\sigma_k}) : l_m^{i_m}\Big|_{\overline{\mathbf{L}}(\mathbf{A}^{\sigma_k})} = \begin{cases} l_m^{i_m}\Big|_{\overline{\mathbf{L}}(\mathbf{A})}, & k \neq m, \\ \mathrm{I}, & k = m \wedge s_m^{i_m} \in \sigma_k, \\ \mathrm{O}, & k = m \wedge s_m^{i_{m,n}} \notin \sigma_k. \end{cases}$$

$$(7)$$

The decomposition operation can be applied iteratively. To simplify the notation, let $\overline{\mathbf{N}}_0$ be an initial contribution vector, decomposed into $\{\overline{\mathbf{N}}_1^{\sigma_0}, \overline{\mathbf{N}}_1\}$, where $\overline{\mathbf{N}}_1^{\sigma_0}$ has the Cartesian property. Then, $\overline{\mathbf{N}}_1$ is decomposed into $\{\overline{\mathbf{N}}_2^{\sigma_1}, \overline{\mathbf{N}}_2\}$, where $\overline{\mathbf{N}}_2^{\sigma_1}$ has the Cartesian property and so forth. Application of the decomposition operation over i iterations replaces $\overline{\mathbf{N}}_0$ with the contribution vector partition $\overline{\overline{\mathbf{N}}}_i = \{\overline{\mathbf{N}}_1^{\sigma_0}, \overline{\mathbf{N}}_2^{\sigma_1}, \overline{\mathbf{N}}_3^{\sigma_2}, \ldots, \overline{\mathbf{N}}_i^{\sigma_{i-1}}, \overline{\mathbf{N}}_i\}$. The iterations terminate when $\overline{\mathbf{N}}_i$ has the Cartesian property, therefore $\overline{\overline{\mathbf{N}}}_i$ is a Cartesian partition. Note that, the $\overline{\mathbf{N}}_i^{\sigma_{i-1}}$ vectors can be derived as $\overline{\mathbf{N}}_i^{\sigma_{i-1}} = \Xi[\overline{\mathbf{L}}_i^{\sigma_{i-1}}]$, using (7) and $\overline{\mathbf{L}}_{i-1} = \Lambda[\overline{\mathbf{N}}_{i-1}]$.

4.1.1. Decomposition Example. To illustrate the decomposition operation consider the set $\mathbf{q}_{r_1}^T(\mathbf{E})$ of Table 1 and its contribution vector $\overline{\mathbf{N}}_0 = [5, 8; 7, 0, 6, 0; 3, 2, 4, 4]$.

Starting from $\overline{\mathbf{N}}_0$, the bicontribution vector is $\overline{\mathbf{L}}_0 = [\mathrm{I}, \mathrm{I}; \mathrm{I}, \mathrm{O}, \mathrm{I}, \mathrm{O}; \mathrm{I}, \mathrm{I}, \mathrm{I}, \mathrm{I}]$ and the Cartesian contribution vector is $\overline{\mathbf{N}}_{\sim 0} = [8, 8; 8, 0, 8, 0; 4, 4, 4, 4]$. Clearly, $\overline{\mathbf{N}}_{\sim 0}$ has three Cartesian entries at s_1^2, s_3^3 and s_3^4. Let $\sigma_0 = \{s_3^3\}$:

(i) $\overline{\mathbf{L}}_1^{\sigma_0} = [\mathrm{I}, \mathrm{I}; \mathrm{I}, \mathrm{O}, \mathrm{I}, \mathrm{O}; \mathrm{O}, \mathrm{O}, \mathbf{I}, \mathrm{O}]$

   $\Rightarrow \overline{\mathbf{N}}_1^{\sigma_0} = [2, 2; 2, 0, 2, 0; 0, 0, \mathbf{4}, 0]$,

(ii) $\overline{\mathbf{N}}_1 = \overline{\mathbf{N}}_0 - \overline{\mathbf{N}}_1^{\sigma_0} = [3, 6; 5, 0, 4, 0; 3, 2, \mathbf{0}, 4]$,

   $\overline{\mathbf{L}}_1 = [\mathrm{I}, \mathrm{I}; \mathrm{I}, \mathrm{O}, \mathrm{I}, \mathrm{O}; \mathrm{I}, \mathrm{I}, \mathbf{O}, \mathrm{I}]$.

Alternatively, the decomposition could be applied *simultaneously* at $\sigma_0^* = \{s_3^3, s_3^4\}$ as follows:

(i) $\overline{\mathbf{L}}_1^{\sigma_0^*} = [\mathrm{I}, \mathrm{I}; \mathrm{I}, \mathrm{O}, \mathrm{I}, \mathrm{O}; \mathrm{O}, \mathrm{O}, \mathbf{I}, \mathbf{I}]$

   $\Rightarrow \overline{\mathbf{N}}_1^{\sigma_0^*} = [4, 4; 4, 0, 4, 0; 0, 0, \mathbf{4}, \mathbf{4}]$,

(ii) $\overline{\mathbf{N}}_1^* = \overline{\mathbf{N}}_0 - \overline{\mathbf{N}}_1^{\sigma_0^*} = [1, 4; 3, 0, 2, 0; 3, 2, \mathbf{0}, \mathbf{0}]$,

   $\overline{\mathbf{L}}_1^* = [\mathrm{I}, \mathrm{I}; \mathrm{I}, \mathrm{O}, \mathrm{I}, \mathrm{O}; \mathrm{I}, \mathrm{I}, \mathbf{O}, \mathbf{O}]$.

In both cases, the vectors $\overline{\mathbf{N}}_1$ and $\overline{\mathbf{N}}_1^*$ do not satisfy the Cartesian property, but they contain Cartesian entries. For instance, $\overline{\mathbf{N}}_1^*$ can be decomposed at $\sigma_1^* = \{s_1^2\}$ as follows:

(i) $\overline{\mathbf{L}}_2^{\sigma_1^*} = [0, \mathbf{1}; 1, 0, 1, 0; 1, 1, 0, 0]$

$\Rightarrow \overline{\mathbf{N}}_2^{\sigma_1^*} = [0, \mathbf{4}; 2, 0, 2, 0; 2, 2, 0, 0],$

(ii) $\overline{\mathbf{N}}_2^* = \overline{\mathbf{N}}_1^* - \overline{\mathbf{N}}_2^{\sigma_1^*} = [1, \mathbf{0}; 1, 0, 0, 0; 1, 0, 0, 0].$

Both $\overline{\mathbf{N}}_2^*$ and $\overline{\mathbf{N}}_2^{\sigma_1^*}$ satisfy the Cartesian property and no further decomposition is necessary. So, the decomposition of $\overline{\mathbf{N}}_0$ yields the Cartesian contribution vector partition $\{\overline{\mathbf{N}}_1^{\sigma_0}, \overline{\mathbf{N}}_2^*, \overline{\mathbf{N}}_2^{\sigma_1^*}\}$.

The set vectors may contain complete blocks, like the first block in $\overline{\mathbf{N}}_1^{\sigma_0}$. The vectors $\overline{\mathbf{N}}_1^{\sigma_0}$ and $\overline{\mathbf{N}}_2^{\sigma_1}$ can be further decomposed until they give implicit contribution vector partitions:

(i)

$$\overline{\mathbf{N}}_1^{\sigma_0} = [4, 4; 4, 0, 4, 0; 0, 0, 4, 4]$$

$$\sim \left\{ \begin{array}{l} [2, 2; 4, 0, 0, 0; 0, 0, 2, 2] \\ [2, 2; 0, 0, 4, 0; 0, 0, 2, 2] \end{array} \right\}$$

$$\sim \left\{ \begin{array}{l} [1, 1; 2, 0, 0, 0; 0, 0, 2, 0] \\ [1, 1; 2, 0, 0, 0; 0, 0, 0, 2] \\ [1, 1; 0, 0, 2, 0; 0, 0, 2, 0] \\ [1, 1; 0, 0, 2, 0; 0, 0, 0, 2] \end{array} \right\},$$

(8)

(ii)

$$\overline{\mathbf{N}}_2^{\sigma_1} = [0, 4; 2, 0, 2, 0; 2, 2, 0, 0]$$

$$\sim \left\{ \begin{array}{l} [0, 2; 0, 0, 2, 0; 1, 1, 0, 0] \\ [0, 2; 2, 0, 0, 0; 1, 1, 0, 0] \end{array} \right\}$$

$$\sim \left\{ \begin{array}{l} [0, 1; 0, 0, 1, 0; 0, 1, 0, 0] \\ [0, 1; 0, 0, 1, 0; 1, 0, 0, 0] \\ [0, 1; 1, 0, 0, 0; 0, 1, 0, 0] \\ [0, 1; 1, 0, 0, 0; 1, 0, 0, 0] \end{array} \right\}.$$

(9)

In the previous example, the cardinality of $\mathbf{q}_{r_1}^T(\mathbf{E})$ was 13, while the cardinality of the implicit partition is equal to $4 + 4 + 1 = 9$. The latter cardinality could be even smaller if a more intelligent decomposition strategy were applied. For instance, $\overline{\mathbf{N}}_0$ has two complete blocks (first and third block) and three Cartesian entries (one in the first block and two in the third). The choice of decomposition order, that is, the sequence of σ_i's is crucial, as discussed in the algorithm implementation section.

4.2. Recomposition of Contribution Vectors.

The outcome of the decomposition operation is an implicit contribution vector partition. Given this, we can seek merging opportunities to create unions that have complete blocks, thus reduce the partition cardinality. Since the contribution vectors are all

implicit, the recomposition operation is hereby discussed in terms of bicontribution vectors.

Consider a block b_m and its set of states $\mathbf{S}_m$. Let $\overline{\overline{\mathbf{L}}}$ be a set of $|\mathbf{S}_m|$ bicontribution vectors, $\overline{\mathbf{L}}_i = [l_{i_m,k}^{i_k}]$, $i_m \in \{1, 2, \ldots, |\mathbf{S}_m|\}$, $i_k \in \{1, 2, \ldots, |\mathbf{S}_k|\}$, $k \in \{1, 2, \ldots, K\}$, such that $l_{i_m,k}^{i_k} = \lambda_k^{i_k} \in \{O, I\}$, for all $k \neq m$, i_m, i_k. Then, the set $\overline{\overline{\mathbf{L}}}$ can be replaced by the single bicontribution vector $\overline{\mathbf{L}} = [\lambda_k^{i_k}]$, where $\lambda_m^{i_m} = I$, for all i_m. Note that, $\overline{\mathbf{L}} = \overline{\mathbf{L}}(\mathbf{S}_1^{\overline{\mathbf{L}}_1} \times \mathbf{S}_2^{\overline{\mathbf{L}}_1} \times \cdots \times \mathbf{S}_m \times \cdots \times \mathbf{S}_K^{\overline{\mathbf{L}}_1})$.

4.2.1. Recomposition Example. Consider the set $\overline{\overline{\mathbf{L}}} = \left\{ \begin{array}{l} [I, O; I, O, O, O; I, O, O, O] \\ [O, I; I, O, O, O; I, O, O, O] \end{array} \right\}$.

From the decomposition example discussed earlier. The two bicomposition vectors have exactly the same entries in the second and third block but different entries in the first block. In addition, the first block has exactly two state instances. The two vectors can be recomposed into a single equivalent vector $[I, I; I, O, O, O; I, O, O, O]$.

5. Algorithm Implementation

Given the table associated with an event tree, the proposed algorithm launches an iterative process of decomposition and recomposition operations until we obtain an implicit partition of minimal cardinality for each one of the tree outcomes. Working with the vectors defined above rather than sets of events decreases significantly the amount of information being stored and the computational effort for manipulating this information.

The algorithm applies the vector decomposition and recomposition operations using a set of heuristic rules. These rules help in the identification of more promising entries to apply the operations. Sections 5.1 and 5.2 describe the heuristic rules, and Section 5.3 discusses how these procedures work together within the proposed algorithm.

5.1. Heuristic Rules for Decomposition Order.

The decomposition operations proceed iteratively, replacing the original $|\mathbf{R}|$ vectors of the outcome partition with $|\mathbf{R}|$ contribution vector sets. Decompositions are applied locally, based on the features (e.g., the Cartesian entries) of each contribution vector.

Given a composition vector, the choice of decomposition order is crucial in preserving the maximum possible of the initial complete blocks. The set $\mathbf{q}_{r_1}^T(\mathbf{E})$ of Table 1 has a complete block of 2 states and a complete block of 4 states. If the decomposition order is different the final implicit partition could be different. In effect, applying the decomposition on entries $\{s_3^3, s_3^4\}$ of $\overline{\mathbf{N}}_0$ results in a reduction by $4 = 13 - 9$ on the cardinality of the final implicit partition of $\mathbf{q}_{r_1}^T(\mathbf{E})$. Applying the decomposition on $\{s_1^2\}$—followed by $\{s_3^3, s_3^4\}$, $\{s_3^1\}$, $\{s_4^3, s_4^4\}$, and so forth, in any order—preserves the third block, and the final reduction is by $13 - 7 = 6$:

$$\overline{\mathbf{N}}_0 = [5, 8; 7, 0, 6, 0; 3, 2, 4, 4] \sim \left\{ \begin{array}{l} [0, 8; 4, 0, 4, 0; 2, 2, 2, 2] \\ [5, 0; 3, 0, 2, 0; 1, 0, 2, 2] \end{array} \right\}$$

$$\sim \left\{ \begin{array}{l} [0, 4; 4, 0, 0, 0; 1, 1, 1, 1] \\ [0, 4; 0, 0, 4, 0; 1, 1, 1, 1] \\ [4, 0; 2, 0, 2, 0; 0, 0, 2, 2] \\ [1, 0; 1, 0, 0, 0; 1, 0, 0, 0] \end{array} \right\}$$

$$\sim \left\{ \begin{array}{ll} [0, 4; 4, 0, 0, 0; 1, 1, 1, 1] & [1, 0; 1, 0, 0, 0; 0, 0, 1, 0] \\ [0, 4; 0, 0, 4, 0; 1, 1, 1, 1] & [1, 0; 1, 0, 0, 0; 0, 0, 0, 1] \\ [1, 0; 1, 0, 0, 0; 1, 0, 0, 0] & [1, 0; 0, 0, 1, 0; 0, 0, 1, 0] \\ & [1, 0; 0, 0, 1, 0; 0, 0, 0, 1] \end{array} \right\}.$$

(10)

The following decomposition rules support the generation of the smallest possible implicit partitions at the minimum possible execution time, and they are applied on each contribution vector that is not Cartesian.

(a) If there are Cartesian entries in the current contribution vector:

   (i) it is prefered to decompose at Cartesian entries in incomplete blocks, to avoid breaking the complete blocks. Note that, in this case, the decomposition order makes no difference to the final partitions,

   (ii) if the only Cartesian entries are within complete blocks, start decomposing the complete blocks with the largest span between their contribution values, relatively to their number of states, that is, sort blocks according to $(\max_k\{n_k^{i_k}\} - \min_k\{n_k^{i_k}\})/|\mathbf{S}_k|$.

(b) If there are no Cartesian entries in the current contribution vector, then:

   (i) if no complete blocks are present, prefer incomplete blocks with more states being present, and decompose them into as many vectors as the number of states present,

   (ii) if only complete blocks are present, decompose the block featuring the entry with the maximum departure from its Cartesian value, that is, $\max_k\{\underline{n}_k^{i_k} - n_k^{i_k}\}$.

Before the application of these rules it is essential to recognize which of the complete blocks make sense, according to the values of the contribution vector entries. For instance, the vector $[5, 8; 7, 0, 6, 0; 3, 2, 4, 4]$ could lead to a partition that includes a bicontribution vector of 2 complete blocks, but this not possible for the vector $[3, 8; 5, 0, 6, 0; 2, 2, 4, 3]$, since the first entry in the compartment of the first block is lower than the number of states in the third block. Similarly, in the vector $[0, 4; 2, 0, 2, 0; 1, 1, 1, 1]$ the third block cannot remain complete under any decomposition order since there is no

single entry in the second block compartment greater or equal to $|\mathbf{S}_3| = 4$.

The final Cartesian partition includes vectors and may need to be further decomposed to derive implicit partitions. In this case, the vectors are decomposed in all their incomplete block entries (which are all Cartesian), and the decomposition order makes no difference.

5.2. Heuristic Rules for Recomposition Order. Once the decomposition stage is completed, the final implicit partitions may have vectors that can be merged. This reduces the partition cardinalities. The choice of recomposition order is crucial, since the application of decomposition operations on large event tables can produce numerous vectors as candidates for recomposition.

Let $\mathbf{Q}_0(\mathbf{q}_{r_i}^T(\mathbf{E}))$, $r_i \in \mathbf{R}$, denote the implicit partitions output from the decomposition stage. The recomposition algorithm proceeds according to the following steps applied on each vector partition $\overline{\overline{\mathbf{N}}}_0 = \overline{\overline{\mathbf{N}}}(\mathbf{Q}_0(\mathbf{q}_{r_i}^T(\mathbf{E})))$ (or the associated bicontribution vectors).

   (i) Partition $\overline{\overline{\mathbf{N}}}$ into sets having the same complete blocks. Let $\overline{\overline{\mathbf{N}}}_a$ be such a subset of $\overline{\overline{\mathbf{N}}}$.

   (ii) The set $\overline{\overline{\mathbf{N}}}_a$ is a candidate for creating an additional complete block b_m, if there is at least one subset $\overline{\overline{\mathbf{N}}}_a^{m_i} \subseteq \overline{\overline{\mathbf{N}}}_a$ of cardinality $|\mathbf{S}_m|$, where $i \in I_m$ is the index of the particular subset for b_m, such that: (i) its overall contribution vector features b_m as a complete block and (ii) the remaining blocks have the same state sets $\mathbf{S}_k^{m_i}$, for all $k \neq m$.

   (iii) If $\overline{\overline{\mathbf{N}}}_a^{m_i}$ is replaced by its composite contribution vector, the resulting vector partition $\overline{\overline{\mathbf{N}}}_1 = \overline{\overline{\mathbf{N}}}(\mathbf{Q}_1(\mathbf{q}_{r_i}^T(\mathbf{E})))$ is implicit.

   (iv) The process is repeated on the new set $\overline{\overline{\mathbf{N}}}_1$ to reduce it to $\overline{\overline{\mathbf{N}}}_2$ and so forth and terminates when there are no more recomposition candidates left.

The set $\overline{\overline{\mathbf{N}}}_a$ may have several subsets $\overline{\overline{\mathbf{N}}}_a^{m_i}$ that relate to the creation of different complete blocks and/or to different ways of creating a particular complete block. The recomposition procedure is supported by intelligent selection biases to ensure that the merging opportunities are properly exploited. During the iterative recomposition process, the following recomposition rules are applied on every set $\overline{\overline{\mathbf{N}}}_a$ of Cartesian contribution vectors featuring the same complete blocks.

   (i) Find all the candidate sets $\overline{\overline{\mathbf{N}}}_a^{m_i} \subseteq \overline{\overline{\mathbf{N}}}_a$, $i \in I_m$, for creating each new complete block at b_m, for all $m \in \{1, 2, \ldots, K\}$.

   (ii) Associate each candidate to a score $w_a^{m_i}$ proportional to $|\mathbf{S}_m|$ and I_m and incorporate the potential to create two complete blocks in one go.

   (iii) Select among conflicting sets according to their $w_a^{m_i}$.

Quality Assurance and Reliability Engineering

TABLE 2: Recomposition example: initial set of bicontribution vectors (Step 1).

$\bar{L}_i$	b_1				b_2			b_3		b_4			b_5		b_6		b_7		b_8		b_9		b_{10}
24	O	O	O	I	I	O	O	I	O	O	I	O	I	I	O	I	O	I	O	**I**	**O**	I	O
26	O	O	O	I	I	O	O	I	O	O	I	O	I	I	O	I	O	I	O	**O**	**I**	I	O
28	O	O	O	I	I	O	O	I	O	O	I	O	I	O	I	I	O	I	O	I	I	I	O
21	I	O	O	O	I	O	I	I	I	O	I	O	I	I	O	I	O	I	O	I	O	I	O
25	I	O	O	O	I	O	I	I	I	O	I	O	I	I	O	I	O	I	O	O	I	I	O
11	O	O	O	I	I	O	O	I	O	I	O	I	O	I	I	I	O	I	O	I	I	I	O
12	O	O	O	I	I	O	O	I	O	O	I	I	O	I	I	I	O	I	O	I	I	I	O
18	O	O	O	I	I	O	O	I	O	I	O	O	I	I	I	I	O	I	O	I	I	I	O
22	O	O	O	I	I	O	I	O	O	O	I	O	I	I	O	I	I	I	I	I	O	I	O
23	O	O	O	I	I	O	O	O	I	O	I	O	I	I	O	I	I	I	I	I	O	I	O
27	I	O	O	O	I	O	I	I	I	O	I	O	I	O	I	I	O	I	O	I	I	I	O
5	I	O	O	O	I	O	I	I	I	I	I	O	I	O	I	I	I	O	I	O	I	I	O
6	I	O	O	O	I	O	I	I	I	O	I	I	O	I	I	I	O	I	O	I	I	I	O
14	I	O	O	O	I	O	I	I	I	I	O	O	I	I	I	I	O	I	O	I	I	I	O
15	O	O	I	O	I	O	I	I	I	I	O	O	I	I	O	I	I	I	I	I	O	I	O
19	O	I	O	O	I	O	I	I	I	O	I	O	I	I	O	I	I	I	I	I	O	I	O
20	O	O	I	O	I	O	I	I	I	O	I	O	I	I	O	I	I	I	I	I	O	I	O
7	O	O	O	I	I	O	I	O	O	O	I	O	I	O	I	I	I	I	I	I	I	I	O
8	O	O	O	I	I	O	I	O	O	O	O	I	I	O	I	I	I	I	I	I	I	I	O
9	O	O	O	I	I	O	O	O	I	I	O	I	O	I	I	I	I	I	I	I	I	I	O
10	O	O	O	I	I	O	O	O	I	O	I	I	O	I	I	I	I	I	I	I	I	I	O
16	O	O	O	I	I	O	I	O	O	I	O	O	I	I	I	I	I	I	I	I	I	I	O
17	O	O	O	I	I	O	O	O	I	I	O	O	I	I	I	I	I	I	I	I	I	I	O
29	O	O	O	I	I	O	I	O	O	I	O	I	O	I	I	I	I	I	I	I	I	O	I
30	O	O	O	I	I	O	I	O	O	I	O	O	I	I	I	I	I	I	I	I	I	O	I
1	O	I	O	O	I	O	I	I	I	I	O	I	O	I	I	I	I	I	I	I	I	I	O
2	O	I	O	O	I	O	I	I	I	O	I	I	O	I	I	I	I	I	I	I	I	I	O
3	O	O	I	O	I	O	I	I	I	I	O	I	O	I	I	I	I	I	I	I	I	I	O
4	O	O	I	O	I	O	I	I	I	O	I	I	O	I	I	I	I	I	I	I	I	I	O
13	O	I	O	O	I	O	I	I	I	I	O	O	I	I	I	I	I	I	I	I	I	I	O

(iv) Apply the recomposition operation on the selected sets.

Note that, in the new partition, the composite vectors should be removed form the subset $\overline{\overline{\mathbf{N}}}_a$ and possibly included in other subsets including the complete blocks of $\overline{\overline{\mathbf{N}}}_a$ and complete blocks created during the recomposition. The reason why the value of I_m is taken into account is that larger I_m's increase the probability of finding "recomposable" sets in the next iteration of the recomposition process.

Consider, for example, the vectors of Table 2 representing the implicit partition for the fifth outcome of the BWR example solved in Section 6.1. As explained above, recomposition operations are applied on bicontribution vectors. The vectors are sorted and divided into different sets according to their complete blocks. The procedure starts from the subset with the fewer complete blocks. This includes the vectors $\overline{\mathbf{L}}_{24}$ and $\overline{\mathbf{L}}_{26}$, which are merged to give $\overline{\mathbf{L}}_{24+26}$. Table 3 shows the Cartesian set updated with $\overline{\mathbf{L}}_{24+26}$, which can now be

merged with $\overline{\mathbf{L}}_{28}$ and so forth. The recomposition choices are not always so few and they can be conflicting. The subset of Table 4 appears after a few iterations. A crude analysis indicates the potential of creating new complete blocks at:

(i) block b_4 by merging $\overline{\mathbf{L}}_i$ & $\overline{\mathbf{L}}_{iii}$ and/or $\overline{\mathbf{L}}_{iv}$ & $\overline{\mathbf{L}}_{viii}$,

(ii) block b_5 by merging $\overline{\mathbf{L}}_{ii}$ & $\overline{\mathbf{L}}_{iii}$, $\overline{\mathbf{L}}_v$ & $\overline{\mathbf{L}}_{vii}$ and/or $\overline{\mathbf{L}}_{vi}$ & $\overline{\mathbf{L}}_{viii}$,

(iii) block b_{10} by merging $\overline{\mathbf{L}}_v$ & $\overline{\mathbf{L}}_{vi}$ and/or $\overline{\mathbf{L}}_{vii}$ & $\overline{\mathbf{L}}_{viii}$.

The above seven merging actions involve binary state blocks, so their initial scores are equal to 2. This score is multiplied by the number of candidate merges per block (I_m), giving a score of 4 for the 4 merges in blocks b_4 and b_{10} and 6 for the 3 merges in block b_5. The possible merging actions are sorted according to their score and an action can take place if it does not conflict with any of the higher score actions. In this sense the proposed actions are $\overline{\mathbf{L}}_{ii}$ & $\overline{\mathbf{L}}_{iii}$, $\overline{\mathbf{L}}_v$ & $\overline{\mathbf{L}}_{vii}$ and $\overline{\mathbf{L}}_{vi}$ & $\overline{\mathbf{L}}_{viii}$.

This involvement of I_m in the score calculation stems from the observation that the resulting vectors have many common entries so the possibility of these vectors being treated as candidates for subsequent merges is very high. In effect, amongst the merged vectors of Table 5, vectors $\overline{L}_{v+vii}$ and $\overline{L}_{vi+viii}$ are candidates to be merged. However, this potential should be examined along with other vectors featuring the same complete blocks.

5.3. Algorithm Implementation.

The decomposition and recomposition operations discussed above are each implemented into an iterative procedure. Following is a step-by-step description of the algorithm, using the motor-operated valve (MOV) example of Table 1 to illustrate the different procedures.

Step 1. Acquire event outcome data. This step returns a table of K columns for the K component blocks and 1 column for the outcomes. Following is the data array for the MOV example:

$$
\begin{bmatrix}
1 & 1 & 1 & \\
1 & & & \\
1 & 1 & 2 & 4 \\
1 & 1 & 3 & 1 \\
1 & 1 & 4 & 1 \\
1 & 2 & 1 & 3 \\
1 & 2 & 2 & 3 \\
1 & & & \\
2 & 3 & 2 & \\
1 & 2 & 4 & 3 \\
1 & 3 & 1 & 4 \\
1 & 3 & 2 & 4 \\
1 & 3 & 3 & 1 \\
1 & 3 & 4 & \\
1 & & & \\
1 & 4 & 1 & 3 \\
1 & 4 & 2 & 3 \\
1 & 4 & 3 & 2 \\
1 & 4 & 4 & 2 \\
2 & 1 & 1 & 1 \\
2 & & & \\
1 & 2 & 1 & \\
2 & 1 & 3 & 1 \\
2 & 1 & 4 & 1 \\
2 & 2 & 1 & 2 \\
2 & 2 & 2 & 2 \\
2 & 2 & 3 & \\
2 & & & \\
2 & 2 & 4 & 2 \\
2 & 3 & 1 & 1 \\
2 & 3 & 2 & 1 \\
2 & 3 & 3 & 1 \\
2 & 3 & 4 & 1 \\
2 & & & \\
4 & 1 & 2 & \\
2 & 4 & 2 & 2 \\
2 & 4 & 3 & 2 \\
2 & 4 & 4 & \\
2 & & &
\end{bmatrix}
$$

Step 2. Get the contribution vectors referring to the outcome partition. The 32 events in columns 1 to 3 of the above array can be divided into 13, 11, 5, and 3 events according to the

outcome they yield. The following arrays can be derived for the 4 outcomes:

$$
\begin{bmatrix}
1 & 1 & 1 & 1 \\
1 & 1 & 3 & 1 \\
1 & 1 & 4 & 1 \\
1 & 3 & 3 & 1 \\
1 & 3 & 4 & 1 \\
2 & 1 & 1 & 1 \\
2 & 1 & 2 & 1 \\
2 & 1 & 3 & 1 \\
2 & 1 & 4 & 1 \\
2 & 3 & 1 & 1 \\
2 & 3 & 2 & 1 \\
2 & 3 & 3 & 1 \\
2 & 3 & 4 & 1
\end{bmatrix}
\begin{bmatrix}
1 & 2 & 3 & 2 \\
1 & 4 & 3 & 2 \\
1 & 4 & 4 & 2 \\
2 & 2 & 1 & 2 \\
2 & 2 & 2 & 2 \\
2 & 2 & 3 & 2 \\
2 & 2 & 4 & 2 \\
2 & 4 & 1 & 2 \\
2 & 4 & 2 & 2 \\
2 & 4 & 3 & 2 \\
2 & 4 & 4 & 2
\end{bmatrix}
\begin{bmatrix}
1 & 2 & 1 & 3 \\
1 & 2 & 2 & 3 \\
1 & 2 & 4 & 3 \\
1 & 4 & 1 & 3 \\
1 & 4 & 2 & 3
\end{bmatrix}
\begin{bmatrix}
1 & 1 & 2 & 4 \\
1 & 3 & 1 & 4 \\
1 & 3 & 2 & 4
\end{bmatrix}
$$

Using composition vectors, each of these arrays gives a row of the following array:

$$
\overline{\overline{N^R}} =
\begin{bmatrix}
5 & 8 & 7 & 0 & 6 & 0 & 3 & 2 & 4 & 4 & 1 \\
3 & 8 & 0 & 5 & 0 & 6 & 2 & 2 & 4 & 3 & 2 \\
5 & 0 & 0 & 3 & 0 & 2 & 2 & 2 & 0 & 1 & 3 \\
3 & 0 & 1 & 0 & 2 & 0 & 1 & 2 & 0 & 0 & 4
\end{bmatrix}.
\tag{11}
$$

The $\overline{\overline{N^R}}$ array now stores the contribution vector partition according to the system outcome. The superscript R is used to indicate that the last column of $\overline{\overline{N^R}}$ also stores the number of the outcome associated with each contribution vector.

Step 3. Check data consistency. This step exposes any irregularities present in the original data, by checking that all the possible complete events are present in the event table, only these events are present, and the table has no duplicate entries.

A fist check involves the columns of the contribution vector array. Based on the number of block states, we estimate the exact number of occurrences for each one of them and see if the sums of entries in the array satisfy this constraint. The MOV example has 3 blocks of 2, 4, and 4 states each. Therefore, each one of the 2 states of block 1 should occur 16 times; each one of the 4 states of blocks 2 and 3 should occur 8 times each. In effect, $16 = 5 + 3 + 5 + 3 = 8 + 8 + 0 + 0$; $8 = 7 + 0 + 0 + 1 = 0 + 5 + 3 + 0 = \cdots = 4 + 3 + 1 + 0$.

The second check involves the rows of the contribution vector array. Each entry refers to a specific state of a specific block. In each row, the total number of occurrences of the block 1 states should be equal to the number of occurrences of blocks 2 and 3. Looking at the first row of the contribution vector array $5 + 8 = 7 + 0 + 6 + 0 = 3 + 2 + 4 + 4$.

Step 3 is repeated for the other rows of $\overline{\overline{N^R}}$.

Step 4. Contribution vector decomposition. This step applies the decomposition operation (Section 4.1) and the decomposition rules (Section 5.1) to the first row of array $\overline{\overline{N^R}}$. Let, for instance, the contribution vector be [5 8 7 0 6 0 3 2 4 4]. The vector exhibits complete state blocks in block 1 (2nd state) and block 3 (3rd,

TABLE 3: Recomposition example: set of bicontribution vectors.

$\overline{L}_i$	b_1			b_2			b_3			b_4		b_5		b_6		b_7		b_8		b_9			b_{10}
24 + 26	O	O	O	I	I	O	O	I	O	O	I	O	I	**I**	**O**	I	O	I	O	I	I	I	O
28	O	O	O	I	I	O	O	I	O	O	I	O	I	**O**	**I**	I	O	I	O	I	I	I	O
21	I	O	O	O	I	O	I	I	I	O	I	O	I	I	O	I	O	I	O	I	O	I	O
25	I	O	O	O	I	O	I	I	I	O	I	O	I	I	O	I	O	I	O	O	O	I	I
11	O	O	O	I	I	O	O	I	O	I	O	I	O	I	I	I	O	I	O	I	I	I	O
...																							

TABLE 4: Recomposition example: subset of bicontribution vectors including only b_6, b_7, b_8, and b_9 as complete blocks.

$\overline{L}_i$	b_1			b_2			b_3			b_4		b_5		b_6		b_7		b_8		b_9			b_{10}
i	O	O	O	I	I	O	O	O	I	**O**	I	**I**	**O**	I	I	I	I	I	I	I	I	I	O
ii	O	O	O	I	I	O	O	O	I	**I**	O	**O**	I	I	I	I	I	I	I	I	I	I	O
iii	O	O	O	I	I	O	O	O	I	**I**	O	I	**O**	I	I	I	I	I	I	I	I	I	O
iv	O	O	O	I	I	O	I	O	O	**O**	I	**I**	**O**	I	I	I	I	I	I	I	I	I	O
v	O	O	O	I	I	O	I	O	O	**I**	O	**O**	I	I	I	I	I	I	I	I	I	O	I
vi	O	O	O	I	I	O	I	O	O	**I**	O	**O**	I	I	I	I	I	I	I	I	I	I	O
vii	O	O	O	I	I	O	I	O	O	**I**	O	I	**O**	I	I	I	I	I	I	I	I	O	I
viii	O	O	O	I	I	O	I	O	O	**I**	O	I	**O**	I	I	I	I	I	I	I	I	I	O

4th states). The rules indicate that the decomposition is first applied to block 1 and later to block 3. According to Section 5.1, the vector [5 8 7 0 6 0 3 2 4 4] is decomposed into the Cartesian contribution vector [0 8 4 0 4 0 2 2 2 2] and the remaining contribution vector [5 0 3 0 2 0 1 0 2 2].

Step 5. Update system arrays. There are $1 + R$ working arrays where contribution vectors (and their outcome) are stored. The first one is $\overline{\overline{N^R}}$. The others are denoted by $\overline{\overline{\underset{\sim}{N}_i}}$ and store the Cartesian vectors generated during Step 4 for each response m, starting from $i = 1$. Each row of $\overline{\overline{N^R}}$ treated during Step 4 is replaced by the remaining contribution vector. In our example, during the first iteration $\overline{\overline{N^R}}$ and $\overline{\overline{\underset{\sim}{N}_1}}$ become

$$\overline{\overline{N^R}} = \begin{bmatrix} 5 & 0 & 3 & 0 & 2 & 0 & 1 & 0 & 2 & 2 & 1 \\ 3 & 8 & 0 & 5 & 0 & 6 & 2 & 2 & 4 & 3 & 2 \\ 5 & 0 & 0 & 3 & 0 & 2 & 2 & 2 & 0 & 1 & 3 \\ 3 & 0 & 1 & 0 & 2 & 0 & 1 & 2 & 0 & 0 & 4 \end{bmatrix}, \tag{12}$$

$$\overline{\overline{\underset{\sim}{N}_1}} = [0 \ 8 \ 4 \ 0 \ 4 \ 0 \ 2 \ 2 \ 2 \ 2].$$

Note that, there is no need to store outcome information in the arrays $\overline{\overline{\underset{\sim}{N}_i}}$.

Steps 4 and 5 are repeated until a decomposition operation leads to two Cartesian vectors. Then, both vectors are added to $\overline{\overline{\underset{\sim}{N}_i}}$, the first row of $\overline{\overline{N^R}}$ is removed, and i is replaced by $i + 1$.

Step 6. Decompose $\overline{\overline{\underset{\sim}{N}_i}}$. The array $\overline{\overline{\underset{\sim}{N}_i}}$ is further decomposed to get an array $\overline{\overline{\underset{\approx}{N}_i}}$ of implicit contribution vectors. Since the vectors have only Cartesian entries, the operations can be easily applied on the bicontribution vectors of $\overline{\overline{\underset{\sim}{N}_i}}$ and $\overline{\overline{\underset{\approx}{N}_i}}$, denoted by $\overline{\overline{\underset{\sim}{L}_i}}$ and $\overline{\overline{\underset{\approx}{L}_i}}$, respectively. The arrays referring to the contribution vector [5 8 7 0 6 0 3 2 4 4] are

$$\overline{\overline{\underset{\sim}{L}_1}} = \begin{bmatrix} 0 & 1 & 1 & 0 & 1 & 0 & 1 & 1 & 1 & 1 \\ 1 & 1 & 0 & 1 & 0 & 1 & 0 & 0 & 0 & 1 \\ 1 & 0 & 1 & 0 & 0 & 0 & 1 & 0 & 0 & 0 \end{bmatrix} \tag{13}$$

and, after further decomposition,

$$\overline{\overline{\underset{\approx}{L}_1}} = \begin{bmatrix} 0 & 1 & 0 & 0 & 0 & 1 & 1 & 1 & 1 & 1 \\ 0 & 1 & 0 & 0 & 1 & 0 & 1 & 1 & 1 & 1 \\ 1 & 0 & 1 & 0 & 0 & 0 & 0 & 0 & 1 & 0 \\ 1 & 0 & 0 & 0 & 1 & 0 & 0 & 0 & 1 & 0 \\ 1 & 0 & 1 & 0 & 0 & 0 & 0 & 0 & 0 & 1 \\ 1 & 0 & 0 & 0 & 1 & 0 & 0 & 0 & 0 & 1 \\ 1 & 0 & 1 & 0 & 0 & 0 & 1 & 0 & 0 & 0 \end{bmatrix}. \tag{14}$$

Step 6 supports the following recomposition actions.

Step 7. Apply recomposition actions. This step applies the recomposition operation (Section 4.2) and the recomposition rules (Section 5.2) to the parts of $\overline{\overline{\underset{\approx}{L}_i}}$ sharing the same outcome. Note that, the example considered here is too small and simple to offer potential for recomposition.

Steps 4–7 are repeated until the array $\overline{\overline{N^R}}$ is empty.

Step 8. Algorithm termination. The final output of the procedures described here is the arrays $\overline{\overline{\underset{\approx}{L}_m}}$, which represent implicit partitions of significantly reduced cardinality compared to the size of the system event table. Note that, the decomposition and the recomposition operations developed here ensure that the consistency of the data is preserved throughout the vector processing.

TABLE 5: Recomposition example: subset of bicontribution vectors including only the vectors appearing in Table 4.

$\bar{L}_i$		b_1			b_2		b_3			b_4		b_5		b_6		b_7		b_8			b_9	b_{10}
i	O	O	O	I	I	O	O	O	I	O	I	I	O	I	I	I	I	I	I	I	I	O
iv	O	O	O	I	I	O	I	O	O	O	I	I	O	I	I	I	I	I	I	I	I	O
v + vii	O	O	O	I	I	O	I	O	O	I	O	I	I	I	I	I	I	I	I	I	**O**	**I**
vi + viii	O	O	O	I	I	O	I	O	O	I	O	I	I	I	I	I	I	I	I	I	**I**	**O**
ii + iii	O	O	O	I	I	O	O	O	I	I	O	I	I	I	I	I	I	I	I	I	I	O

The Matlab environment is chosen as suitable for the fast manipulation of matrices using the built-in matrix operations. For instance, the decomposition when there are no Cartesian entries requires knowledge of the exact event subspace that corresponds to the processed contribution vector. The algorithm can either keep track of the events contributing in each vector or go through the original event table to isolate the subspace relating to each vector. The former, though it is more sophisticated, takes up a lot of memory even for relatively small problems. Matlab takes advantage of its built-in matrix operations for sort and find, to reduce significantly the execution time.

6. Case Studies

6.1. Case Study 1. The first case study is taken from Papazoglou [10] and concerns the development of an event tree for a boiling water nuclear reactor. The system involves 10 state blocks with 2 to 4 block states each. The event space consists of 3072 complete events and the system has 5 outcomes. Papazoglou [10] provided a set of Boolean equations and developed functional block diagrams that embedded information on the dependencies between the blocks. He finally presented a reduced event tree of 41 branches. Note that, if the reactor system is treated in BowTieBuilder [12, 13] without providing dependency information, the resulting event tree has 110 branches. This confirms that the efficiency of functional block diagram applications in reducing the size of event trees depends on the structure of the Boolean model, that dictates the dependencies between the blocks.

The methodology proposed here takes as input the original 3072 × 11 event table and produces the results reported in Table 6. Note that

(i) the states $\{1, 2, 3, 4\}$ of block I correspond to $\{L, M, N, T\}$, and

(ii) the outcomes $\{1, 2, 3, 4, 5\}$ correspond to $\{CI, CII, CIII, CIV, Success\}$

of Papazoglou [10]. Each row of Table 6 gives a Cartesian vector (or an implicant) corresponding to a branch of the event tree. In this sense, the reduced tree described here has only 38 branches. The proposed algorithm identifies an inconsistency in the partition of block C of the original data. Resolving it leads to different results for outcome CIV, and this explains the three branches difference between 41 and 38. The rest of the branches/implicants are notably the same, with a single exception, involving the choice to expand block U rather than block Q (see bold cells of Table 6, lines

12–15). While both choices yield four branches/implicants, this differentiation shows that the procedure proposed here is not biased by the order of the blocks in the event table data.

6.2. Case Study 2. The proposed methodology is tested against a large problem involving 16 blocks, including one block having four state instances, four blocks with three states and eleven binary. The original event table has 663552 × 17 cells. The system has 5 possible outcomes. The initial event table is constructed via recursive partitions of the event subspace.

The resulting implicit partition has totally 273 vectors; in particular 86, 115, 34, 28, and 71 for the five outcomes. The recomposition stage requires 1.08 CPU seconds. Then, the final implicit partitions have totally 178 vectors; in particular 31, 54, 16, 20, and 57 for the respective partitions of the five outcomes.

CPU times can give an idea of the relative effort invested in the different activities taking place during a run. In this relatively large problem, the preparatory Steps 1–3 of Section 5.3 require 1.27 CPU seconds. The decomposition steps require only 0.0469 CPU seconds for a total of 86 decompositions using rule (a) and 124 CPU seconds for a total of 52 decompositions using rule (b) of Section 5.1. Therefore, the application of decompositions on the basis of Cartesian entries reduces the computational effort by almost 4 orders of magnitude. Clearly, an intelligent reduction of the frequency of visiting the event table would bring significant benefits in the computational times. Note that, CPU times refer to an Intel Core Quad 2.50 GHz processor with 1.95 GB RAM.

Finally, the proposed procedure manages to reduce the expanded event tree to just 0.0268% of its original size. The final partitions are easily translated into a set of implicants. There is no proof that this is a prime set, since there is lack of theoretical background on sufficient and necessary minimality conditions. In any case, the proposed methodology is a fast, effective, and intelligent way to reduce substantially a large event tree and facilitate the quantification of risk.

7. Conclusions

This work presents a new methodology for the reduction of event trees without the use of structural or functional information on the system. The work applies a holistic approach based on the concept of contribution vectors, to generate a minimal set of implicants representative of the system behavior. The method inherits the advantages and limitations of event tree representation. In this sense, the method is not hindered by component interdependences

TABLE 6: Final reduced event table for case study 1.

No. of event tree branch (or implicant)	Blocks states										Outcome r_i
	I	C	M	Q	U	X	E	I	V	W	
1	—	2	—	—	—	—	—	—	—	—	4
2	1	1	—	—	—	—	1	1	—	1	5
3	1	1	—	—	—	—	1	1	—	2	2
4	1	1	—	—	—	—	1	2	—	—	3
5	1	1	—	—	—	—	2	—	—	—	3
6	2	1	—	1	—	—	—	—	—	1	5
7	2	1	—	1	—	—	—	—	—	2	2
8	2	1	—	2	1	—	—	—	—	1	5
9	2	1	—	2	1	—	—	—	—	2	2
10	2	1	—	2	2	1	—	—	1	1	5
11	2	1	—	2	2	1	—	—	1	2	2
12	2	1	—	2	2	1	—	—	2	—	1
13	2	1	—	2	2	2	—	—	—	—	1
14	3	1	—	—	1	—	—	—	—	1	5
15	3	1	—	—	1	—	—	—	—	2	2
16	3	1	—	—	2	1	—	—	1	1	5
17	3	1	—	—	2	1	—	—	1	2	2
18	3	1	—	—	2	1	—	—	2	—	1
19	3	1	—	—	2	2	—	—	—	—	1
20	4	1	1	1	—	—	—	—	—	—	5
21	4	1	1	2	1	—	—	—	—	1	5
22	4	1	1	2	1	—	—	—	—	2	2
23	4	1	1	2	2	1	—	—	1	1	5
24	4	1	1	2	2	1	—	—	1	2	2
25	4	1	1	2	2	1	—	—	2	—	1
26	4	1	1	2	2	2	—	—	—	—	1
27	4	1	2	—	—	—	1	1	—	1	5
28	4	1	2	—	—	—	1	1	—	2	2
29	4	1	2	—	—	—	1	2	—	—	3
30	4	1	2	—	—	—	2	—	—	—	3
31	4	1	3	—	1	—	—	—	—	1	5
32	4	1	3	—	1	—	—	—	—	2	2
33	4	1	3	1	2	—	—	—	—	1	5
34	4	1	3	1	2	—	—	—	—	2	2
35	4	1	3	2	2	1	—	—	1	1	5
36	4	1	3	2	2	1	—	—	1	2	2
37	4	1	3	2	2	1	—	—	2	—	1
38	4	1	3	2	2	2	—	—	—	—	1

and noncoherent behavior in the considered systems. The proposed representation framework stems from Cartesian products to define partitions using composition vectors. The representation provides the basis for the application of decomposition and recomposition operations on single composition vectors and composition vector partitions. Implementation issues for the efficient use of these operations within an iterative algorithmic framework are discussed thoroughly.

The proposed method is tested against two case studies, one found in the literature and a fictitious large scale problem. In the former, the method provides a set of prime implicants very similar to the one reported in the literature. The latter illustrates the efficiency of the method in handling large-scale problems and proves the computational advantages from the proposed representation and operations.

Future work considers the use of the theoretical background presented here to develop necessary and sufficient conditions for the minimality of the final set of implicants. These conditions could then be incorporated in the recomposition stage to guide an optimal search algorithm towards the set of prime implicants.

Nomenclature

K: Number of system blocks

b_k: System block, $k \in \{1, 2, \dots, K\}$

$\mathbf{S}_k$:   Set of internal states of block b_k, $\mathbf{S}_k = \{s_k^1, s_k^2, \dots, s_k^{K_{S_k}}\}$

$s_k^{i_k}$: i_kth internal state of block b_k, $i_k \in \{1, 2, \dots, K_{s_k^{i_k}}\}$

$\mathbf{E}$:   System event space

$\hat{e}$:   Complete joint event, $\hat{e} \in \mathbf{E}$

$\mathbf{R}$:   Set of the all possible system outcomes

r_j: System outcome, $j \in \{1, 2, \dots, |\mathbf{R}|\}$

T: Event table mapping $T : \hat{e} \mapsto r = T(\hat{e})$, where $\hat{e} \in \mathbf{E}$ and $r \in \mathbf{R}$

$\mathbf{A}, \mathbf{B}$:   Nonempty subspaces of $\mathbf{E}$

$|\mathbf{X}|,\ K_\mathbf{X}$:   Number of elements (cardinality) of set $\mathbf{X}$

$\mathbf{Q}(\mathbf{A})$:   Partition applied over $\mathbf{A}$

$\mathbf{q}_i^Q(\mathbf{A})$: ith subset of $\mathbf{A}$ according to $\mathbf{Q}(\mathbf{A})$, $i \in \{1, 2, \dots, K_{\mathbf{Q}(\mathbf{A})}\}$

$\mathbf{Q}^T(\mathbf{A})$:   Outcome-based partition of $\mathbf{A}$ (according to mapping $\mathbf{T}$)

$\mathbf{S}_k^\mathbf{A}$:   Set of b_k-block states, $k \in \{1, 2, \dots, K\}$, in the events comprising $\mathbf{A}$

$C(\mathbf{A})$:   Cartesian product $\mathbf{S}_1^\mathbf{A} \times \mathbf{S}_2^\mathbf{A} \times \cdots \times \mathbf{S}_K^\mathbf{A}$

$\underline{\mathbf{A}}$:   Cartesian subspace

$\underset{\approx}{\mathbf{A}}$:   Implicit subspace

$\underline{\mathbf{Q}}(\mathbf{A})$:   Cartesian partition over subspace $\mathbf{A}$

$\underline{\mathbf{q}}_i^{\mathbf{Q}}(\mathbf{A})$: ith subset of $\underline{\mathbf{Q}}(\mathbf{A})$, $i \in \{1, 2, \dots, K_{\underline{\mathbf{Q}}(\mathbf{A})}\}$

$\underset{\approx}{\mathbf{Q}}(\mathbf{A})$:   Implicit partition over $\mathbf{A}$

$\underset{\approx}{\mathbf{q}}_i^{\mathbf{Q}}(\mathbf{A})$: ith subset of $\underset{\approx}{\mathbf{Q}}(\mathbf{A})$, $i \in \{1, 2, \dots, K_{\underset{\approx}{\mathbf{Q}}(\mathbf{A})}\}$

$\overline{\mathbf{N}}(\mathbf{A})$:   Contribution vector of $\mathbf{A}$

$n_k^{i_k}(\mathbf{A})$:   Entry of $\overline{\mathbf{N}}(\mathbf{A})$, $n_k^{i_k}(\mathbf{A}) \in \mathbb{Z}_+$ and $k \in \{1, 2, \dots, K\}$, $i_k \in \{1, 2, \dots, K_{s_k^{i_k}}\}$

$\overline{\mathbf{L}}(\mathbf{A})$:   Bicontribution vector of subspace $\mathbf{A}$

$l_k^{i_k}(\mathbf{A})$:   Entry of $\overline{\mathbf{L}}(\mathbf{A})$, $l_k^{i_k}(\mathbf{A}) \in \{\mathrm{O, I}\}$ and $k \in \{1, 2, \dots, K\}$, $i_k \in \{1, 2, \dots, K_{s_k^{i_k}}\}$

$\underline{\overline{\mathbf{N}}}(\mathbf{A})$:   Cartesian contribution vector of $\mathbf{A}$

$\underset{\approx}{n}_k^{i_k}(\mathbf{A})$:   Entry of $\underset{\approx}{\overline{\mathbf{N}}}(\mathbf{A})$, $\underset{\approx}{n}_k^{i_k}(\mathbf{A}) \in \mathbb{Z}_+$ and $k \in \{1, 2, \dots, K\}$, $i_k \in \{1, 2, \dots, K_{s_k^{i_k}}\}$

$\overline{\overline{\mathbf{N}}}(\mathbf{Q}(\mathbf{A}))$:   Contribution vector partition of $\mathbf{Q}(\mathbf{A})$

$\overline{\mathbf{N}}(\mathbf{q}_i^Q(\mathbf{A}))$: ith member of $\overline{\overline{\mathbf{N}}}(\mathbf{Q}(\mathbf{A}))$, $i \in \{1, 2, \dots, K_{\mathbf{Q}(\mathbf{A})}\}$

λ: Vector length

$\overline{Y}$:   Vector

$\overline{\overline{Y}}$:   Vector partition (i.e., set of vectors)

$\underline{\mathcal{X}}$:   Entity obeying the Cartesian property

$\underset{\approx}{Y}$:   Entity obeying the property of implicitness

$\Lambda[\overline{\mathbf{N}}(\mathbf{A})]$:   Operation applied on $\overline{\mathbf{N}}(\mathbf{A})$ to obtain $\overline{\mathbf{L}}(\mathbf{A})$

$\Xi[\overline{\mathbf{L}}(\mathbf{A})]$:   Operation applied on $\overline{\mathbf{L}}(\mathbf{A})$ to obtain $\underline{\overline{\mathbf{N}}}(\mathbf{A})$

σ_k: Subset of $\mathbf{S}_k$

$\mathbf{A}^{\sigma_k}$:   Subset of $\mathbf{A}$ such that $\mathbf{A}^{\sigma_k} = \{\hat{e} \in \mathbf{A} \cap \Delta^{\sigma_k}\}$

$\Delta_\mathbf{A}^{\sigma_k}$:   Event subspace such that $$\Delta_\mathbf{A}^{\sigma_k} = \{\mathbf{S}_1^\mathbf{A} \times \mathbf{S}_2^\mathbf{A} \times \cdots \times \sigma_k \times \cdots \times \mathbf{S}_K^\mathbf{A}\}.$$

Glossary

Complete set of block states:	Set of all the possible states of a certain system block
Complete joint event:	Joint event containing an instance of each one of the system blocks
Cartesian property:	Event subspaces and contribution vectors that can be generated by a Cartesian product. Also, subspaces partitions and partition contribution vectors that can be generated by a set of Cartesian products
Property of implicitness:	Cartesian entities (i.e., subspaces, vectors, partitions) whose associated Cartesian products contain only complete or singleton sets of block states
Cartesian entries:	Nonzero entries of a contribution vector equal to their relative Cartesian contribution vector entries.

References

[1] E. Zio, "Reliability engineering: old problems and new challenges," *Reliability Engineering and System Safety*, vol. 94, no. 2, pp. 125–141, 2009.

[2] J. Andrews, "System reliability modelling: the current capability and potential future developments," *Proceedings of the Institution of Mechanical Engineers C*, vol. 223, no. 12, pp. 2881–2897, 2009.

[3] I. A. Papazoglou and B. J. M. Ale, "A logical model for quantification of occupational risk," *Reliability Engineering and System Safety*, vol. 92, no. 6, pp. 785–803, 2007.

[4] R. Remenyte-Prescott and J. D. Andrews, "An efficient real-time method of analysis for non-coherent fault trees," *Quality and Reliability Engineering International*, vol. 25, no. 2, pp. 129–150, 2009.

[5] A. Lisnianski and G. Levitin, *Multi-State System Reliability: Assessment, Optimization and Applications*, World Scientific, 2003.

[6] S. C. M. Rocco and M. Muselli, "Approximate multi-state reliability expressions using a new machine learning technique," *Reliability Engineering and System Safety*, vol. 89, no. 3, pp. 261–270, 2005.

[7] Y. Dutuit and A. Rauzy, "Approximate estimation of system reliability via fault trees," *Reliability Engineering and System Safety*, vol. 87, no. 2, pp. 163–172, 2005.

[8] W. S. Jung, J.-E. Yang, and J. Ha, "Development of measures to estimate truncation error in fault tree analysis," *Reliability Engineering and System Safety*, vol. 90, no. 1, pp. 30–36, 2005.

[9] I. A. Papazoglou, "Mathematical foundations of event trees," *Reliability Engineering and System Safety*, vol. 61, no. 3, pp. 169–183, 1998.

[10] I. A. Papazoglou, "Functional block diagrams and automated construction of event trees," *Reliability Engineering and System Safety*, vol. 61, no. 3, pp. 185–214, 1998.

[11] A. Rauzy and Y. Dutuit, "Exact and truncated computations of prime implicants of coherent and non-coherent fault trees within Aralia," *Reliability Engineering and System Safety*, vol. 58, no. 2, pp. 127–144, 1997.

[12] B. J. M. Ale, H. Baksteen, L. J. Bellamy et al., "Quantifying occupational risk: the development of an occupational risk model," *Safety Science*, no. 2, pp. 176–185, 2008.

[13] A. R. Hale, B. J. M. Ale, L. H. J. Goossens et al., "Modeling accidents for prioritizing prevention," *Reliability Engineering & System Safety*, vol. 92, no. 12, pp. 1701–1715, 2007.

16

Robust Control Charts for Monitoring Process Mean of Phase-I Multivariate Individual Observations

Asokan Mulayath Variyath and Jayasankar Vattathoor

Department of Mathematics and Statistics, Memorial University of Newfoundland, St. John's, NL, Canada A1C 5S7

Correspondence should be addressed to Asokan Mulayath Variyath; variyath@mun.ca

Academic Editor: Adiel Teixeira de Almeida

Hoteling's T^2 control charts are widely used in industries to monitor multivariate processes. The classical estimators, sample mean, and the sample covariance used in T^2 control charts are highly sensitive to the outliers in the data. In Phase-I monitoring, control limits are arrived at using historical data after identifying and removing the multivariate outliers. We propose Hoteling's T^2 control charts with high-breakdown robust estimators based on the reweighted minimum covariance determinant (RMCD) and the reweighted minimum volume ellipsoid (RMVE) to monitor multivariate observations in Phase-I data. We assessed the performance of these robust control charts based on a large number of Monte Carlo simulations by considering different data scenarios and found that the proposed control charts have better performance compared to existing methods.

1. Introduction

Control charts are widely used in industries to monitor/control processes. Generally, the construction of a control chart is carried out in two phases. The Phase-I data is analyzed to determine whether the data indicates a stable (or in-control) process and to estimate the process parameters and thereby the construction of control limits. The Phase-II data analysis consists of monitoring future observations based on control limits derived from the Phase-I estimates to determine whether the process continues to be in control or not. But trends, step changes, outliers, and other unusual data points in the Phase-I data can have an adverse effect on the estimation of parameters and the resulting control limits. That is, any deviation from the main assumption (in our case, identically and independently distributed from normal distribution) may lead to an out-of-control situation. Therefore, it becomes very important to identify and eliminate these data points prior to calculating the control limits. In this paper, all these unusual data points are referred to as "outliers."

Multivariate quality characteristics are often correlated, and to monitor the multivariate process mean Hoteling's T^2 control chart [1, 2] is widely used. To implement Hoteling's T^2 control chart for individual observations in Phase-I, for each observation $\mathbf{x}_j$ we calculate

$$T^2\left(\mathbf{x}_j\right) = \left(\mathbf{x}_j - \bar{\mathbf{x}}\right)' \mathbf{S}_1^{-1} \left(\mathbf{x}_j - \bar{\mathbf{x}}\right), \quad (1)$$

where $\mathbf{x}_j = (x_{j1}, x_{j2}, \ldots, x_{jp})'$ is the jth p-variate observation, $(j = 1, 2, \ldots, m)$ and the sample mean $\bar{\mathbf{x}}$, sample covariance matrix $\mathbf{S}_1$ are based on m Phase-I observations. In Phase-I monitoring, the $T^2(\mathbf{x}_j)$ values are compared with the T^2 control limit derived by assuming that the $\mathbf{x}_j$'s are multivariate normal so that the T^2 control limits are based on the beta distribution with the parameters $p/2$ and $(m - p - 1)/2$. However, the classical estimators, sample mean, and sample covariance are highly sensitive to the outliers, and hence robust estimation methods are preferred as they have the advantage of not being unduly influenced by the outliers. The use of robust estimation methods is well suited to detect multivariate outliers because of their high breakdown points which ensure that the control limits are reasonably accurate. Sullivan and Woodall [3] proposed a T^2 chart with an estimate of the covariance matrix based on the successive differences of observations and showed that it is effective in detecting process shift. However, these charts are not effective in detecting multiple multivariate outliers because of their low breakdown point.

Vargas [4] introduced two robust T^2 control charts based on robust estimators of location and scatter, namely, the minimum covariance determinant (MCD) and minimum volume ellipsoid (MVE) for identifying the outliers in Phase-I multivariate individual observations. Jensen et al. [5] showed that T^2_{MCD} and T^2_{MVE} control charts have better performance when outliers are present in the Phase-I data. Chenouri et al. [6] used reweighted MCD estimators for monitoring the Phase-II data, without constructing Phase-I control charts. However, in many situations Phase-I control charts are necessary to assess the performance of the process and also to identify the outliers. We propose T^2 control charts based on the reweighted minimum covariance determinant (RMCD)/reweighted minimum volume ellipsoid (RMVE) ($T^2_{\text{RMCD}}/T^2_{\text{RMVE}}$) for monitoring Phase-I multivariate individual observations. RMCD/RMVE estimators are statistically more efficient than MCD/MVE estimators and have a manageable asymptotic distribution. We empirically arrive at Phase-I control limits for the $T^2_{\text{RMCD}}/T^2_{\text{RMVE}}$ control chart for some specific sample sizes and fitted a nonlinear model to determine control limits for any sample size for dimensions 2 to 10. Our simulation studies show that $T^2_{\text{RMCD}}/T^2_{\text{RMVE}}$ control charts are performing well compared to $T^2_{\text{MCD}}/T^2_{\text{MVE}}$ control charts for monitoring the Phase-I data.

The organization of the remaining part of the paper is as follows. In Section 2, we discuss the properties of a good robust estimator and we briefly explain the MCD/MVE estimators and their reweighted versions. The proposed $T^2_{\text{RMCD}}/T^2_{\text{RMVE}}$ control charts are given in Section 3 along with the control limits arrived at based on Monte Carlo simulations. We assess the performance of the proposed control charts in Section 4, and the implementation of the proposed methods is illustrated in a case example in Section 5. Our conclusions are given in Section 6.

2. Robust Estimators

The affine equivariance property of the estimator is important because it makes the analysis independent of the measurement scale of the variables as well as the transformations or rotations of the data. The breakdown point concept introduced by Donoho and Huber [7] is often used to assess the robustness. The breakdown point is the smallest proportion of the observations which can render an estimator meaningless. A higher breakdown point implies a more robust estimator, and the highest attainable breakdown point is 1/2 in the case of median in the univariate case. For more details on affine equivariance and breakdown points one may refer to Chenouri et al. [6] or Jensen et al. [5].

An estimator is said to be relatively efficient compared to any other estimator if the mean square error for the estimator is the least for at least some values of the parameter compared to others. A robust estimator is considered to be good if it carries the property of affine equivariance along with a higher breakdown point and greater efficiency. In addition to the

above three properties of a good robust estimator, it should be possible to calculate the estimator in a reasonable amount of time to make it computationally efficient.

It is difficult to get an affine equivariant and robust estimator as affine equivariance and high breakdown will not come simultaneously. Lopuhaä and Rousseeuw [8] and Donoho and Gasko [9] showed that the finite sample breakdown point of $(m - p + 1)/(2m - p + 1)$ is difficult for an affine equivariant estimator. The largest attainable finite sample breakdown point of any affine equivariant estimator of the location and scatter matrix with a sample size m and dimension p is $(m - p + 1)/2m$ [10]. Therefore relaxing the affine equivariance condition of the estimators to invariance under the orthogonal transformation makes it easy to find an estimator with the highest breakdown point.

The classical estimators, sample mean vector, and covariance matrix of location and scatter parameters are affine equivariant but their sample breakdown point is as low as $1/m$. The MCD and MVE estimators have the highest possible finite sample breakdown point $(m - p + 1)/2m$. However, both of these estimators have very low asymptotic efficiency under normality. But the reweighted versions of MCD and MVE estimators have better efficiency without compromising on the breakdown point and rate of convergence compared to MCD and MVE. In the next two subsections, we discuss in detail about the MCD and MVE estimators and their reweighted versions.

2.1. MCD and RMCD Estimators. The MCD estimators of location and scatter parameters of the distribution are determined by a two-step procedure. In step 1, all possible subsets of observations of size $h = (m * \gamma)$, where $0.5 \leq \gamma \leq 1$ are obtained. In step 2, the subset whose covariance matrix has the smallest possible determinant is selected. The MCD location estimator $\bar{x}_{\text{MCD}}$ is defined as the average of this selected subset of h points, and the MCD scatter estimator is given by $\mathbf{S}_{\text{MCD}} = a_{\gamma,p} * b^m_{\gamma,p} * \mathbf{C}_{\text{MCD}}$, where $\mathbf{C}_{\text{MCD}}$ is the covariance matrix of the selected subset, the constant $a_{\gamma,p}$ is the multiplication factor for consistency [11], and $b^m_{\gamma,p}$ is the finite sample correction factor [12]. Here $(1 - \gamma)$ represents the breakdown point of the MCD estimators. The MCD estimator has its highest possible finite sample breakdown point when $h = (m + p + 1)/2$ and has an $m^{-1/2}$ rate of convergence but has a very low asymptotic efficiency under normality. Computing the exact MCD estimators ($\bar{x}_{\text{MCD}}$, $\mathbf{S}_{\text{MCD}}$) is computationally expensive or even impossible for large sample sizes in high dimensions [13], and hence various algorithms have been suggested for approximating the MCD. Hawkins and Olive [14] and Rousseeuw and van Driessen [15] independently proposed a fast algorithm for approximating MCD. The FAST-MCD algorithm of Rousseeuw and van Driessen finds the exact MCD for small datasets and gives a good approximation for larger datasets, which is available in the standard statistical software SPLUS, R, SAS, and Matlab.

MCD estimators are highly robust, carry equivariance properties, and can be calculated in a reasonable time using the FAST-MCD algorithm; however, they are statistically not efficient. The reweighted procedure will help to carry both

TABLE 1: Estimates of the model parameters $a_{1(p,\alpha)}$, $a_{2(p,\alpha)}$, $a_{3(p,\alpha)}$ for $T^2_{\text{RMCD}}/T^2_{\text{RMVE}}$ control charts.

p	$\alpha = 0.05$			$\alpha = 0.01$			$\alpha = 0.001$		
	$\hat{a}_1$	$\hat{a}_2$	$\hat{a}_3$	$\hat{a}_1$	$\hat{a}_2$	$\hat{a}_3$	$\hat{a}_1$	$\hat{a}_2$	$\hat{a}_3$
					T^2_{RMCD}				
2	17.223	41102	2.647	21.134	38170	2.329	27.051	192909	2.508
3	20.134	35844	2.209	24.287	128924	2.344	31.350	1144947	2.718
4	23.152	269357	2.548	28.181	1272773	2.773	35.575	5989325	2.973
5	24.685	467949	2.524	28.437	1417059	2.632	31.013	2666196	2.593
6	26.962	1762051	2.746	29.654	3061216	2.711	31.662	5414248	2.669
7	24.892	1099128	2.493	22.882	1585224	2.416	19.058	3465278	2.444
8	27.236	2908821	2.667	27.245	4922576	2.644	28.326	12134778	2.710
9	23.974	2447649	2.534	21.420	4726835	2.554	18.772	14096595	2.676
10	31.894	12572909	2.914	37.085	34375654	3.033	56.573	172176786	3.301
					T^2_{RMVE}				
2	17.442	29553	2.494	21.365	31571	2.244	27.594	148747	2.434
3	20.286	22497	2.066	24.387	59096	2.13	31.326	338665	2.402
4	23.095	108855	2.286	27.549	291064	2.372	35.109	1255429	2.576
5	24.796	238966	2.334	28.302	508097	2.367	32.008	1063783	2.377
6	27.585	1041090	2.606	31.126	1882888	2.601	37.136	4714353	2.671
7	28.151	1541634	2.598	30.936	3183762	2.635	39.357	12199414	2.827
8	34.917	14798692	3.127	45.767	75616029	3.419	70.875	840512379	3.904
9	39.191	59094377	3.415	50.271	275604839	3.679	72.768	1960966919	4.039
10	50.733	950607720	4.099	68.154	4696452032	4.379	110.587	56398461817	4.881

robustness and efficiency. That is, first a highly robust but perhaps an inefficient estimator is computed, which is used as a starting point to find a local solution for detecting outliers and computing the sample mean and covariance of the cleaned data set as in Rousseeuw and van Zomeren [16]. This consists of discarding those observations whose Mahalanobis distances exceed a certain fixed threshold value. MCD is the current best choice for the initial estimator of a two-step procedure as it contains the robustness, equivariance, and computational efficiency properties along with its $m^{-1/2}$ rate of convergence. Hence RMCD estimators are the weighted mean vector

$$\bar{\mathbf{x}}_{\text{RMCD}} = \frac{\left(\sum_{j=1}^{m} w_j \mathbf{x}_j\right)}{\left(\sum_{j=1}^{m} w_j\right)}, \qquad (2)$$

and the weighted covariance matrix

$$\mathbf{S}_{\text{RMCD}} = c_{\alpha,p} * d_{\gamma,\alpha}^{m,p}$$

$$* \frac{\sum_{j=1}^{m} w_j \left(\mathbf{x}_j - \bar{\mathbf{x}}_{\text{RMCD}}\right) \left(\mathbf{x}_j - \bar{\mathbf{x}}_{\text{RMCD}}\right)'}{\sum_{j=1}^{m} w_j}, \qquad (3)$$

where $c_{\alpha,p}$ is the multiplication factors for consistency [11], $d_{\gamma,\alpha}^{m,p}$ is the finite sample correction factor [12], and the weights w_j are defined as

$$w_j = \begin{cases} 1 & \text{if } \text{RD}(\mathbf{x}_j) \leq \sqrt{q_\alpha}, \\ 0 & \text{otherwise}, \end{cases} \qquad (4)$$

where the robust distance $\text{RD}(\mathbf{x}_j) = \sqrt{(\mathbf{x}_j - \bar{\mathbf{x}}_{\text{MCD}})' \mathbf{S}_{\text{MCD}}^{-1} (\mathbf{x}_j - \bar{\mathbf{x}}_{\text{MCD}})}$ and q_α is $(1 - \alpha)100\%$ quantile of the chi-square distribution with p degrees of freedom.

This reweighting technique improves the efficiency of the initial MCD estimator while retaining (most of) its robustness. Hence the RMCD estimator inherits the affine equivariance, robustness, and asymptotic normality properties of the MCD estimators with an improved efficiency.

2.2. MVE and RMVE Estimators. Determining the MVE estimators of location and scatter parameters of the distribution is almost in line with that of the MCD estimator. As in the case of MCD, all the possible subsets of data points with size $h = (m * \gamma)$ (where $0.5 \leq \gamma \leq 1$) is obtained first. Then the ellipsoid of minimum volume that covers the subsets are obtained to determine the MVE estimators. The MVE location estimator is the geometrical center of the ellipsoid, and the MVE scatter estimator is the matrix defining the ellipsoid itself, multiplied by an appropriate constant to ensure consistency [13, 16]. Thus MVE estimator does not correspond to the sample mean vector and the sample covariance matrix as in the case of the MCD estimator. Here $(1 - \gamma)$ represents the breakdown point of the MVE estimators, as in the case of MCD, and it has the highest possible finite sample breakdown point when $h = (m + p + 1)/2m$ [8, 17]. The MVE estimator has an $m^{-1/3}$ rate of convergence and a nonnormal asymptotic distribution [17].

As in the case for MCD estimators, MVE estimators are also not efficient. Hence, a reweighted version similar to that

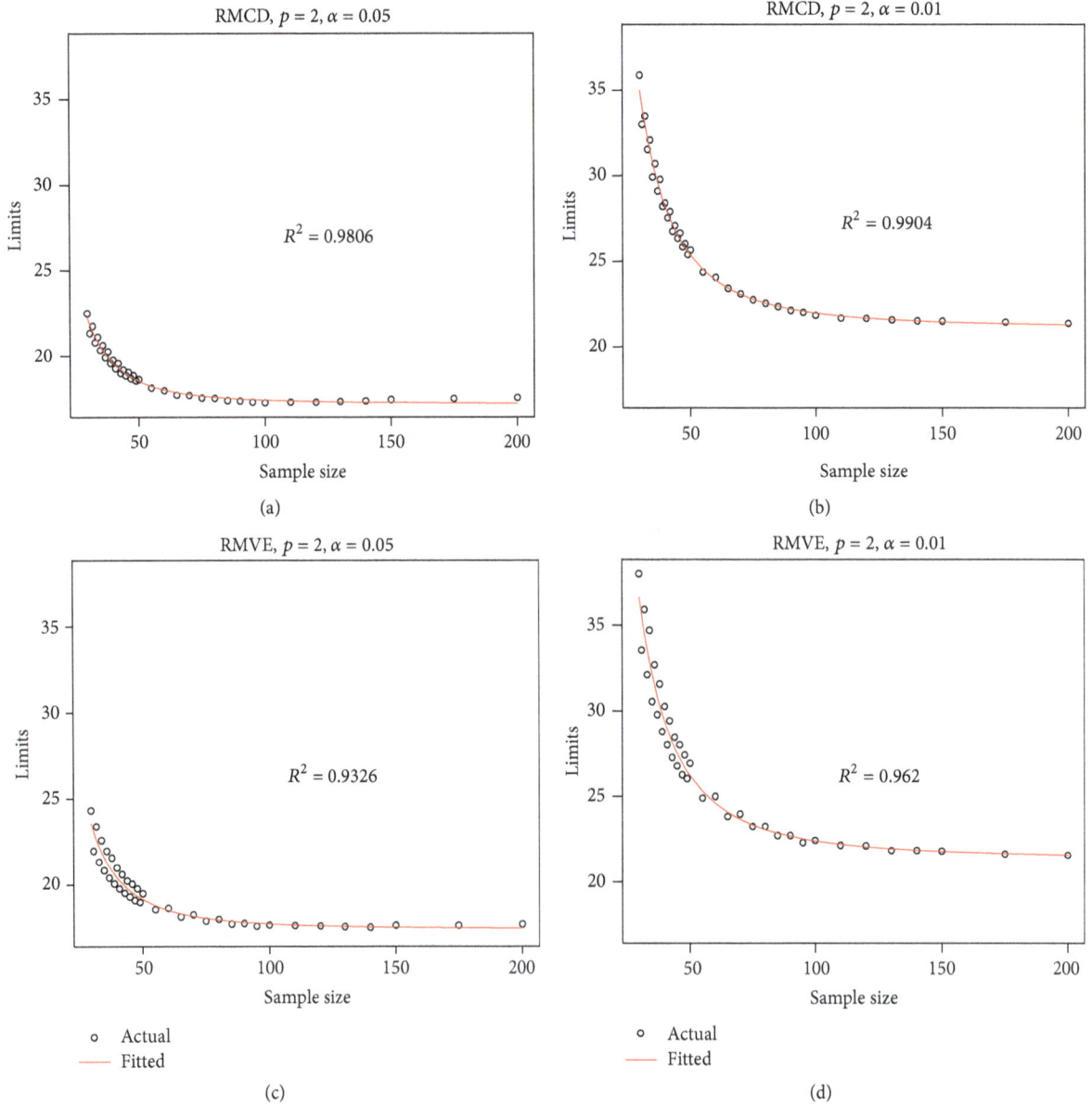

FIGURE 1: Scatter plot of $T^2_{\text{RMCD}}/T^2_{\text{RMVE}}$ control limits and the fitted curve for $p = 2$.

for MCD has been proposed by Rousseeuw and van Zomeren [16]. Note that it has been shown more recently that the RMVE estimators do not improve on the convergence rate (and thus the 0% asymptotic efficiency) of the initial MVE estimator [8, 12]. Therefore, as an alternative, a one-step M-estimator can be calculated with the MVE estimators as the initial solution [13, 18] which results in an estimator with the standard $m^{-1/2}$ convergence rate to a normal asymptotic distribution. For more details on MCD/MVE estimators one may refer to Chenouri et al. [6] or Jensen et al. [5]. The algorithm to determine the MVE/RMVE estimators is available in the statistical software SPLUS, R, SAS, and Matlab.

3. Robust Control Charts

We propose to use T^2 charts with robust estimators of location and dispersion parameters based on RMCD/RMVE

for monitoring the process mean of Phase-I multivariate individual observations. RMCD/RMVE estimators inherit the nice properties of initial MCD estimators such as affine equivariance, robustness, and asymptotic normality while achieving a higher efficiency. We now define a robust T^2 control chart with RMCD and RMVE estimators for ith multivariate observation as

$$
\begin{aligned}
T^2_{\text{RMCD}}\left(\mathbf{x}_i\right) &= \left(\mathbf{x}_i - \overline{\mathbf{x}}_{\text{RMCD}}\right)' \mathbf{S}^{-1}_{\text{RMCD}} \left(\mathbf{x}_i - \overline{\mathbf{x}}_{\text{RMCD}}\right), \\
T^2_{\text{RMVE}}\left(\mathbf{x}_i\right) &= \left(\mathbf{x}_i - \overline{\mathbf{x}}_{\text{RMVE}}\right)' \mathbf{S}^{-1}_{\text{RMVE}} \left(\mathbf{x}_i - \overline{\mathbf{x}}_{\text{RMVE}}\right),
\end{aligned}
\tag{5}
$$

where $\overline{\mathbf{x}}_{\text{RMCD}}$, $\overline{\mathbf{x}}_{\text{RMVE}}$ are the mean vectors and $\mathbf{S}_{\text{RMCD}}$, $\mathbf{S}_{\text{RMVE}}$ are the dispersion matrices under the RMCD/RMVE methods based on m multivariate observations.

The exact distribution of $T^2_{\text{RMCD}}/T^2_{\text{RMVE}}$ estimators not available, hence the control limits for Phase-I data are

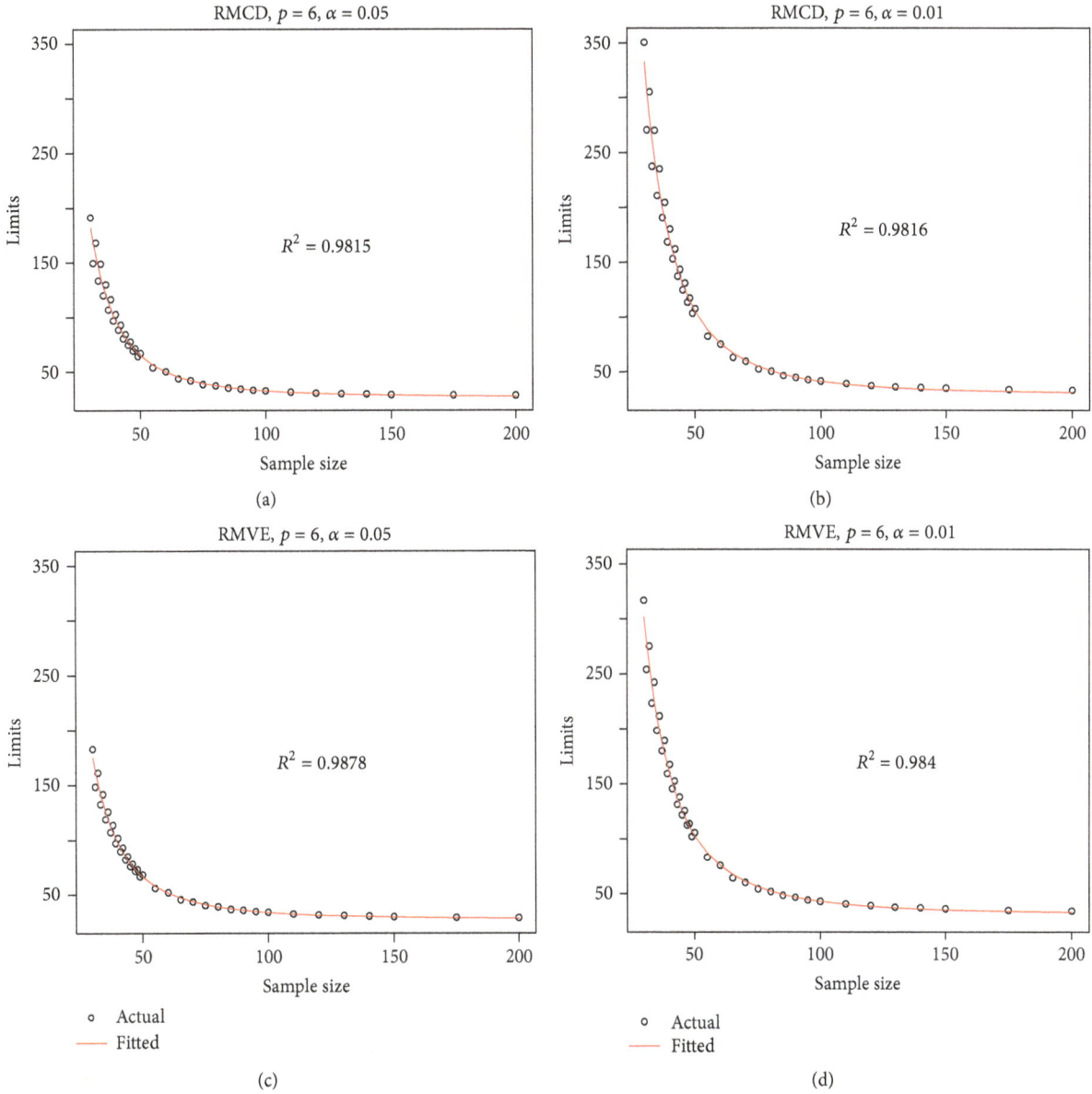

FIGURE 2: Scatter plot of $T^2_{\mathrm{RMCD}}/T^2_{\mathrm{RMVE}}$ control limits and the fitted curve for $p = 6$.

obtained empirically. In the next subsection we apply Monte Carlo simulation to estimate quantiles of the distribution of T^2_{RMCD} and T^2_{RMVE} for several combinations of sample sizes and dimensions. For each dimension, we further introduce a method to fit a smooth nonlinear model to arrive, the control limits for any given sample size.

3.1. Computation of Control Limits. We performed a large number of Monte Carlo simulations to obtain the control limits. We generated $n = 200,000$ samples of size m from a standard multivariate normal distribution MVN(0, I_p) with dimension p. Due to the invariance of the T^2_{RMCD} and T^2_{RMVE} statistics, these limits will be applicable for any values of μ and Σ. Using the reweighted MCD/MVE estimators $\bar{\mathbf{x}}_{\mathrm{RMCD}}$, $\mathbf{S}_{\mathrm{RMCD}}$, $\bar{\mathbf{x}}_{\mathrm{RMVE}}$, and $\mathbf{S}_{\mathrm{RMVE}}$ with a breakdown value

of $\gamma = 0.50$, $T^2_{\mathrm{RMCD}}/T^2_{\mathrm{RMVE}}$ statistics for each observation in the data set were calculated using (5), and the maximum value attained for each data set of size m was recorded. The empirical distribution of maximum of T^2_{RMCD} and T^2_{RMVE} was inverted to determine the $(1 - \alpha)100\%$ quantiles. We used the R-function "CovMcd()" in the "rrcov" package written by Torodov [19] to ascertain the RMCD/RMVE estimators.

We have constructed the empirical distribution of $T^2_R\mathrm{MCD}/T^2_R\mathrm{MVE}$ as above for $m = [30(1)50, 55(5)100, 110(10)200]$, $p = (2, 3, \ldots 10)$ when $\gamma = 0.50$ and arrived at the control limits for $\alpha = (0.05, 0.01, \text{and } 0.001)$. The scatter plots of the quantiles and sample sizes for different dimensions suggest a family of nonlinear models of the form

$$f_{p,\alpha,\gamma,m} = a_{1,(p,\alpha,\gamma)} + \frac{a_{2,(p,\alpha,\gamma)}}{m^{a_{3,(p,\alpha,\gamma)}}}, \quad (6)$$

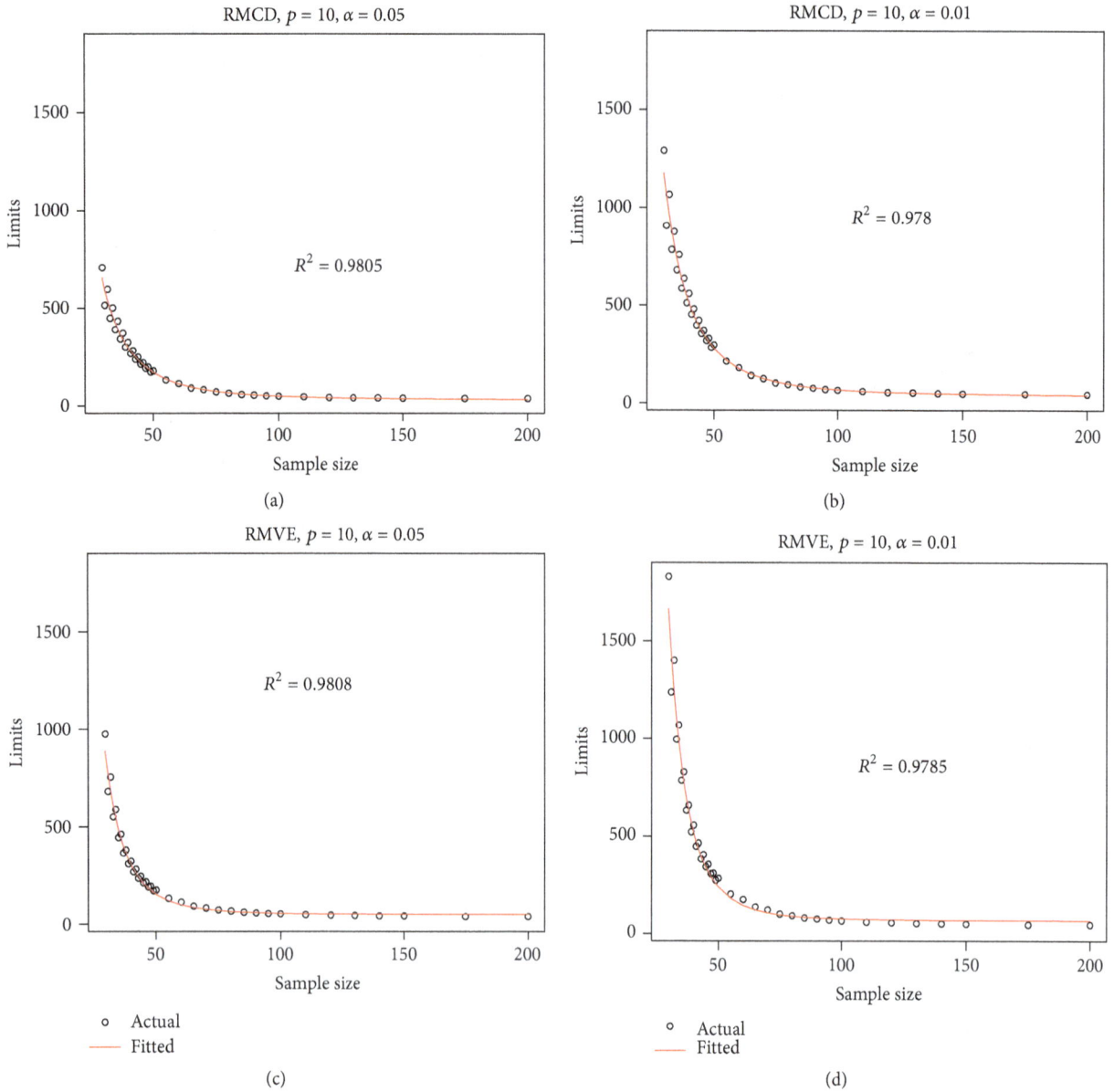

FIGURE 3: Scatter plot of $T^2_{\text{RMCD}}/T^2_{\text{RMVE}}$ control limits and the fitted curve for $p = 10$.

where $a_{1(p,\alpha,\gamma)}$, $a_{2(p,\alpha,\gamma)}$, and $a_{3(p,\alpha,\gamma)}$ are the model parameters. For clarity, the scatter plot of the actual and the fitted values of the quantiles of T^2_{RMCD} and T^2_{RMVE} for $p = 2$, 6, and 10 are given in Figures 1, 2, and 3; other plots are omitted to save space.

From Figures 1, 2, and 3, we can see that the nonlinear fit is very well supported by the high R^2 values, which help us to determine the T^2_{RMCD} and T^2_{RMVE} control limits for any given sample size. The least square estimates of the parameters $a_{1(p,\alpha)}$, $a_{2(p,\alpha)}$, and $a_{3(p,\alpha)}$ when $\gamma = 0.50$ for dimensions $p = (2, 3, \ldots, 10)$ and $\alpha = (0.05, 0.01$ and $0.001)$ for $T^2_{\text{RMCD}}/T^2_{\text{RMCD}}$ control charts are given in Table 1. Using these estimates, the

control limits for T^2_{RMCD} and T^2_{RMVE} can be found using (6) for any sample size.

For the implementation of a robust control chart, first collect a sample of m multivariate individual observations with dimension p. Compute robust estimates of mean and covariance matrix using R or any other software with $\gamma = 0.50$, and determine $T^2_{\text{RMCD}}/T^2_{\text{RMVE}}$. Outliers can be determined by comparing the $T^2_{\text{RMCD}}/T^2_{\text{RMVE}}$ values with control limits obtained using (6) for specific values of α, m, p, and the constants given in Table 1. The outlier free data can be used to construct the standard T^2 control chart for monitoring the Phase-II observations.

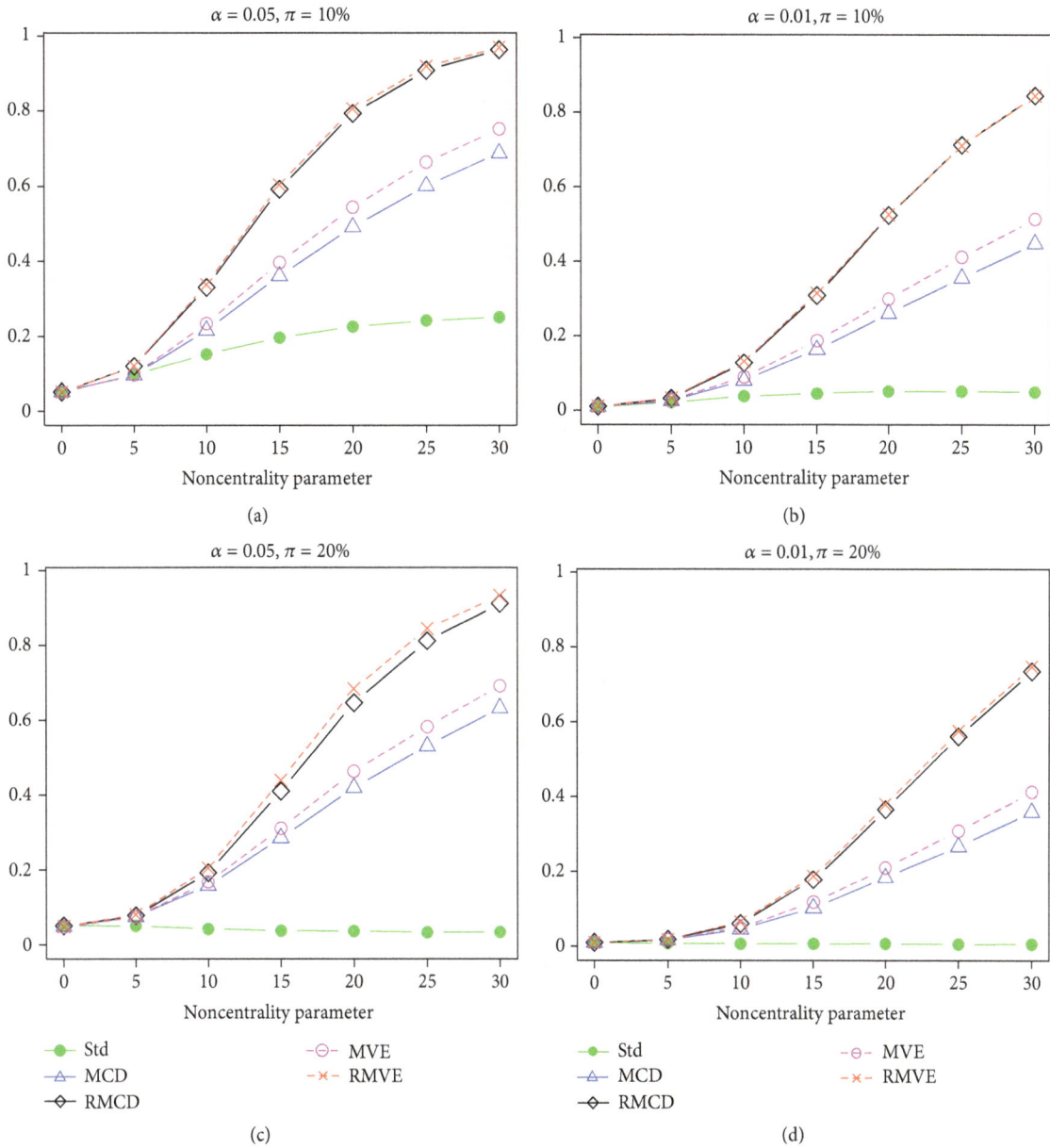

FIGURE 4: Probability of signal for T^2 control chart with different estimation methods for $p = 2$, $m = 50$.

4. Performance Analysis

We assess the performance of the proposed charts when outliers are present due to the shift in the process mean. In their study, Jensen et al. [5] concluded that the $T^2_{\text{MCD}}/T^2_{\text{MVE}}$ control charts had better performance in terms of probability of signal. Hence, we compare the performance of our proposed method with $T^2_{\text{MCD}}/T^2_{\text{MVE}}$ charts as well as the standard T^2 charts based on classical estimators. Our study compares more combinations of dimension p, sample size m, and π. For a particular combination of p, m, and π, a number of datasets are generated. Out of the m observations generated, $m * \pi$ of them are random data points generated from the out-of-control distribution, and the remaining $m * (1 - \pi)$ observations are generated from the in-control distribution so that the sample of m data points may contain some

outliers. We set $\pi = 0.10$ and 0.20 to ensure that the sample contains few outliers. Without loss of generality, we consider the in-control distribution as $N(0, I_p)$. The out-of-control distribution is a multivariate normal with a small shift in the mean vector with same covariance matrix. The amount of mean shift is defined through a noncentrality parameter (δ), which is given by

$$\delta = (\mu_1 - \mu)' \Sigma^{-1} (\mu_1 - \mu), \qquad (7)$$

where $(\mu_1 - \mu)$ is the shift in the mean vector. The larger the value of δ is, the more extreme the outliers are. The proportion of datasets that had at least one T^2_{RMCD} or T^2_{RMVE} statistic greater than the control limit was calculated, and this proportion becomes the estimated probability of signal. We compared the performance of these charts with standard T^2

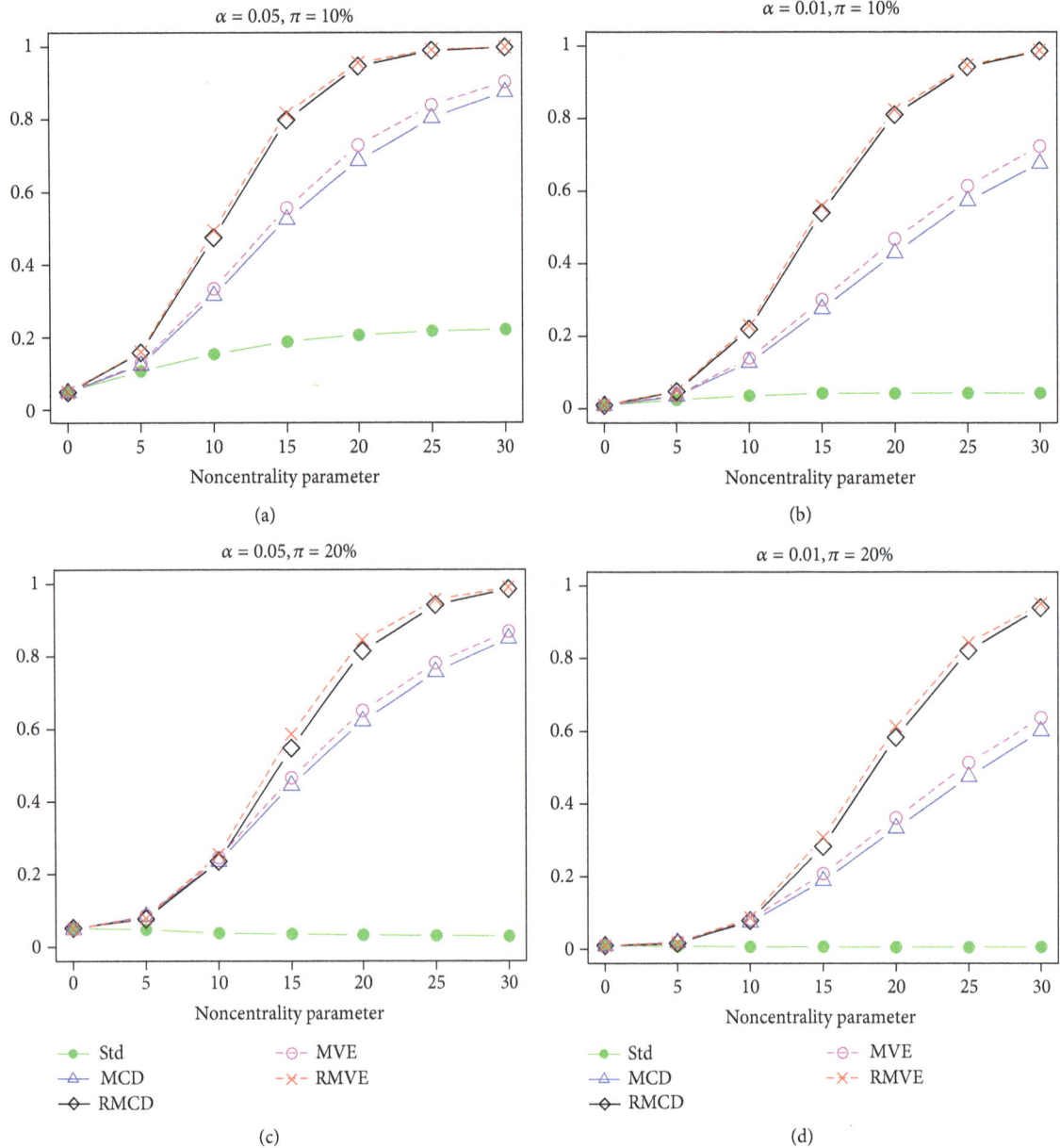

FIGURE 5: Probability of signal for T^2 control chart with different estimation methods for $p = 2$, $m = 100$.

charts, T^2_{MCD}, and T^2_{MVE} charts. The standard T^2 chart was included in our performance study as a reference because of its common usage.

The probability of a signal for different values of $\delta = (0, 5, 10, 15, 20, 25, 30)$ and for some of the values of $m = (30, 50, 100, 150)$, $p = (2, 6, 10)$ and $\pi = (10\%, 20\%)$ was considered in our study. Fifty thousand datasets of size m were generated for each combination of p, π, and δ, and the probability of signal was estimated for $\alpha = 0.05, 0.01$, and 0.001. We considered various combinations of μ_1, μ_2, and ρ which determine δ as per (7) and found that the probability of signal is the same irrespective of the combination of μ_1, μ_2 and ρ. Hence we have considered $\mu_1 = \mu_2$ and $\rho = 0$ for various

values of δ. We have presented only a selected set of plots to save space. The plots of probability of signal for $\alpha = 0.05$ and 0.01, $p = 2$ and 6, and $m = 50$ and 100 are given in Figures 4, 5, 6, and 7 for easier understanding. For dimension $p = 10$, we used $m = 100$ and 150, and the plots of probability of signal are given in Figures 8 and 9.

From Figures 4–9, we can see that when the value of the noncentrality parameter is zero or close to zero, the probability of signal is close to α which is expected for an in-control process. As the value of the noncentrality parameter increases the probability of signals also increases. Using this criterion, we select the best method for identifying the outliers. If the probability of signal does not increase for

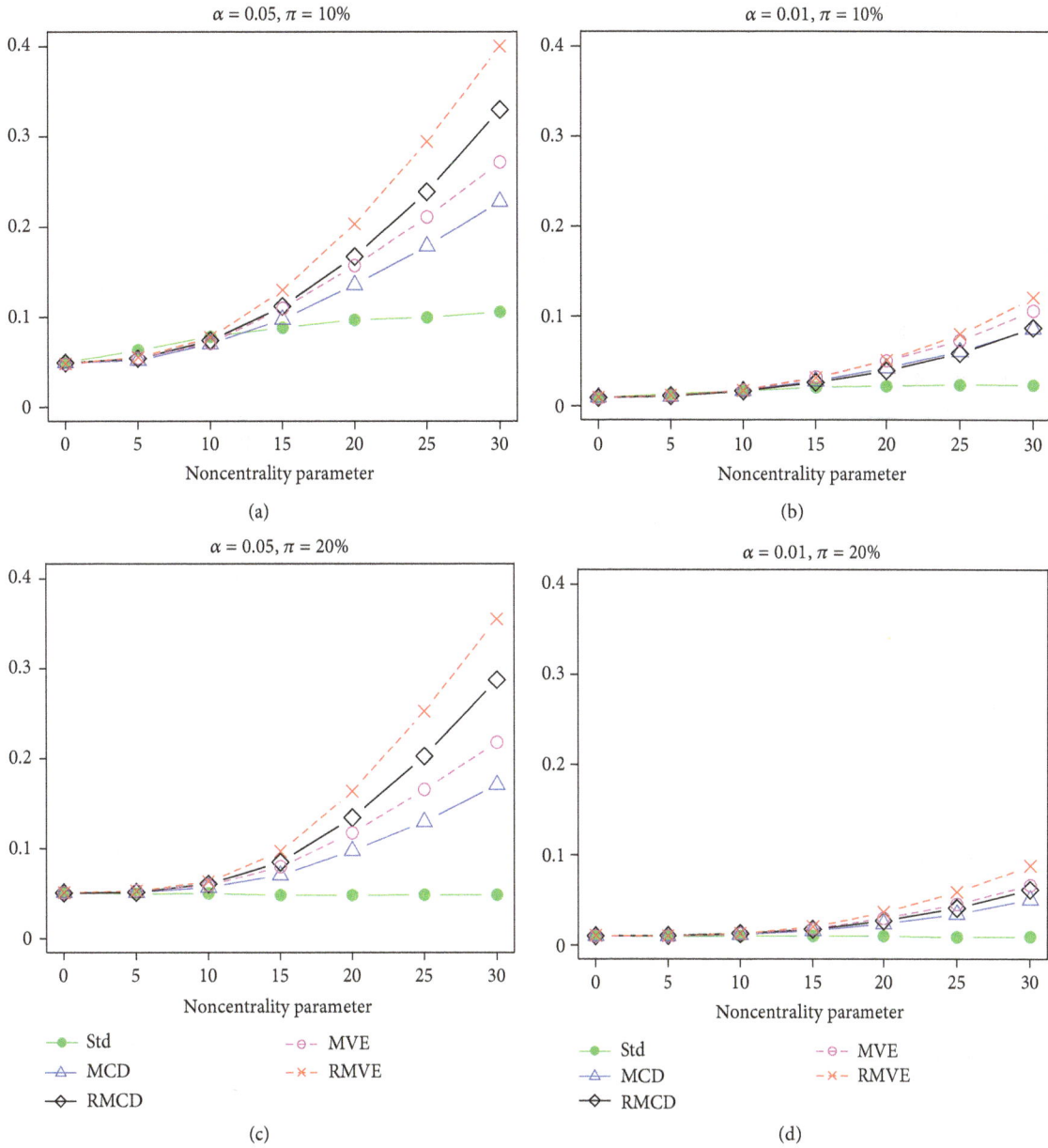

FIGURE 6: Probability of signal for T^2 control chart with different estimation methods for $p = 6$, $m = 50$.

increase in noncentrality parameter, then it is clear that the estimator has broken down and is not capable of detecting the outliers.

A careful examination of these plots of probability of signals corresponding to various values of p, m, and π indicates that for small values of p and m, T^2_{RMVE} performs well. As m and p increase, T^2_{RMCD} chart is superior. For example, from Figures 4 and 5 we see that T^2_{RMVE} has slight advantage over T^2_{RMCD}. But compared to $T^2_{\text{MCD}}/T^2_{\text{MVE}}$ charts, $T^2_{\text{RMCD}}/T^2_{\text{RMVE}}$ charts are performing well which is evident from all the plots presented here. When p is large (see Figures 8 and 9), the T^2_{RMCD} has clear advantage compared to T^2_{RMVE}. From these figures, we see that standard T^2 control chart

possesses little ability to detect the outliers and the T^2_{MVE}, and T^2_{MVE} stands below the $T^2_{\text{RMCD}}/T^2_{\text{RMVE}}$ charts throughout all the values of δ.

As p increases for a fixed value of m, the breakdown points of RMCD and RMVE get smaller as the breakdown value is given by $(m - p + 1)/2m$. This suggests that the larger p is, the larger m will need to be in order to maintain the breakdown point, which is very well demonstrated in Figures 8 and 9. In general, there was always one estimator, RMCD or RMVE, that was found to be superior across all the values of the noncentrality parameter as long as the proportion of outliers was not so big as to cause the estimators to break down. This greatly simplifies the conclusions that

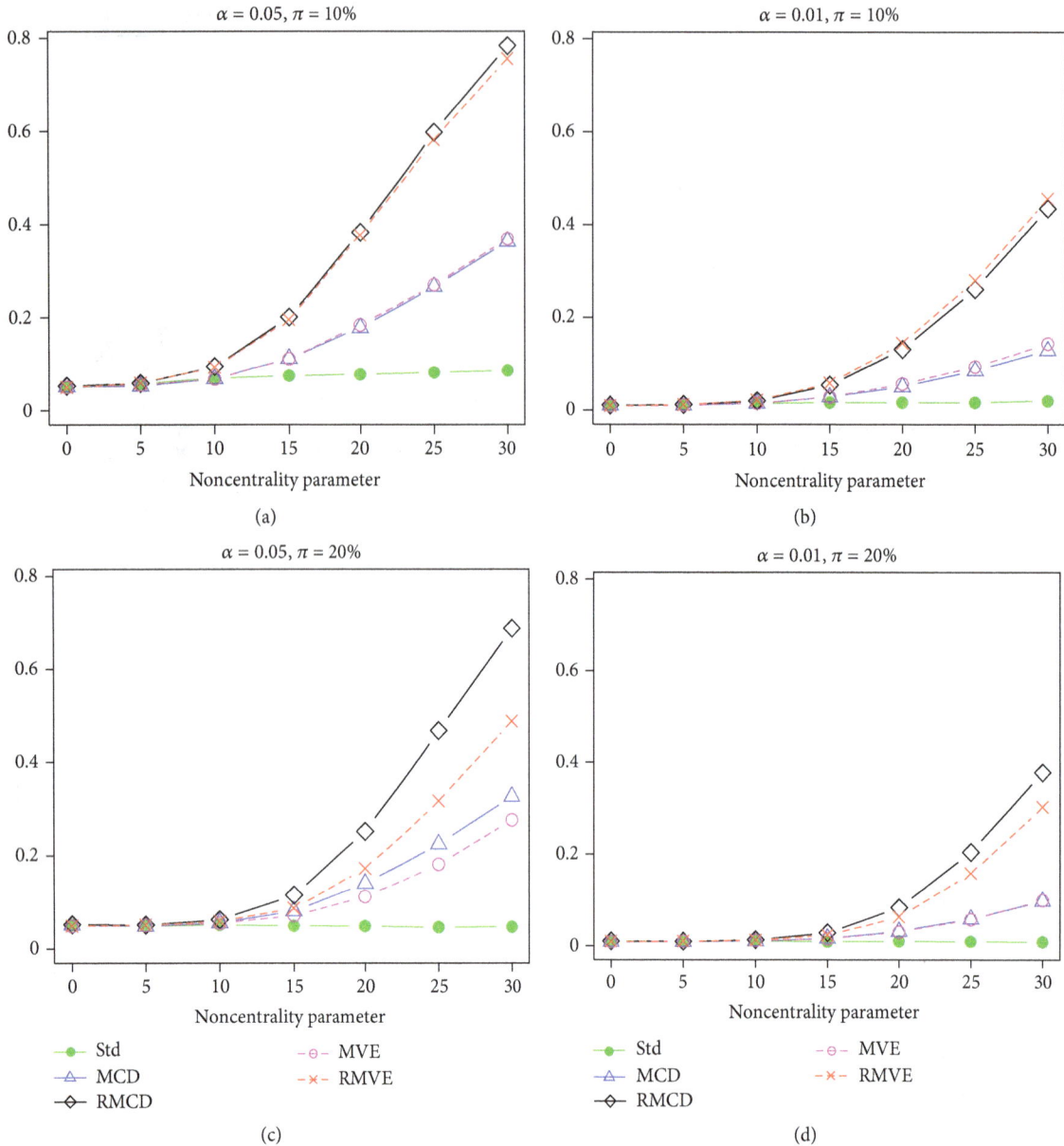

FIGURE 7: Probability of signal for for T^2 control chart with different estimation methods for $p = 6$, $m = 100$.

can be made about when the RMCD or RMVE estimators are preferred to the MCD and MVE estimators.

Nevertheless, T^2_{RMCD} and T^2_{RMCD} charts are preferred for the various combinations of m, p, and π, and some broad recommendations can be made on the selection among these two charts. When $m < 100$, the T^2_{RMVE} will be the best for small dimension. When $m \geq 100$, the T^2_{RMCD} is preferred. As p increases, then the percentage of outliers that can be detected by the T^2_{RMVE} chart decreases. It is true for both the charts that when p is higher, the number of outliers that can be detected decreases for smaller sample sizes. Thus

for Phase-I applications where the number of outliers is unknown, T^2_{RMVE} should be used only for smaller sample sizes, and it is also computationally feasible. T^2_{RMCD} should be used for larger sample sizes or when it is believed that there is a large number of outliers. When the dimension is large, larger sample sizes are needed to ensure that the estimator does not break down and lose its ability to detect outliers. Hence for larger dimension cases, T^2_{RMCD} is preferred with large sample sizes. For very small samples ($m < 30$), one may opt for higher values of γ, for which control limits need to be developed.

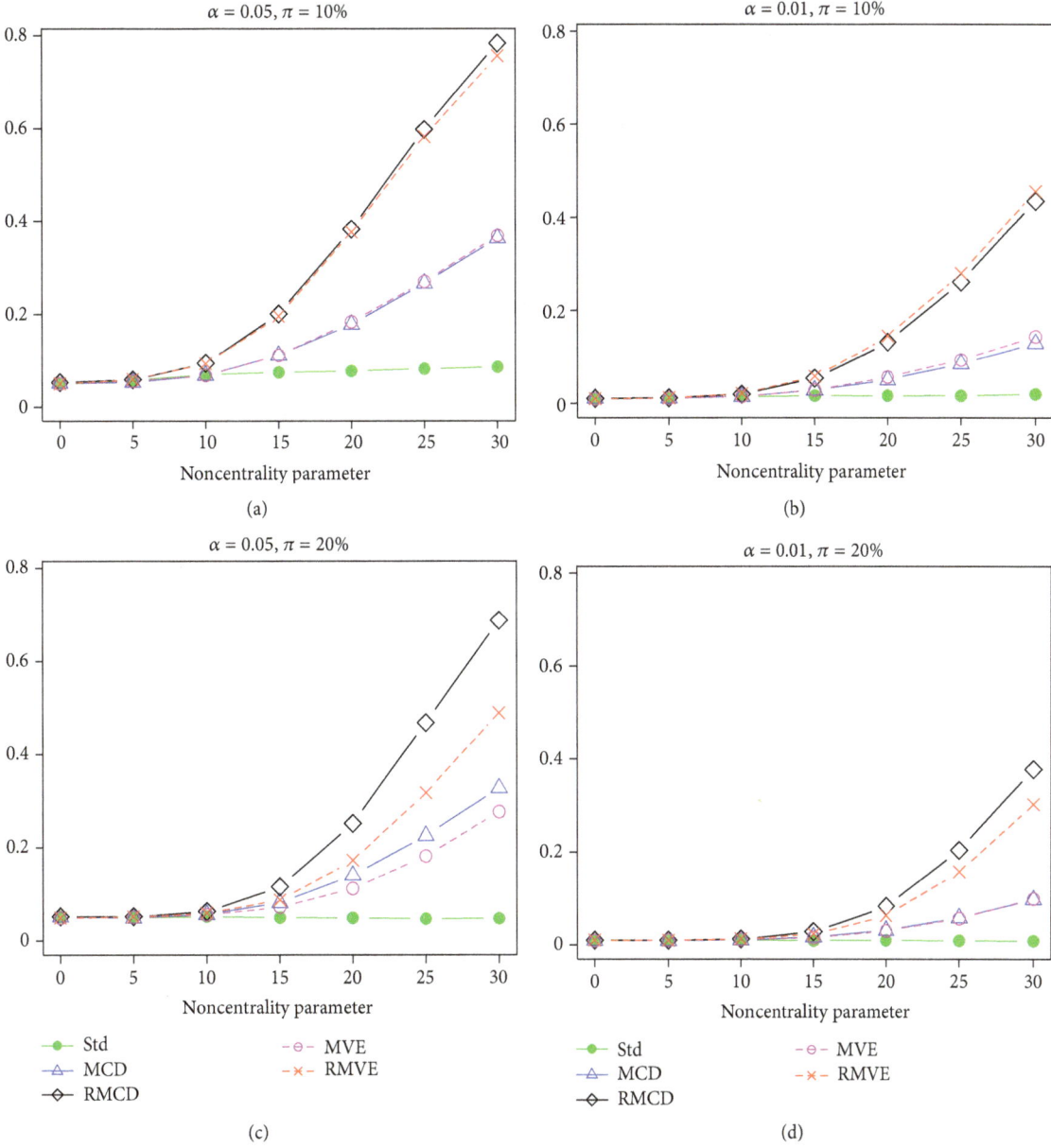

FIGURE 8: Probability of signal for T^2 control chart with different estimation methods for $p = 10$, $m = 100$.

5. Case Example

To illustrate the applicability of the proposed control chart method, we discuss a real case example taken from an electronic industry. The data gives 105 measurements of 3 axial components of acceleration measured by accelerometer on a e-compass unit fixed on the objects. The mean vector and covariance matrix under the classical, RMCD, and RMVE methods of the sample data considered are given by

$$\overline{X} = \begin{pmatrix} 6.3143 \\ 5.7339 \\ 5.7527 \end{pmatrix},$$

$$S = \begin{pmatrix} 3.4022 & -1.1524 & -1.0746 \\ -1.1524 & 1.9249 & 1.1209 \\ -1.0746 & 1.1209 & 2.2004 \end{pmatrix},$$

$$\overline{X}_{\text{RMCD}} = \begin{pmatrix} 5.7125 \\ 6.2643 \\ 6.0837 \end{pmatrix},$$

$$S_{\text{RMCD}} = \begin{pmatrix} 2.8549 & 0.1901 & -0.2926 \\ 0.1901 & 1.1175 & 0.4433 \\ -0.2926 & 0.4433 & 2.3115 \end{pmatrix},$$

$$\overline{X}_{\text{RMVE}} = \begin{pmatrix} 5.7790 \\ 6.1379 \\ 5.9894 \end{pmatrix},$$

$$S_{\text{RMVE}} = \begin{pmatrix} 2.7837 & -0.2642 & -0.5745 \\ -0.2642 & 1.2778 & 0.6080 \\ -0.5745 & 0.6080 & 2.2909 \end{pmatrix}.$$

$$(8)$$

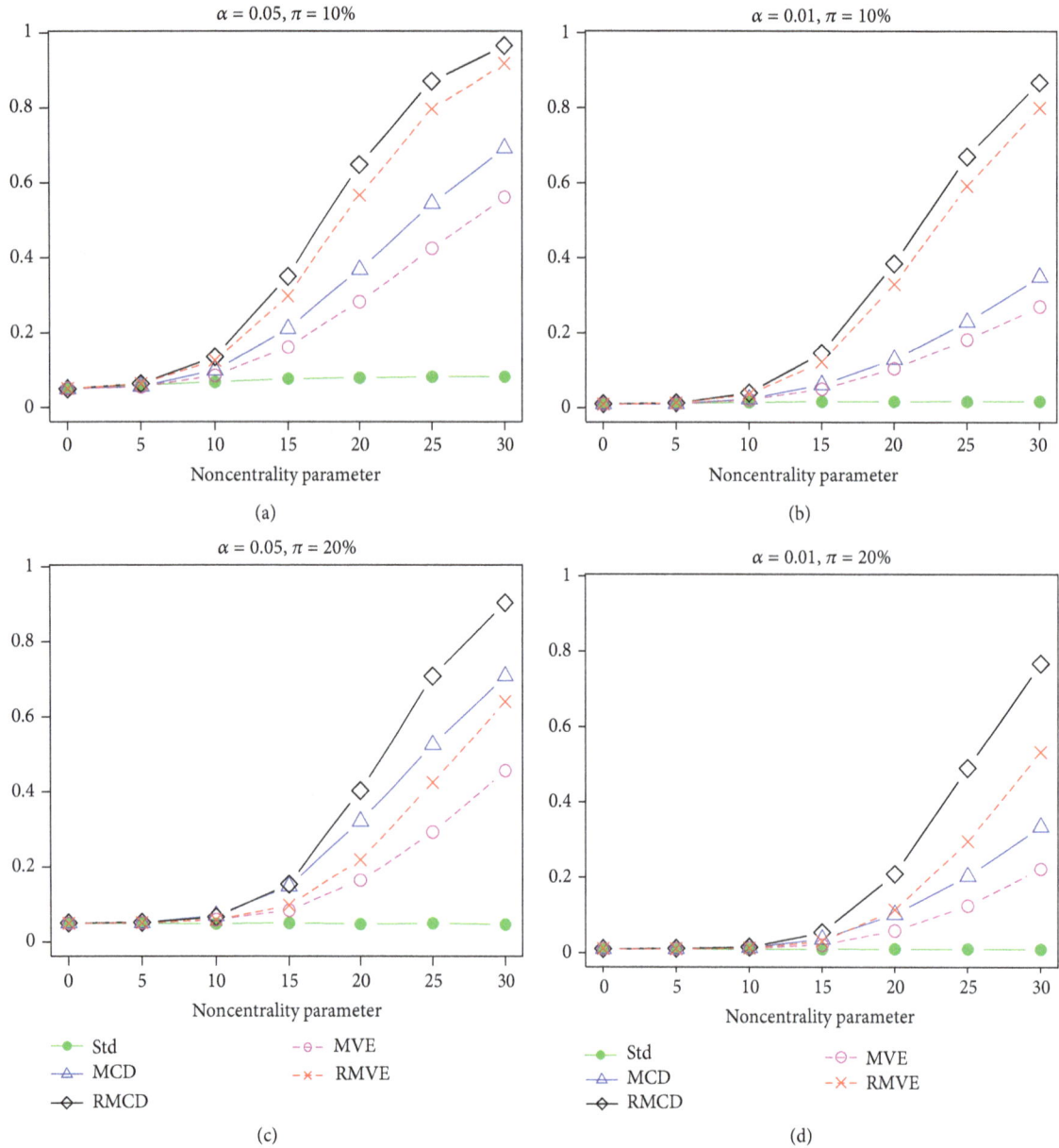

FIGURE 9: Probability of signal for T^2 control chart with different estimation methods for $p = 10$, $m = 150$.

A simple comparison of these estimators indicates that there are outliers in the Phase-I data. The plots of T^2, T^2_{RMCD}, and T^2_{RMVE} values along with the respective control limits at 99% confidence level for the sample data are given in Figure 10.

The control limits for T^2 are arrived at based on beta distribution, and $T^2_{\text{RMCD}}/T^2_{\text{RMVE}}$ are calculated using (6) for $p = 3$ and $m = 105$. From Figure 10, it is very clear that both T^2_{RMCD} and T^2_{RMVE} control chart alarms signal for 3 outliers whereas the standard T^2 control chart alarm signals for none even though all the charts are having the same pattern. This indicates the effectiveness of the proposed robust control charts in identifying the outliers.

6. Conclusions

Use of robust control chart in Phase-I monitoring is very important to assess the performance of the process as well as detecting outliers. We propose $T^2_{\text{RMCD}}/T^2_{\text{RMVE}}$ control charts for Phase-I monitoring of multivariate individual observations. The control limits for these charts are arrived empirically and a non-linear regression model is used for arriving control limits for any sample size. The performance of the proposed charts were compared under various data scenarios using large number of Monte Carlo simulations. Our simulation studies indicate that T^2_{RMVE} control charts are performing well for smaller sample sizes and smaller

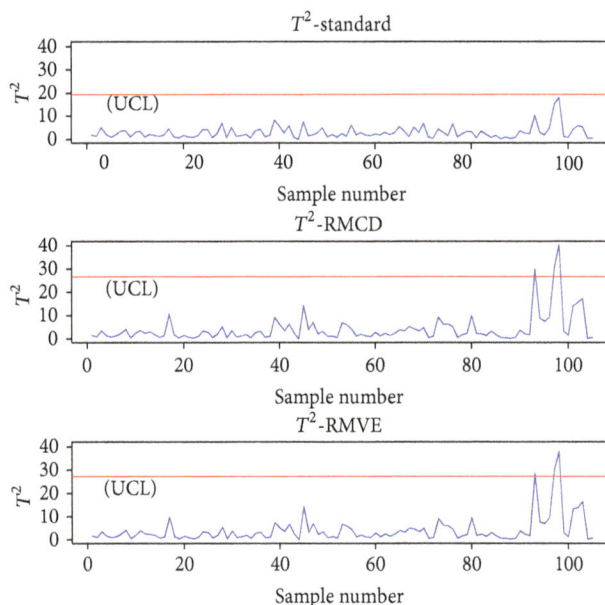

FIGURE 10: T^2, T^2_{RMCD}, and T^2_{RMVE} control charts for the sample data.

dimension where as T^2_{RMCD} control charts are performing well for larger sample sizes and larger dimensions. We illustrated our proposed robust control chart methodology using a case study from the electronic industry.

Acknowledgments

The authors would like to thank the editor and two anonymous referees for their valuable comments and suggestions that substantially improved the overall quality of an earlier version of this paper. The research is supported by a grant from the Natural Science and Engineering Council of Canada.

References

[1] H. Hotelling, "Multivariate quality control," in *Techniques of Statistical Analysis*, C. Eisenhart, H. Hastay, and W. A. Wallis, Eds., pp. 111–184, McGrawHill, New York, NY, USA, 1947.

[2] N. D. Tracy, J. C. Young, and R. L. Mason, "Multivariate control charts for individual observations," *Journal of Quality Technology*, vol. 24, pp. 88–95, 1992.

[3] J. H. Sullivan and W. H. Woodall, "A comparison of multivariate control charts for individual observations," *Journal of Quality Technology*, vol. 28, no. 4, pp. 398–408, 1996.

[4] J. A. N. Vargas, "Robust estimation in multivariate control charts for individual observations," *Journal of Quality Technology*, vol. 35, no. 4, pp. 367–376, 2003.

[5] W. A. Jensen, J. B. Birch, and W. H. Woodall, "High breakdown estimation methods for phase I multivariate control charts," *Quality and Reliability Engineering International*, vol. 23, no. 5, pp. 615–629, 2007.

[6] S. Chenouri, S. H. Steiner, and A. M. Variyath, "A multivariate robust control chart for individual observations," *Journal of Quality Technology*, vol. 41, no. 3, pp. 259–271, 2009.

[7] D. L. Donoho and P. J. Huber, "The notion of breakdown point," in *A Festschrift for Erich Lehmann*, P. Bickel, K. Doksum, and J. Hodges, Eds., pp. 157–184, Wadsworth, Belmont, Calif, USA, 1983.

[8] H. P. Lopuhaä and P. J. Rousseeuw, "Breakdown points of affine equivariant estimators of multivariate location and covariance matrices," *The Annals of Statistics*, vol. 19, pp. 229–248, 1991.

[9] D. L. Donoho and M. Gasko, "Breakdown properties of location estimates based on halfspace depth and projected outlyingness," *The Annals of Statistics*, vol. 20, pp. 1803–1827, 1992.

[10] P. L. Davies, "Asymptotic behavior of S-estimates of multivariate location parameters and dispersion matrices," *The Annals of Statistics*, vol. 15, pp. 1269–1292, 1987.

[11] C. Croux and G. Haesbroeck, "Influence function and efficiency of the minimum covariance determinant scatter matrix estimator," *Journal of Multivariate Analysis*, vol. 71, no. 2, pp. 161–190, 1999.

[12] G. Pison, S. van Aelst, and G. Willems, "Small sample corrections for LTS and MCD," *Metrika*, vol. 55, no. 1-2, pp. 111–123, 2002.

[13] D. L. Woodru and D. M. Rocke, "Computable robust estimation of multivariate location and shape in high dimension using compound estimators," *Journal of American Statistical Association*, vol. 89, pp. 888–896, 1994.

[14] D. M. Hawkins and D. J. Olive, "Improved feasible solution algorithms for high breakdown estimation," *Computational Statistics and Data Analysis*, vol. 30, no. 1, pp. 1–11, 1999.

[15] P. J. Rousseeuw and K. van Driessen, "A fast algorithm for the minimum covariance determinant estimator," *Technometrics*, vol. 41, no. 3, pp. 212–223, 1999.

[16] P. J. Rousseeuw and B. C. van Zomeren, "Unmasking multivariate outliers and leverage points," *Journal of American Statistical Association*, vol. 85, pp. 633–639, 1990.

[17] P. L. Davies, "The asymptotics of Rousseeuw's minimum volume ellipsoid estimator," *The Annals of Statistics*, vol. 20, no. 4, pp. 1828–1843, 1992.

[18] C. Croux and G. Haesbroeck, "An easy way to increase the finite-sample efficiency of the resampled minimum volume ellipsoid estimator," *Computational Statistics and Data Analysis*, vol. 25, no. 2, pp. 125–141, 1997.

[19] V. Torodov, "rrcov: Scalable Robust Estimators with High Breakdown point," R package version 0.5-03, 2009, http://cran.r-project.org/web/packages/rrcov/.

17

Epistemic Uncertainty Analysis: An Approach Using Expert Judgment and Evidential Credibility

Patrick Hester

Engineering Management and Systems Engineering Department, Old Dominion University, Kaufman Hall, Room 241, Norfolk, VA 23529, USA

Correspondence should be addressed to Patrick Hester, pthester@odu.edu

Academic Editor: Tadashi Dohi

When dealing with complex systems, all decision making occurs under some level of uncertainty. This is due to the physical attributes of the system being analyzed, the environment in which the system operates, and the individuals which operate the system. Techniques for decision making that rely on traditional probability theory have been extensively pursued to incorporate these inherent aleatory uncertainties. However, complex problems also typically include epistemic uncertainties that result from lack of knowledge. These problems are fundamentally different and cannot be addressed in the same fashion. In these instances, decision makers typically use subject matter expert judgment to assist in the analysis of uncertainty. The difficulty with expert analysis, however, is in assessing the accuracy of the expert's input. The credibility of different information can vary widely depending on the expert's familiarity with the subject matter and their intentional (i.e., a preference for one alternative over another) and unintentional biases (heuristics, anchoring, etc.). This paper proposes the metric of evidential credibility to deal with this issue. The proposed approach is ultimately demonstrated on an example problem concerned with the estimation of aircraft maintenance times for the Turkish Air Force.

1. Introduction

Real-world decision making is always performed under uncertainty. This uncertainty is present in the physical attributes of the system being analyzed, the environment in which it operates, and the individuals which operate the system. Decision makers must make decisions which best incorporate these uncertainties. With some problems, such as determining the probability of a terrorist attack on a given target, assigning probabilistic estimations to uncertain parameters is impossible due to the lack of statistical evidence upon which to base probabilistic estimates. Given these complex problems, decision makers often solicit subject matter expert opinion to provide estimates on uncertain parameters within a model. While this is a valid approach, soliciting expert opinions introduces additional uncertainty due to the varying degree of knowledge of the expert about the subject matter (i.e., one individual may truly be the world renowned expert in a field whereas others are merely

seasoned practitioners). Additionally, as human beings, they have the potential for intentional and unintentional biases.

The challenge when performing this type of analysis, in which expert judgment is essential to address uncertainty, is in assigning "weights" to the information provided by different experts in accordance with the level of expertise the expert provides. The credibility of different experts can vary widely depending on the expert's familiarity with the subject matter and their intentional (i.e., a preference for one alternative over another) and unintentional biases (e.g., heuristics, and anchoring). While expert opinion in an area that is of little familiarity to the expert may be not be entirely correct, there is no reason to believe that the information should be ignored completely, as the expert may have a particular insight to bring to the analysis. Further, the principle of complementarity [1] indicates that no one individual has complete knowledge of a complex system; thus, additional perspectives are value-added. Additionally, even though human beings have inherent biases and prejudices, the information they provide should not be completely discounted. This paper develops an approach to address these problems.

This paper begins with a background discussion about uncertainty analysis, expert judgment elicitation, evidence combination, and expert biases. It then develops an approach which allows the decision maker to determine a level of credibility to use in incorporating each expert's evidence. The proposed approach is demonstrated on an example problem concerned with the estimation of aircraft maintenance times for the Turkish Air Force. Finally, conclusions and recommendations for future work are presented.

2. Background

Uncertainty is typically separated into aleatory uncertainty and epistemic uncertainty (see, e.g., [3, 4]). "Aleatory uncertainty is also referred to as variability, irreducible uncertainty, inherent uncertainty, stochastic uncertainty, and uncertainty due to chance. Epistemic uncertainty is also referred to as reducible uncertainty, subjective uncertainty, and uncertainty due to lack of knowledge" ([5], p. 10-2). Aleatory uncertainty refers to variation which is inherent to a given system, typically as a result of the random nature of model inputs. Aleatory uncertainties are typically modeled as random variables described by probability distributions, where decision makers typically make assumptions about the distribution's descriptive statistics (i.e., its mean and variance). "Epistemic uncertainty as a source of nondeterministic behavior derives from lack of knowledge of the system or the environment" Oberkampf et al. [6]. Oberkampf and Helton [5] elaborate on this definition:

> The key feature stressed in this definition is that the fundamental source of epistemic uncertainty is incomplete information or incomplete knowledge of some characteristic of the system or the environment. As a result, an increase in knowledge or information can lead to a reduction in the predicted uncertainty of the response of the system, all things being equal. Examples of sources of epistemic uncertainty are: little or no experimental data for a fixed (but unknown) physical parameter, a range of possible values of a physical quantity provided by expert opinion, limited understanding of complex physical processes, and the existence of fault sequences or environmental conditions not identified for inclusion in the analysis of a system (p.10-2).

Epistemic uncertainty often becomes an issue when expert opinion is required to solve a problem. In trying to determine the likelihood of a terrorist attack on a given building, a decision maker may solicit many expert opinions due to a lack of sufficient knowledge about the problem. In doing so, the decision maker is introducing additional uncertainty into the analysis, both in the lack of knowledge about the credibility of the experts being solicited and in the experts' own intentional (i.e., a preference for one alternative over another) and unintentional (heuristics, anchoring, etc.)

biases that influence the information they provide. Epistemic uncertainty can be reduced with increased information, but aleatory uncertainty is a function of the problem characteristics itself.

Oberkampf et al. [7] describe various methods for estimating the total uncertainty in a model by identifying all sources of variability and uncertainty. Traditionally, uncertainty has been handled with probability theory, but recent developments maintain that representing all uncertainty information in the same manner is inappropriate and, in order to be analyzed appropriately, several experts believe that aleatory and epistemic uncertainty should be addressed separately (e.g., [8–15]).

Traditional quantification of uncertainty uses probability theory, which represents uncertainties as random variables by utilizing a probability density function which presents the probability information about the variable. Probability theory, however, has problems separating aleatory from epistemic uncertainty [8]. As a result, various techniques including Dempster-Shafer theory [16, 17] have gained increased use in recent years as techniques that can adequately separate differing types of uncertainty. Other theories such as generalized information theory [18] and approximate reasoning [19] have also proven useful in characterizing uncertainty.

Modern approaches to deal with epistemic uncertainty include fuzzy sets [20, 21], Dempster-Shafer theory [16, 17, 22], and possibility theory [23]. Dempster-Shafer theory was chosen for use in this paper due to its strong theoretical basis, large number of recent example problems to draw from, and versatility of Dempster-Shafer theory to represent and combine potentially dissimilar evidence from various sources. A brief discussion of the mathematics of Dempster-Shafer evidence theory is provided in the approach section (Section 3) of this paper.

In addition to dealing with uncertainties present in the problem domain, analysts must also understand what inherent biases are incorporated into an individual's thought process. The following section discusses biases that may influence an expert's judgment.

2.1. Biases. Whenever an expert is utilized as a source of information, their beliefs and experiences bias how they view the problem and what information they choose to provide to help solve the problem. These biases take the form of either intentional or unintentional biases. Intentional biases are a result of the expert's willful decision to bias the results of their assessment. This willful deceit can occur due to preference of one alternative over another. The expert may prefer one alternative over another due to gains that he/she stands to receive as a result of the analysis. An example would be a company that is using expert judgment to assess its building's level of security. If the expert were to have an interest in convincing the company that their security levels were subpar (such as if the expert owned his/her own security company), then the expert may intentionally bias the results. Alternatively, the expert may have a reason not to prefer a particular alternative and may intentionally bias the results accordingly. Typically, these intentional biases are easier for an outside observer to recognize as strong connections between the expert and his/her intentionally biased choice (such as significant financial connections) should emerge. It is important to note that the vast majority of experts will not exhibit this behavior, but it is important for the analyst to be cognizant of the potential for this bias nonetheless.

It is mistakenly assumed that because an individual is an expert in a particular subject matter, he/she is perfectly capable of providing accurate likelihood estimates for particular events. Even without intentional biases to account for, all human beings have unintentional cognitive biases that affect the information that is elicited from them. These cognitive biases include behaviors such as the availability heuristic, conjunction fallacy, representativeness heuristic, and anchoring.

The availability heuristic [24, 25] refers to the practice of basing probabilistic evidence on an available piece of information in one's own set of experiences. That is to say, humans estimate the likelihood of an event based on a similar event that they can remember. Further, since newer events are fresher in our minds, they influence our reasoning in larger proportion than older events. Since experts have a larger set of experiences to draw from, and thus more available data, it is likely that their propensity for the availability heuristic will decrease as their experience level increases. However, a more naïve expert may be able to provide a better result if he/she has experienced a relevant event recently, whereas an expert in the field with many years of relevant experience (none of which are recent), may not be as likely to provide useful information.

Another bias that humans incorporate when providing uncertainty estimates is the conjunction fallacy. This fallacy occurs when individuals identify specific scenarios as being more likely than general ones. Tversky and Kahneman [26] explored this phenomenon and found that this mistake is commonly committed despite the fact that it is mathematically impossible for the joint probability of two events to be more likely than the probability of either of the individual events. Individuals often make this mistake as the specific scenario seems more realistic to them and it is possible that experts can be prone to this type of fallacy as well. While experts are less prone to this type of behavior, the phenomenon is still something that analysts should be aware of when eliciting expert opinion.

The representativeness heuristic [25] occurs when commonalities between objects are assumed. For example, an expert has estimated the probability of attack against a building before and assumes the current building that is being analyzed is similar to his/her previous work, and therefore, estimates the probabilities to be similar. There may, in fact, be a glaring difference between the two problems that the expert is overlooking.

Another bias is the anchoring and adjustment heuristic, observed by Tversky and Kahneman [24]. Humans anchor their judgments and base subsequent observations on the initial value that was provided to them. In other words, if the expert is provided a baseline value, he/she can be influenced to a degree where subsequent probability values will be anchored by the provided baseline value. Even experts can be influenced to provide probabilistic values close to values that the analyst desires by anchoring the questions that are asked when eliciting their opinion.

The biases discussed here are only a few of the possible that may affect experts. The important takeaway with respect to biases, both intentional and unintentional, is that decision makers must be cognizant of their effect on results obtained when eliciting expert judgment. Any approach that incorporates expert judgment must take into the account the presence of biases and adjust its approach accordingly.

An approach is developed in the next section which provides a method for dealing with these biases when using expert judgment to address epistemic uncertainty.

3. Solution Approach

3.1. Dempster-Shafer Theory. Dempster-Shafer theory is a mathematical theory of evidence, defined by three important functions: the basic probability assignmentfunction (BPA or m), the belief function (Bel), and the plausibilityfunction (Pl). The seminal work on the subject is [17], which is an expansion of [16]. In evidence theory, uncertainty is separated into belief and plausibility, whereas traditional probability theory uses only the probability of an event to analyze uncertainty. Belief and plausibility provide bounds on probability. In special cases, they converge on a single value, probability. In other cases, such as in the evidence theory representation of uncertainty, belief, and plausibility represents a range of potential values for a given parameter, without any assumptions on the likelihood of the underlying data.

In evidence theory, for a sample space X, degrees of evidence are assigned to subsets (events) of X. A subset (x) with a nonzero degree of evidence is called a focal element. Based on available information, a basic probability assessment (BPA), denoted by $m(x)$, can be defined as

$$m(x) \geq 0 \quad \text{for } x \subset X \tag{1a}$$

$$\sum_{x \subset X} m(x) = 1. \tag{1b}$$

An BPA is provided by experts in lieu of a traditional probability assessment. Imagine a scenario in which experts are being asked to predict weather occurrences in a given city. Two experts ($E1$ and $E2$) are providing their opinions on the likelihood of three weather occurrences (W1, W2, and W3). The potential weather phenomena are as follows: W1 is sunny, W2 is cloudy and W3 is rainy.

In this case, the objective is to find the likely weather occurrence. As such, $X = \{W1, W2, W3\}$ and the frame of discernment representing all possible categories of evidence, $2^X = \{W1, W2, W3, W1 \cup W2, W2 \cup W3, W1 \cup W3, W1 \cup W2 \cup W3\}$.

Suppose the following information is collected: Expert one ($E1$) says it will be rainy or sunny with 90% probability, while Expert two ($E2$) says it will be sunny or cloudy with 75% probability.

The BPA for expert one can then be stated as

$$m_{E1}(W1 \cup W3) = 0.90, \qquad m_{E1}(W1 \cup W2 \cup W3) = 0.10.$$
(2)

The second piece of evidence ($m_{E1}(W1 \cup W2 \cup W3)$) is due to the fact that nothing is known about the remaining evidence, so it must be allocated to what is termed the remaining frame of discernment. That is, since nothing is known (except that weather must occur), a judgment cannot be made about the unknown frame of reference (the remaining 0.10 is not specified by the expert, so it is specified as being either W1 or W2 or W3).

The BPAs for the second expert can be stated as

$$m_{E2}(W1 \cup W2) = 0.75, \qquad m_{E2}(W1 \cup W2 \cup W3) = 0.25.$$
(3)

In other words, *E1* says the weather phenomena are most probably W1 or W3, while *E2* says the weather phenomena are most probably W1 or W2. The sources of evidence can be combined using the Dempster rule of combination as follows:

$$m_{12}(A) = \frac{\sum\limits_{B \cap C = A} m_1(B)m_2(C)}{1 - \sum\limits_{B \cap C = \varnothing} m_1(B)m_2(C)}.$$
(4)

This equation yields the combined evidence of experts 1 and 2 that support A (which is composed of the intersection of B and C). This can be applied recursively to combining the evidence of more than two experts. For example, the results obtained from combining experts one and two can then be combined with expert three in order to determine the combination of the three experts' evidence. The order in which the experts' evidence is combined is irrelevant.

Dempster's rule of combination has been subject to criticism in that it tends to ignore conflict available within the evidence (as pointed out by [27]) and attributes evidence supporting conflict to the null set [22]. Additional combination rules deal with this complication, but they require that the relative importance (in the case of this approach, the evidential credibility values) of each expert is known. Information on additional rules of combination is provided in Agarwal et al. [8] and Sentz and Ferson [28]. For the purposes of this paper, it will be assumed that the evidence provided does not have a large enough level of conflict that using Dempster's rule will adversely affect the results. The approach presented here can be generalized to other rules of combination if desired.

The evidence of the two weather experts can then be combined as in Table 1.

Given this combined BPA, the evidence can now be used to form belief and plausibility bounds on the uncertainty. Belief in any set is the sum of all probabilities of all subsets of that set. It represents any proof that has been provided (it is believed) that a particular event is true.

Belief values for the individual events in the aforementioned problem are

$$\mathrm{Bel}_{E1,E2}(W1) = m_{E1,E2}(W1) = 0.675,$$

$$\mathrm{Bel}_{E1,E2}(W2) = 0, \qquad\qquad (5)$$

$$\mathrm{Bel}_{E1,E2}(W3) = 0.$$

On the other hand, plausibility is more general. It represents the degree to which it is plausible that a particular event is true. Another way to look at belief is it is a measure of the degree to which an event will happen, whereas plausibility is a measure of the degree to which an event could happen.

Plausibility values for the individual events in the aforementioned problem are

$$\mathrm{Pl}_{E1,E2}(W1) = m_{E1,E2}(W1) + m_{E1,E2}(W1 \cup W3)$$

$$+ m_{E1,E2}(W1 \cup W2)$$

$$+ m_{E1,E2}(W1 \cup W2 \cup W3)$$

$$= 0.675 + 0.225 + 0.075 + 0.025 = 1,$$

$$\mathrm{Pl}_{E1,E2}(W2) = m_{E1,E2}(W1 \cup W2)$$

$$+ m_{E1,E2}(W1 \cup W2 \cup W3)$$

$$= 0.075 + 0.025 = 0.10,$$

$$\mathrm{Pl}_{E1,E2}(W3) = m_{E1,E2}(W1 \cup W3)$$

$$+ m_{E1,E2}(W1 \cup W2 \cup W3)$$

$$= 0.225 + 0.025 = 0.25.$$
(6)

More information on the mathematics and application of evidence theory can be found in Dempster [16] and Shafer [17].

3.2. Evidential Credibility. When expert judgments are elicited, and epistemic uncertainty is introduced due to lack of knowledge about the credibility of the evidence provided by experts, however, a modified version of evidence theory must be developed to deal with this additional layer of uncertainty. In this modified approach, the proposed expert's modified BPA, $m_i^*(x)$, is given as

$$m_i^*(x) = m_i(x)\mathrm{EC}_i(x), \qquad\qquad (7)$$

where $m_i(x)$ is as defined in (1a) and (1b) and $\mathrm{EC}_i(x)$ is the evidential credibility value, with i and x being the indices corresponding to the particular expert and event, respectively.

Evidential credibility is a measure of the analyst's confidence in the expert's estimated likelihood values; it acts as a weight to adjust the likelihood estimate given by the individual expert. An evidential credibility value of one means that the analyst has complete confidence in the expert's estimate and it should be taken into account fully,

TABLE 1: Combined expert evidence.

	$m_{E2}(W1 \cup W2) = 0.75$	$m_{E2}(W1 \cup W2 \cup W3) = 0.25$
$m_{E1}(W1 \cup W3) = 0.90$	$m_{E1,E2}(W1) = 0.675$	$m_{E1,E2}(W1 \cup W3) = 0.225$
$m_{E1}(W1 \cup W2 \cup W3) = 0.10$	$m_{E1,E2}(W1 \cup W2) = 0.075$	$m_{E1,E2}(W1 \cup W2 \cup W3) = 0.025$

whereas an evidential credibility value of less than one demonstrates the analyst's reluctance to place full confidence in the likelihood estimates provided by the expert. The remainder of evidence not attributed to an event by the expert (independent of his/her evidential credibility) is allocated to the remaining frame of discernment as detailed earlier in this section (with this evidence increasing as $EC_i(x)$ approaches zero). If evidential credibility is calculated as one for an expert, the adjusted BPA in (7) reverts to the original form in (1a) and (1b). If the expert's calculated evidential credibility is zero, then his/her likelihood for the specified event reverts to zero. If an expert is deemed to have an evidential credibility of zero for all possible events, all evidence for the given expert is allocated to complete ignorance $(m^*((\cup x)\forall x \in x) = 1)$. This reflects the notion that the analyst has no confidence in the expert's predictions, based on his/her evidential credibility. For the previous weather example, a single expert being polled with no evidential credibility would result in an BPA of $m_{E1}^*(\overline{W1 \cup W2 \cup W3}) = 1.0$.

For the proposed approach, evidential credibility is calculated in a manner derived from the Brier score (1950) approach to evaluating experts, which is straightforward and provides a good basis for the development of an evidential credibility measure. Brier's work is predicated on the existence of verifiable data, whereas there is no established "right" answer in many of the applications for which this paper's proposed approach is intended. There are two options to deal with this complication: (1) develop a Brier score based on the information that is present (e.g., historical data of similar systems) or (2) adjust the Brier score to create a new scoring rule which reflects the lack of available data. The author utilizes the second approach in this paper, as the first approach is valid only in simpler systems where the variance between new and old systems is trivial, thus making the necessity of the method developed in this paper unnecessary. The adjusted Brier score, then, is used to calculate the evidential credibility as follows:

$$EC_i(x) = \left[1 - \left|f_i(x) - \overline{m(x)}\right|^{1/2}\right], \qquad (8)$$

where $f_i(x)$ is the ith expert's estimate of the average of all expert forecasts for event x and $\overline{m(x)}$ is the average of actual expert predictions of event x.

Equation (8) reflects an error-function-based approach to scoring experts' evidence. Evidential credibility is not intended to reflect the expert's individual credibility, but rather it provides a discount factor of the individual's knowledge about what the collective judgment of the experts will be. This point is illustrated in (8), where the difference between the expert's prediction of the average estimation and the actual average of the expert's estimations (the error) is

calculated. The closer the two values are together, the lower the prediction error of the expert is, and, thus, the more knowledgeable the expert is proven to be. A true expert would be able to provide an accurate prediction of an event $(m_i(x))$ and an estimate of what other experts would say $(f_i(x))$. If he/she is not accurate in this regard, the resulting $EC_i(x)$ will be reduced. This makes an intuitive sense as the true expert should understand what knowledge he/she has that would alter his/her prediction relative to other experts. If the expert is not able to do so, the analyst's confidence in his/her predictions should be reduced (as demonstrated by an evidential credibility value of less than one) as they are likely not as knowledgeable as originally predicted. If the expert is not privy to additional information which biases his/her predictions, he/she is likely to assume $f_i(x) = m_i(x)$.

Utilizing the above definition for evidential credibility, individuals are incentivized to report their true predictions for the pool of experts, as shading their predictions would result in their evidential credibility being reduced and their initial prediction, $m_i(x)$, being adjusted significantly by a lower $EC_i(x)$ value. Thus, less evidential credibility means the evidence provided by the expert (in terms of their likelihood values) will be lessened as there is less confidence in their estimates. It should be noted that if an expert is not comfortable with providing a likelihood estimate for a particular event, he/she may abstain from providing one and the average prediction that is calculated will exclude any individuals who choose to omit a response. As each response provided by the experts has its own evidential credibility value, there is no incentive for a particular expert to answer more questions than he/she is comfortable with to artificially inflate their $EC_i(x)$. If an event has only one expert providing an estimate for its likelihood, there is no expert to contrast with and therefore their evidential credibility is taken to be one (since they are incentivized to tell the truth, there is no reason to believe that their answer is anything less than completely truthful, as they have no knowledge of how many other experts will answer this particular question).

Additionally, the evidential credibility metric is useful in combating biases facing experts. If an expert is prone to the availability heuristic, conjunction fallacy, representativeness heuristic, or anchoring, and these biases influence their average predictions, their resultant evidence will be reduced through a reduced $EC_i(x)$ value. Thus, the experts are incentivized to examine their predictions before reporting them and report their true values, free of intentional and unintentional biases.

The approach proposed in this paper is useful for several reasons.

(1) It has the benefit of assigning a separate evidential credibility value for each expert and each event being

estimated so that the opinions of experts with varying levels of experience can be combined without having to ignore experts with less experience.

(2) The evidential credibility measure is flexible and can be utilized regardless of how much information each expert chooses to provide for a particular problem.

(3) There is no need for this approach to have the "right" answer a priori, as experts are ranked relative to one another's estimates and not relative to a correct baseline. This is especially useful for complex problems as the correct answer is often not known until after the analysis occurs, if at all.

(4) This approach incentivizes experts to tell the truth, both in their own estimates and in their predictions of others' estimates, limiting the effect of intentional and unintentional biases on the resulting expert judgments.

Utilizing the proposed approach, two example problems are solved in the next section.

4. Example Discussion

The example demonstration of the proposed method for calculating evidential credibility is drawn from Kudak and Hester [2] and based on an example regarding maintenance time estimation in the Turkish Air Force. Kudak and Hester [2] explain how evidence theory was used to estimate maintenance times for damaged aircraft repairs. Three experts were surveyed and asked to provide their estimates regarding the three major statistically observed failure sources of ignition, fuel, and electrical systems [29].

Hester and Kudak [2] describe the problem:

For wartime operations, Maintenance Time (MT) of each specific failure can be divided into three separate time intervals (x_1, x_2 and x_3) as shown in [Figure 1]. Expected Actual Time, $E_a(t)$, is the exact MT as assigned in [29] for normal operation times. The first time interval, x_1, represents times less than the Best Time (defined by the expert as the shortest maintenance time expected to complete the failure, with a default time given as 5–10 % less than $E_a(t)$, depending on the system). The second interval, x_2, represents the period between Best Time and Worst Time (defined by the expert as the longest maintenance time expected to complete the failure, with a default time given as 5–10% more than $E_a(t)$, depending on the system), and it includes the Expected Actual Time $E_a(t)$. The third time interval, x_3, represents times greater than the Worst Time. (pp. 56-57).

This example will be used to explore the evidential credibility metric proposed in this paper.

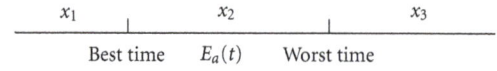

Figure 1: Maintenance time intervals (adapted from Kudak and Hester [2]).

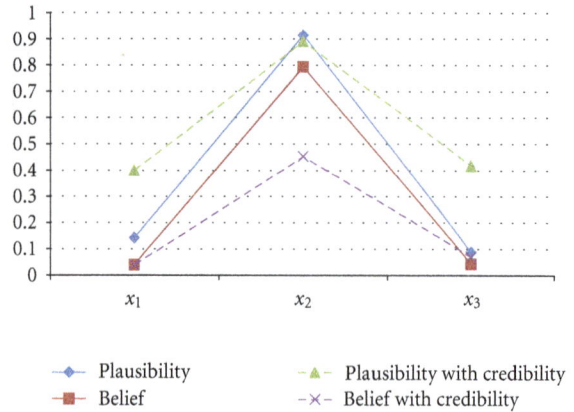

Figure 2: Belief and plausibility measures with and without evidential credibility.

5. Results and Analysis

The three experts discussed in Kudak and Hester [2] were asked to provide their BPAs for the three time intervals shown in Figure 1 and their combinations. Their BPAs are provided in Table 2.

Using (4), these estimates can be combined to provide belief and plausibility estimates as shown in Tables 3 and 4, respectively.

Now, let us suppose that each expert is asked to provide an estimate of the average of all expert forecasts, $f_i(x)$. In the absence of greater knowledge about his or her fellow experts, each individual assumes their estimates are equal to the average, that is, $f_i(x) = m_i(x)$, as discussed in Section 3.2. Using (8), we can then calculate the $EC_i(x)$ for each event and expert as shown in Table 5.

Using (7), modified BPAs can be generated for the experts, as shown in Table 6.

Using (4), these modified BPA assignments can be combined to provide belief and plausibility estimates (including evidential credibility) as shown in Tables 7 and 8.

Further, belief and plausibility values for each of the three time ranges (x_1, x_2, and x_3) can be compared graphically as shown in Figure 2.

It is clear from investigating Figure 2 that evidence supporting x_2 has decreased when evidential credibility is taken into account. While baseline belief and plausibility estimates indicate a narrow band at x_2, incorporation of evidential credibility widens this band and allocates associated evidence to both x_1 and x_3. Further investigation of Tables 5 and 2 to determine the cause of this change reveals that Expert 3 is likely the reason behind the adjusted estimates. His/her low evidential credibility (0.42, as shown in Table 5) with respect to estimating x_2 reduces the credibility of his/her

TABLE 2: BPA Assignments (adapted from Kudak and Hester, [2]).

	$m_i\{x_1\}$	$m_i\{x_2\}$	$m_i\{x_3\}$	$m_i\{x_1,x_2\}$	$m_i\{x_1,x_3\}$	$m_i\{x_2,x_3\}$	$m_i\{x_1,x_2,x_3\}$
Expert 1	0	0	0.2	0.6	0	0	0.2
Expert 2	0	0.3	0	0	0	0.3	0.4
Expert 3	0.1	0.65	0	0	0	0	0.25

TABLE 3: Belief values.

$Bel_{123}\{x_1\}$	$Bel_{123}\{x_2\}$	$Bel_{123}\{x_3\}$	$Bel_{123}\{x_1,x_2\}$	$Bel_{123}\{x_1,x_3\}$	$Bel_{123}\{x_2,x_3\}$
0.04	0.794	0.044	0.911	0.085	0.857

TABLE 4: Plausibility values.

$Pl_{123}\{x_1\}$	$Pl_{123}\{x_2\}$	$Pl_{123}\{x_3\}$	$Pl_{123}\{x_1,x_2\}$	$Pl_{123}\{x_1,x_3\}$	$Pl_{123}\{x_2,x_3\}$
0.142	0.915	0.088	0.955	0.205	0.959

TABLE 5: Evidential credibility values.

	$EC_i\{x_1\}$	$EC_i\{x_2\}$	$EC_i\{x_3\}$	$EC_i\{x_1,x_2\}$	$EC_i\{x_1,x_3\}$	$EC_i\{x_2,x_3\}$
Expert 1	0.82	0.44	0.63	0.37	1.00	0.68
Expert 2	0.82	0.87	0.74	0.55	1.00	0.55
Expert 3	0.74	0.42	0.74	0.55	1.00	0.68

TABLE 6: Modified BPA assignments incorporating evidential credibility.

	$m_i^*\{x_1\}$	$m_i^*\{x_2\}$	$m_i^*\{x_3\}$	$m_i^*\{x_1,x_2\}$	$m_i^*\{x_1,x_3\}$	$m_i^*\{x_2,x_3\}$	$m_i^*\{x_1,x_2,x_3\}$
Expert 1	0.00	0.00	0.13	0.22	0.00	0.00	0.65
Expert 2	0.00	0.26	0.00	0.00	0.00	0.17	0.57
Expert 3	0.07	0.27	0.00	0.00	0.00	0.00	0.65

TABLE 7: Modified belief values incorporating evidential credibility.

$Bel_{123}^*\{x_1\}$	$Bel_{123}^*\{x_2\}$	$Bel_{123}^*\{x_3\}$	$Bel_{123}^*\{x_1,x_2\}$	$Bel_{123}^*\{x_1,x_3\}$	$Bel_{123}^*\{x_2,x_3\}$
0.04	0.45	0.07	0.58	0.11	0.60

TABLE 8: Modified plausibility values incorporating evidential credibility.

$Pl_{123}^*\{x_1\}$	$Pl_{123}^*\{x_2\}$	$Pl_{123}^*\{x_3\}$	$Pl_{123}^*\{x_1,x_2\}$	$Pl_{123}^*\{x_1,x_3\}$	$Pl_{123}^*\{x_2,x_3\}$
0.40	0.89	0.42	0.93	0.55	0.96

strong evidence supporting this event (an BPA of 0.65, as shown in Table 2). Thus, it would be worthwhile for the analyst to seek more information from expert 3 to determine why he/she felt so strongly about x_2 and yet did not provide an accurate estimate with respect to his/her fellow experts ($f_i(x_2)$). Similar analyses can be undertaken to support Kudak and Hester's [2] discussion regarding fuel and electrical systems, or in any other problem where evidence theory is an appropriate candidate for uncertainty quantification.

6. Conclusions

This paper proposed an approach for including a measure of evidential credibility into analysis when eliciting expert opinion to estimate epistemic uncertainty in a problem. It is the hope of the author that this approach can be extended to other evidence theory combination rules in the future in order to further explore its usefulness. Other scoring rules (specifically Prelec [30] and Matheson and Winkler [31]) should also be explored for use in this approach. Further, it is also thought that this approach, while demonstrated on a single case study in this paper, must be further explored to ensure its validity and utility.

References

[1] N. Bohr, "Quantum postulate and the recent development of atomic theory," *Nature*, vol. 121, no. 3050, pp. 580–591, 1928.

[2] H. Kudak and P. Hester, "Application of Dempster-Shafer theory in aircraft maintenance time assessment: a case study," *Engineering Management Journal*, vol. 23, no. 2, pp. 55–62, 2011.

[3] J. C. Helton and D. E. Burmaster, "Treatment of aleatory and epistemic uncertainty in performance assessments for complex systems," *Reliability Engineering and System Safety*, vol. 54, no. 2-3, pp. 91–94, 1996.

[4] W. L. Oberkampf, "Uncertainty quantification using evidence theory," in *Proceedings from the Advanced Simulation & Computing Workshop*, Albuquerque, NM, USA, 2005.

[5] W. L. Oberkampf and J. C. Helton, "Evidence theory for engineering applications," in *Engineering Design Reliability Handbook*, E. Nikolaidis, D. M. Ghiocel, and S. Singhal, Eds., CRC Press, Boca Raton, Fla, USA, 2005.

[6] W. L. Oberkampf, S. M. DeLand, B. M. Rutherford, K. V. Diegert, and Alvin, "Error and uncertainty in modeling and simulation," *Reliability Engineering and System Safety*, vol. 75, no. 3, pp. 333–357, 2002.

[7] W. L. Oberkampf, K. V. Diegert, K. F. Alvin, and B. M. Rutherford, "Variability, uncertainty, and error in computational simulation," in *ASME Proceedings of the 7th AIAA/ASME Joint Thermophysics and Heat Transfer Conference*, vol. 2, pp. 259–272, 1998.

[8] H. Agarwal, J. E. Renaud, E. L. Preston, and D. Padmanabhan, "Uncertainty quantification using evidence theory in multidisciplinary design optimization," *Reliability Engineering and System Safety*, vol. 85, no. 1–3, pp. 281–294, 2004.

[9] J. Darby, *Evaluation of Risk from Acts of Terrorism: The Adversary/Defender Model Using Belief and Fuzzy Sets (SAND2006-5777)*, Sandia National Laboratories, Albuquerque, NM, USA, 2006.

[10] J. C. Helton, "Treatment of uncertainty in performance assessments for complex systems," *Risk Analysis*, vol. 14, no. 4, pp. 483–511, 1994.

[11] J. C. Helton, "Uncertainty and sensitivity analysis in the presence of stochastic and subjective uncertainty," *Journal of Statistical Computation and Simulation*, vol. 57, no. 1–4, pp. 3–76, 1997.

[12] F. O. Hoffman and J. S. Hammonds, "Propagation of uncertainty in risk assessments: the need to distinguish between uncertainty due to lack of knowledge and uncertainty due to variability," *Risk Analysis*, vol. 14, no. 5, pp. 707–712, 1994.

[13] M. E. Paté-Cornell, "Uncertainties in risk analysis: six levels of treatment," *Reliability Engineering and System Safety*, vol. 54, no. 2-3, pp. 95–111, 1996.

[14] S. N. Rai, D. Krewski, and S. Bartlett, "A general framework for the analysis of uncertainty and variability in risk assessment," *Human and Ecological Risk Assessment*, vol. 2, no. 4, pp. 972–989, 1996.

[15] W. D. Rowe, "Understanding uncertainty," *Risk Analysis*, vol. 14, no. 5, pp. 743–750, 1994.

[16] A. P. Dempster, "Upper and lower probabilities induced by a multivalued mapping," *The Annals of Statistics*, vol. 38, no. 2, pp. 325–339, 1967.

[17] G. Shafer, *A Mathematical Theory of Evidence*, Princeton University Press, Princeton, NJ, USA, 1976.

[18] G. J. Klir and M. J. Wierman, *Uncertainty Based Information: Elements of Generalized Information Theory*, Physica, Heidelberg, Germany, 1998.

[19] D. Padmanabhan and S. M. Batill, "Reliability based optimization using approximations with applications to multidisciplinary system design," in *Proceedings of the 40th AIAA Sciences Meeting & Exhibit*, AIAA-2002-0449, Reno, Nev, USA, 2002.

[20] L. A. Zadeh, "Fuzzy sets," *Information and Control*, vol. 8, no. 3, pp. 338–353, 1965.

[21] G. J. Klir and B. Yuan, *Fuzzy Sets and Fuzzy Logic: Theory and Applications*, Prentice Hall, Upper Saddle River, NJ, USA, 1995.

[22] R. R. Yager, "Arithmetic and other operations on Dempster-Shafer structures," *International Journal of Man-Machine Studies*, vol. 25, no. 4, pp. 357–366, 1986.

[23] D. Dubois and H. Prade, *Possibility Theory: An Approach to Computerized Processing of Uncertainty*, Plenum Press, New York, NY, USA, 1988.

[24] A. Tversky and D. Kahneman, "Availability: a heuristic for judging frequency and probability," *Cognitive Psychology*, vol. 5, no. 2, pp. 207–232, 1973.

[25] A. Tversky and D. Kahneman, "Judgment under uncertainty: heuristics and biases," *Science*, vol. 185, no. 4157, pp. 1124–1131, 1974.

[26] A. Tversky and D. Kahneman, "Extensional versus intuitive reasoning: the conjunction fallacy in probability judgment," *Psychological Review*, vol. 90, no. 4, pp. 293–315, 1983.

[27] L. A. Zadeh, "Review of Shafer's mathematical theory of evidence," *Artifical Intelligence Magazine*, vol. 5, pp. 81–83, 1984.

[28] K. Sentz and S. Ferson, *Combination of Evidence in Dempster-Shafer Theory (SAND2002-0835)*, Sandia National Laboratories, Albuquerque, NM, USA, 2002.

[29] T. O. F-4E-1-9, "F-4E aircraft jet engine maintenance technical order manual," U.S. Air Force, 1983.

[30] D. Prelec, "A Bayesian truth serum for subjective data," *Science*, vol. 306, no. 5695, pp. 462–466, 2004.

[31] J. E. Matheson and R. L. Winkler, "Scoring rules for continuous probability distributions," *Management Science*, vol. 22, no. 10, pp. 1087–1096, 1976.

18

A Review of Quality Criteria Supporting Supplier Selection

Mohammad Abdolshah

Engineering Faculty, Islamic Azad University, Semnan Branch, Semnan 35136-93688, Iran

Correspondence should be addressed to Mohammad Abdolshah; abdolshah@gmail.com

Academic Editor: Shey-Huei Sheu

This paper presents a review of decision criteria reported in the literature for supporting the supplier selection process. The review is based on an extensive search in the academic literature. After a literature review of decision criteria, we discuss the most important criteria: quality. Then different methods and factors for assessing the quality of supplier are discussed. Results showed that all methods and factors mentioned in this paper are not appropriate tools for quality evaluation. Moreover, we propose a novel method (using loss functions) in order to assess the quality of suppliers.

1. Introduction

Nowadays companies hope to establish a longer-term working relationship with the suppliers. Therefore, supplier selection is one of the main parts of decisions in supply chain management. Because there are many suppliers with many criteria, so it is impossible to find the best way to evaluate and select suppliers. Therefore, in studies, scholars have used different methods, variables, criteria, and factors in order to select the suppliers.

Evaluation of suppliers is a process that leads companies to select their desired suppliers. This process has two main aims, which are to reduce all costs of purchasing and to increase the overall value of the purchasing [1]. Regarding to the costs of evaluating the suppliers (such as time and travel budget), companies basically evaluate those suppliers that have a good chance of qualifying for purchasing from them. In this process, formally, companies send expert teams to the supplier site, and with evaluating different criteria and factors, they will do an in-depth evaluation.

There are different steps that must be done in the process of supplier evaluation and selection [1]. As it is shown in Figure 1, in order to evaluate and select the suppliers, companies must identify some important things such as methodologies, criteria, and problems (strategies).

This paper focused on the literature review of criteria for supplier evaluation and especially the criteria related to the quality evaluation of suppliers.

2. Decision Criteria Formulation

Regarding to multicriteria decision-making concept of supplier selection problems, one of the main parts of this process is to define related factors and criteria. There are many criteria such as price, quality, and process capability and on time delivery, which can affect selecting the proper supplier. There are many studies since the 1960s about factors, which affect supplier selection. Roa and Kiser [2], Ellram [3], and Stamm and Golhar [4] mentioned 60, 18, and 13 criteria for supplier selection, respectively. Weber at al. [5] reviewed 47 articles in which more than one criterion was considered in supplier selection models. One crucial study, which was done by Dickson [6], identified 23 different criteria evaluated in supplier selection.

This study was on the base of a questionnaire that was sent to 273 purchasing agents and managers from the United States and Canada. Their survey showed that price, delivery, and quality are three main factors, which are important for them. Weber et al. [5] also did a similar survey and concluded that price, delivery, quality, production capacity, and localization are the most important criteria.

FIGURE 1: Supplier evaluation and selection process [1].

Although the evolution of the industrial environment modified the degrees of the relative importance of supplier selection criteria since the 1960s, the 23 ones presented by Dickson [6] still cover the majority of those presented in the literature until today. Ha and Krishnan [8] and Aissaoui et al. [9] investigated 31 main criteria which were related to supplier selection in the literature since 1966. In this study, we added some new studies and summarized them in Table 2: A: Dickson [6]; B: Wind et al. [10]; C: Lehmann and haughnessy [11]; D: Perreault and Russ [12]; E: Abratt [13]; F: Billesbach et al. [14]; G: Weber et al. [5]; I: Min and Galle [15]; J: Stavropolous [16]; K: Pi and Low [17]; L: Pi and Low [18], M: Teeravaraprug [19]; N: Sanayei et al. [20]; O: Parthiban et al. [21]; P: Peng [22]; Q: Bilişik et al. [23]; R: Tektas and Aytekin [24]; S: Li [25]; T: Betül et al. [26]; U: Mehralian et al. [27].

3. Assessing the Quality

Quality is one of the main criteria for supplier evaluation and supplier selection. As it showed in Table 1, among 23 criteria, quality was the main criterion for supplier evaluation [6]. Holjevac [28] defined quality as follows:

(i) quality refers to the ability of a product or service to consistently meet or exceed customer's expectations;

(ii) quality means getting what you have paid for;

(iii) quality is not something that is adopted as a special feature; instead, it is an integral part of a product or service.

Assessing quality as one of the main factors for supplier selection is so important. In order to assess the quality, there are many factors and methods that can be used. There are a lot of literatures which have been accumulated on the subject of vendor evaluation and selection models, and in order to evaluate quality, most of these models have used rate of rejects [20, 29, 30], while rate of rejects cannot present the

quality appropriately. In a recent study, Lee [31] used yield rate in order to evaluate the quality. In the next chapter, it is mentioned why the rate of rejects or yield rate are not appropriate tools for assessing the quality.

Some scholars integrated some factors in order to evaluate the quality. Teng and Jaramillo [32] integrated continuous improvement programs, quality of customer, support services, certifications, and percentage on time shipment. In another study, Hou and Su [33] defined a quality index with integration of technical and design level, ease of repair, and reliability. Xia and Wu [34] used just technical and design level, and reliability in order to evaluate the quality. These factors also cannot present the core of quality level of products.

Process capability indices are appropriate and suitable indices in order to evaluate the quality, but presenting them as cost are so complicated, so some scholars presented an innovative qualitative method with the integration of process capability indices and other factors.

In the other hand, some scholars have used process capability indices in order to evaluate the quality. Tseng and Wu [35] considered the problem of selecting the best manufacturing process from some available processes based on the "precision" capability index C_p and have proposed a modified likelihood ratio (MLR) selection rule. They prepared some tables of the sample size and of the critical values for selecting the best manufacturing by controlling the probability based on the proposed MLR selection rule. Moreover, in case a nonnormal symmetric distribution, Tseng and Wu [35] used simulation to examine the robustness of the selection rule. Their results showed that the proposed modified likelihood ratio selection rule is acceptable for nonnormal symmetric process distributions.

Chou [36] developed an approximate method for selecting a better supplier based on one-sided capability indices C_{pu} and C_{pl} when the sample sizes are the same, which deals with comparing two one-sided processes and selects better one with the higher process yield. Pearn et al. [37] investigated the selection power analysis of the method via simulation and process capability. Huang and Lee (1995) proposed a model for selection a subset of processes containing the best supplier from a given set of processes. Under the circumstances, a search for the larger C_{pm} which are used to provide a measure of the process performance is equivalent to a search for the smaller γ^2 [38].

The selection method proposed by Chou [36] utilized some approximating results but provide no indication on how one could further proceed with selecting the better supplier by testing process capability index C_{pu} or C_{pl}. Pearn et al. [37] investigated the selection power analysis of the method via simulation. The accuracy analysis provides useful information regarding the sample size required for designated selection power. To render this method practical for in-plant applications, a two-phase selection procedure is developed by Pearn et al. [37] to select the better supplier and examine further the magnitude of the difference between the two suppliers [38].

There are some innovative research with the integration of process capability and other factors in order to select the

TABLE 1: 23 supplier evaluation criteria [6].

Rank	Criteria	Mean rating	Evaluation
1	Quality	3.508	Extreme importance
2	Delivery	3.417	
3	Performance history	2.998	
4	Warranties and claim policies	2.849	
5	Production facilities and capacity	2.775	Considerable importance
6	Price	2.758	
7	Technical capability	2.545	
8	Financial position	2.514	
9	Procedural compliance	2.488	
10	Communication system	2.426	
11	Reputation and position in industry	2.412	
12	Desire for business	2.256	
13	Management and organization	2.216	
14	Operating controls	2.211	
15	Repair service	2.187	Average importance
16	Attitude	2.120	
17	Impression	2.054	
18	Packaging ability	2.009	
19	Labor relations record	2.003	
20	Geographical location	1.872	
21	Amount of past business	1.597	
22	Training aids	1.537	
23	Reciprocal arrangements	0.610	Slight importance

best supplier. Linn et al. [7] presented CPC chart for supplier selection problem (Figure 2). The CPC chart integrates the process capability and price information of multiple suppliers and presents them in a single chart. It provides a simple and useful method to consider quality and price simultaneously in the supplier selection process. However, this chart uses C_{pk} index, while we know that this process capability index is not a proper index to evaluate quality.

In order to evaluate supplier, Linn et al. [7] considered some factors such as quality, on-time delivery, price, and service, but the strategy is not clear if the purchase department wants to buy many kinds of materials from one supplier. It means in this case that some other factors must be considered, because purchase department always wants to buy all materials from one or two suppliers, so the shipment cost also will be decreased. Similarly Zhu [39] established a suppliers capability and price information chart (SCPIC) (Figure 3) focused on the case where the specification limits are symmetric about the target for evaluating supplier performance which applies the process incapability index to measure supplier quality performance.

Assessing quality as one of the main factors for supplier selecting is so important. In order to assess the quality there are many factors and methods that can be used. There are a lot of literatures accumulated on the subject of vendor evaluation and selection models, and in order to evaluate quality, most of these models have used rate of rejects [20, 29, 30], while rate of rejects cannot present the quality appropriately. In a recent study, Lee [31] used yield rate in order to evaluate the quality.

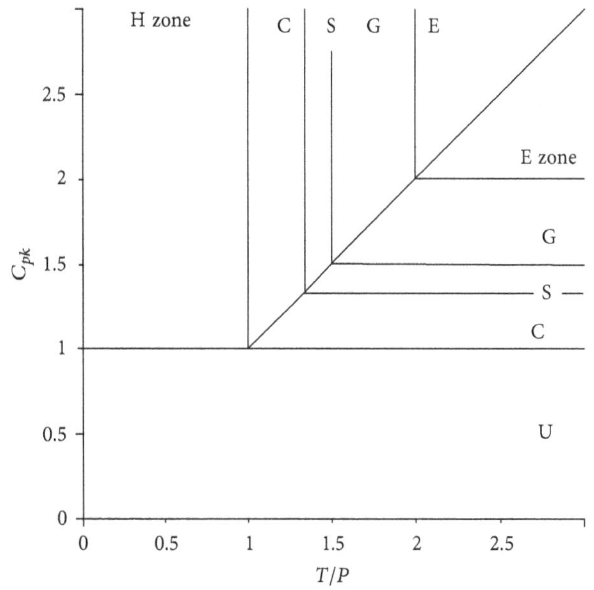

FIGURE 2: CPC chart [7].

There are few studies about the usage of loss functions in order to selection the suppliers. In these studies, scholars just have used the concepts of loss function in order to weigh the criteria. Pi and Low [17] used Taguchi loss [42] and assumed 3% defective products as the standard rate of rejects, and the quality could be calculated with Taguchi loss function and

TABLE 2: Various selection criteria that have emerged in the literature.

Selection criteria	A	B	C	D	E	F	G	H	I	J	K	L	M	N	O	P	Q	R	S	T	U
Price (cost)	√	√	√	√	√	√	√		√	√	√	√	√		√	√			√	√	√
Quality	√	√	√	√		√	√		√		√	√	√		√	√	√	√	√		√
Delivery	√	√	√	√		√	√		√		√	√	√	√	√		√	√	√	√	√
Warranties and claims	√		√																		
After sales service	√		√	√			√					√			√				√		
Technical support			√		√	√															
Training aids	√		√				√														
Attitude	√					√	√														
Performance history	√						√														
Financial position	√		√				√											√			
Geographical location	√	√		√			√														
Management and organization	√			√			√						√							√	
Labor relations	√						√							√							
Communication system	√						√										√				
Response to customer request		√				√							√								
E-commerce capability								√	√	√											√
JIT capability						√		√													
Technical capability	√	√					√	√								√	√				√
Production facilities and capacity	√						√							√					√		
Packaging ability	√						√														
Operational controls	√						√														
Ease of use				√	√																
Maintainability				√	√																
Amount of past business	√	√	√				√														
Reputation and position in industry	√	√	√	√			√						√								√
Reciprocal arrangements	√	√		√			√														
Impression	√		√		√		√														
Environmentally friendly products								√						√				√			√
Product appearance									√												
Catalog technology									√												
Dependability[1]											√	√									
Flexibility[2]														√							√
Payment terms														√							
Productivity															√	√	√				
Applicable of conceptual Manufacturing															√						
Manufacturing challenges															√						
Driving Power															√						
To match the lead times																	√				
Personnel capability																	√				
To be solution oriented																	√				
Global factors																			√		
Environmental risk																					√

[1]Dependability refers to the ability to supply items as promised. Dependability sometimes includes delivery due date, delivery accuracy, and delivery completeness.
[2]Ability to change or revise in the production operations.
A: Dickson [6]; B: Wind et al. [10]; C: Lehmann and haughnessy [11]; D: Perreault and Russ [12]; E: Abratt [13]; F: Billesbach et al. [14]; G: Weber et al. [5]; I: Min and Galle [15]; J: Stavropolous [16]; K: Pi and Low [17]; L: Pi and Low [18], M: Teeravaraprug [19]; N: Sanayei et al. [20]; O: Parthiban et al. [21]; P: Peng [22]; Q: Bilişik et al. [23]; R: Tektas and Aytekin [24]; S: Li [25]; T: Betül et al. [26]; U: Mehralian et al. [27].

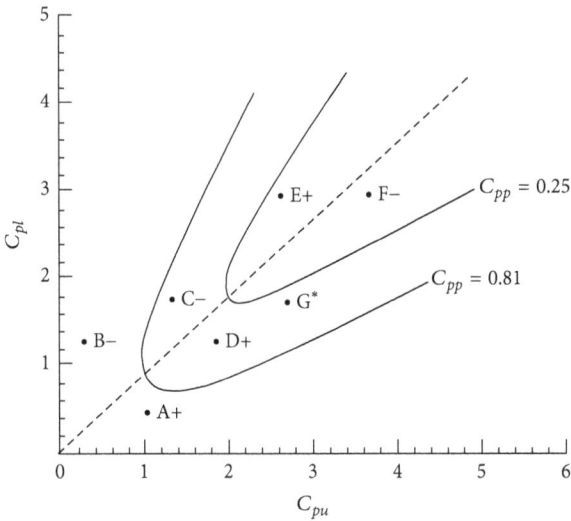

FIGURE 3: Suppliers' capability and price information chart [39].

this target value. In another study, they [18] continued their research and proposed an AHP method in order to select the final supplier.

Teeravaraprug [19] proposed a new model for outsourcing and vendor selection based on Taguchi loss function. In this interesting study loss, the function was used in order to calculate the quality rate, but in its methodology Taguchi loss function was used for just weighting the different rates, and again in order to calculate the quality, just rate of rejects assumed. Summary of the main studies in this area is shown in Table 3.

Regarding to fierce worldwide competition, quality turned to be one of the main factorsdwhich directly affects the supplier selection decision. As stated in the literature review, quality can be assessed by methods categorized in two different groups. The first group consists of qualitative methods such as continuous improvement programs, quality of customer and support services, certifications, technical and design level, capability of handling abnormal quality, and ease of repair. The second group consist of quantitative methods including reliability, rate of rejects, yield rate, process capability indices, and loss functions.

Since qualitative methods can evaluate just one aspect of an organization but cannot evaluate the whole production process, they are not suitable methods for assessing the quality of a process. For example, one product of an organization may have a bad quality, but the recent top manager of the organization has taken the basic decision to implement some continuous improvement programs. Continuous improvement programs help to enhance the quality, but they cannot guarantee the quality of the current products. There are some quality certifications, such as ISO 9000, focused on the quality management in organizations. These quality systems can be chosen in order to assess the quality of an organization, but they cannot be an appropriate representative of the quality of the products. For example, Reimann and Hertz [44] stated that ISO registration does not necessarily mean the following words: (1) good or improving product quality, (2) satisfaction

of customer's needs, (3) comparable levels of product quality among registered companies, (4) better quality than nonregistered companies, and (5) good or improving productivity, responsiveness, competitiveness, or workforce development. Moreover, Juran [41] acknowledged that a comprehensive quality system defined by ISO standards has a degree of merit, but the certification alone will not enable companies to attain world-class quality.

Quality of customer and support services is another criterion in order to evaluate the quality of suppliers. This criterion is a sign of implementing customer-based systems in organizations, but customers just with considering this factor cannot assure the quality of the products. Some organizations may show off their responsibility and hide their weak points with a flashy customer and support service.

Technical and design level is another criteria, which is used for assessing the quality in some studies. Technical and design level helps organizations to produce products with better quality, but for multiproduction organization, it is hard to assess whether high technical and design level has an effect on the quality of a specific product or not. Also, in assembly organizations, the technical and design level cannot guarantee the product quality. Since nowadays customers often tend to replace the defective products or defective parts, the product, or one part of it, repairing has missed its position in the term of quality. So the item "ease of repair" is not an appropriate criterion for quality assessment.

Capability of handling abnormal quality products is another qualitative method, which is important but cannot guarantee the quality of products. Nowadays, some new approaches such as zero defects which rely on the zero rates of rejects are emerging, so organizations, instead of focusing on problems of abnormal quality products such as handling of them, focus on low rate of rejects. On the other hand, a supplier can have a bad quality of products, but a good service for handling abnormal quality products.

The second group of quality assessment methods consists of quantitative methods. The quantitative methods also have some weak points. For example, some studies evaluated the reliability of products instead of assessing the quality. Reliability is defined as the probability that a product will successfully perform without any failure, under specified environmental conditions, for a specified period of time [45]. In this definition, there is not a distinction between a product, which its quality specifications are really close to the target, and an accepted product, which its quality performance is far from the target. Ramakrishnan et al. [43] presented an example of solder bump failures in assembly process (Figure 4). They showed a high process capability index, but actually the rate of rejects is high and consequently the reliability is low.

The example of Ramakrishnan et al. [43] showed that process capability indices also have some weak points in order to evaluate the quality of products. Moreover, Perakis and Xekalaki [46] illustrated that process capability indices such as C_p, C_{pk}, and C_{pm} do not have a direct relationship with the conformance proportion of the process.

Rate of rejects is one of the main methods for assessing the quality. This method selects the supplier with least rate

TABLE 3: The literature review about the methods of assessing quality for supplier selection.

ID	Article	Year	Rate of rejects	Continuous improvement programs	Quality of customer and support services	Certifications	Percentage on time shipment	Technical and design level	Ease of repair	Reliability	Capability of handling abnormal quality	Yield rate	Process capability indices	Loss functions	Customer satisfaction	Culture compatibility	Enterprise credit
1	Tseng and Wu [35]	1991											✓				
2	Chou [36]	1994											✓				
3	Huang and Lee	1995											✓				
4	Pearn et al. [37]	2004											✓				
5	Teng and Jaramillo [32]	2005		✓	✓	✓	✓										
6	Liu and Hai [40]	2005	✓	✓													
7	Pi and Low [17]	2005												✓			
8	Pearn and Kotz [38]	2006											✓				
9	Linn et al. [7]	2006											✓				
10	Wang and Guo [29]	2007	✓														
11	Hou and Su [33]	2007						✓	✓	✓							
12	Xia and Wu [34]	2007	✓					✓	✓	✓							
13	Zhu [39]	2007											✓				
14	Sanayei et al. [20]	2008	✓														
15	Teeravaraprug [19]	2008												✓			
16	Kokangul and Susuz [30]	2009	✓														
17	Lee [31]	2009										✓					
18	Li [25]	2010												✓			
19	Tektas and Aytekin [24]	2011	✓	✓		✓											
20	Parthiban et al. [21]	2012	✓										✓				
21	Peng [22]	2012													✓		
22	Bilişik et al. [23]	2012	✓													✓	✓
23	Mehralian et al. [41]	2012	✓			✓									✓		

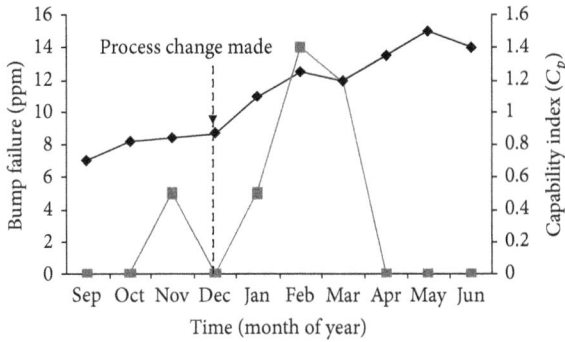

FIGURE 4: An example about relation of process capability index and bump failures (ppm) [43].

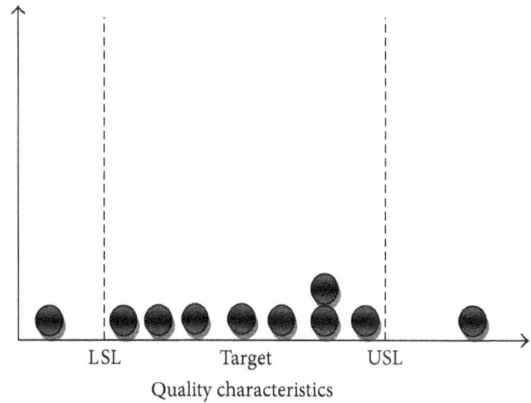

FIGURE 6: Samples from process B.

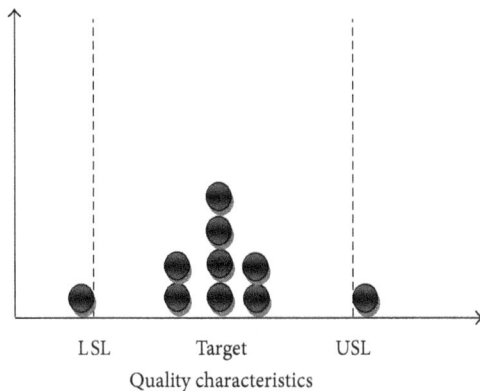

FIGURE 5: Samples from process A.

of rejects. This is a good quantitative method for assessing the quality, but it cannot be useful in case of similar rate of rejects. For instance, suppose that two samples were selected from suppliers A and B. Each Sample consisted of 10 parts. Suppose that for both samples, eight quality characteristics fall within specification limits (LSL, USL), as stated in Figure 5. Regarding to rate of rejects method, there is no any differences between them because rate of rejects of supplier A = rate of reject of supplier B = 0.2, but Figures 5 and 6 show that there is a significant difference between them and process A has a better quality.

Yield rate is an opposite concept to rate of rejects and behaves the same. Similarly based on Figures 5 and 6, it is concluded that yield rate of both processes is the same (80%), but there is a significant difference between them.

On time shipment percentage similarly presents a quality factor, which is indirectly related to the quality of products. On time shipment deals with delivery, and it does not take into account the quality. Finally, in some studies, scholars in order to assess the quality have used the loss functions. Since loss functions take into account all samples and also have a significant relation with loss, they seem to be appropriate and reliable functions for evaluating the quality of suppliers. However, as stated in the literature review, the studies which are based on loss functions also have some weak points. For

example, Pi and Low [17] who used Taguchi loss function actually used the rate of rejects and assumed 3% defective products as the standard rate of rejects. Moreover, in the methodology of Teeravaraprug [19], Taguchi loss functions were used for just weighting the different rates, and again in order to calculate the quality, just rate of rejects, assumed.

4. Conclusion and Suggestions for Future Research

There are many supplier selection methods based on different criteria that were employed for solving the supplier selection problems. This paper presented a review of decision criteria reported in the literature for supporting the supplier selection process. The review was based on an extensive search in the academic literature. Therefore, all different criteria related to supplier selection were reviewed.

Quality is the most important criteria for supplier selection [6]. The methods for assessing the quality can be divided to two main categories: qualitative methods and quantitative methods. In this paper it was shown that all qualitative methods and quantitative methods can evaluate just one aspect of an organization, but cannot evaluate the whole production process; they are not suitable methods for assessing the quality of a process. Therefore, these methods have some weak points and are not appropriate tools to assess the quality. In fact, the objective functions of these methods are not realistic objects.

The suggestion of this paper for quality evaluation is the use of loss functions. Among the numerous methods that have been proposed for assessing the supplier, loss functions such as Taguchi loss function without any range are considered one of the most effective techniques for identifying quality parts. Quality loss functions are more reliable and precise functions in order to assess the quality. There are few studies such as Teeravaraprug [19] and Pi and Low [17] that used loss functions in order to evaluate the quality, but actually they have used some definitions such as weighting or ranges. This study proposes the use of loss functions without any range and weight in order to evaluate the quality of the suppliers.

References

[1] R. M. Monczka, R. Trent, and R. Handfield, *Purchasing and Supply Chain Management*, International Thomson publishing, 1998.

[2] C. P. Roa and G. E. Kiser, "Educational buyer's perception of vendor attributes," *Journal of Purchasing Material Management*, vol. 16, pp. 25–30, 1980.

[3] L. Ellram, "The supplier selection decision in strategic partnerships," *Journal of Purchasing Material Management*, vol. 26, no. 4, pp. 8–14, 1990.

[4] C. L. Stamm and D. Y. Golhar, "JIT purchasing attribute classification and literature review," *Production Planning & Control*, vol. 4, no. 3, pp. 273–282, 1993.

[5] C. A. Weber, J. R. Current, and W. C. Benton, "Vendor selection criteria and methods," *European Journal of Operational Research*, vol. 50, no. 1, pp. 2–18, 1991.

[6] G. W. Dickson, "An analysis of vendor selection: systems and decisions," *Journal of Purchasing*, vol. 1, pp. 5–17, 1966.

[7] R. J. Linn, F. Tsung, and L. W. C. Ellis, "Supplier selection based on process capability and price analysis," *Quality Engineering*, vol. 18, no. 2, pp. 123–129, 2006.

[8] S. H. Ha and R. Krishnan, "A hybrid approach to supplier selection for the maintenance of a competitive supply chain," *Expert Systems with Applications*, vol. 34, no. 2, pp. 1303–1311, 2008.

[9] N. Aissaoui, M. Haouari, and E. Hassini, "Supplier selection and order lot sizing modeling: a review," *Computers & Operations Research*, vol. 34, no. 12, pp. 3516–3540, 2007.

[10] Y. Wind, P. E. Green, and P. J. Robinson, "The determinants of vendor selection: the evaluation function approach," *Journal of Purchasing*, vol. 4, no. 3, pp. 29–42, 1968.

[11] D. R. Lehmann and J. O'Shaughnessy, "Difference in attribute importance for different industrial products," *Journal of Marketing*, vol. 38, no. 2, pp. 36–42, 1974.

[12] W. D. Perreault and F. A. Russ, "Physical distribution service in industrial purchase decisions," *Journal of Marketing*, vol. 40, no. 1, pp. 3–10, 1976.

[13] R. Abratt, "Industrial buying in high-tech markets," *Industrial Marketing Management*, vol. 15, no. 4, pp. 293–298, 1986.

[14] T. J. Billesbach, A. Harrison, and S. Croom-Morgan, "Supplier performance measures and practices in JIT companies in the US and UK," *International Journal of Purchasing and Materials Management*, vol. 21, no. 4, pp. 24–28, 1991.

[15] H. Min and W. P. Galle, "Electronic commerce usage in business-to-business purchasing," *International Journal of Operations & Production Management*, vol. 19, no. 9, pp. 909–921, 1999.

[16] N. Stavropolous, "Suppliers in the new economy," *Telecommunication Journal of Australia*, vol. 50, no. 4, pp. 27–29, 2000.

[17] W. N. Pi and C. Low, "Supplier evaluation and selection using Taguchi loss functions," *International Journal of Advanced Manufacturing Technology*, vol. 26, no. 1-2, pp. 155–160, 2005.

[18] W. N. Pi and C. Low, "Supplier evaluation and selection via Taguchi loss functions and an AHP," *International Journal of Advanced Manufacturing Technology*, vol. 27, no. 5-6, pp. 625–630, 2006.

[19] J. Teeravaraprug, "Outsourcing and vendor selection model based on Taguchi loss function," *Songklanakarin Journal of Science and Technology*, vol. 30, no. 4, pp. 523–530, 2008.

[20] A. Sanayei, S. Farid Mousavi, M. R. Abdi, and A. Mohaghar, "An integrated group decision-making process for supplier selection and order allocation using multi-attribute utility theory and linear programming," *Journal of the Franklin Institute*, vol. 345, no. 7, pp. 731–747, 2008.

[21] P. Parthiban, H. A. Zubar, and C. P. Garge, "A multi criteria decision making approach for suppliers selection," *Procedia Engineering*, vol. 38, pp. 2312–2328, 2012.

[22] J. Peng, "Selection of logistics outsourcing service suppliers based on AHP," *Energy Procedia*, vol. 17, pp. 595–601, 2012.

[23] M. E. Bilişik, N. Çağlar, and O. N. A. Bilişik, "A comparative performance analyze model and supplier positioning in performance maps for supplier selection and evaluation," *Procedia—Social and Behavioral Sciences*, vol. 58, pp. 1434–1442, 2012.

[24] A. Tektas and A. Aytekin, "Supplier selection in the international environment: a comparative case of a Turkish and an Australian company," *IBIMA Business Review*, vol. 2011, Article ID 598845, 14 pages, 2011.

[25] C.-N. Li, "Supplier selection project using an integrated Delphi, AHP and Taguchi loss function," *ProbStat Forum*, vol. 3, pp. 118–134, 2010.

[26] O. Betül, H. Başlıgil, and N. Şahin, "Supplier selection using analytic hierarchy process: an application from Turkey," in *Proceedings of the World Congress on Engineering (WCE '11)*, vol. 2, London, UK, July 2011.

[27] G. Mehralian, A. Rajabzadeh Ghatari, H. Morakabzti, and H. Vatanpour, "Developing a suitable model for supplier selection based on supply chain risks: an empirical study from Iranian pharmaceutical companies," *Iranian Journal of Pharmaceutical Research*, vol. 11, no. 1, pp. 209–219, 2012.

[28] I. A. Holjevac, "Business ethics in tourism—as a dimension of TQM," *Total Quality Management and Business Excellence*, vol. 19, no. 10, pp. 1029–1041, 2008.

[29] Q. E. Wang and X. M. Guo, "A study of supplier selection and quality strategy based on quality costs theory," in *International Conference on Management Science and Engineering (ICMSE '07)*, pp. 838–843, August 2007.

[30] A. Kokangul and Z. Susuz, "Integrated analytical hierarch process and mathematical programming to supplier selection problem with quantity discount," *Applied Mathematical Modelling*, vol. 33, no. 3, pp. 1417–1429, 2009.

[31] A. H. I. Lee, "A fuzzy supplier selection model with the consideration of benefits, opportunities, costs and risks," *Expert Systems with Applications*, vol. 36, no. 2, pp. 2879–2893, 2009.

[32] S. G. Teng and H. Jaramillo, "A model for evaluation and selection of suppliers in global textile and apparel supply chains," *International Journal of Physical Distribution and Logistics Management*, vol. 35, no. 7, pp. 503–523, 2005.

[33] J. Hou and D. Su, "EJB-MVC oriented supplier selection system for mass customization," *Journal of Manufacturing Technology Management*, vol. 18, no. 1, pp. 54–71, 2007.

[34] W. Xia and Z. Wu, "Supplier selection with multiple criteria in volume discount environments," *Omega*, vol. 35, no. 5, pp. 494–504, 2007.

[35] S. T. Tseng and T. Y. Wu, "Selecting the best manufacturing process," *Journal of Quality Technology*, vol. 23, pp. 53–62, 1991.

[36] Y. M. Chou, "Selecting a better supplier by testing process capability indices," *Quality Engineering*, vol. 6, no. 3, pp. 427–438, 1994.

[37] W. L. Pearn, C. W. Wu, and H. C. Lin, "Procedure for supplier selection based on Cpm applied to super twisted nematic liquid crystal display processes," *International Journal of Production Research*, vol. 42, no. 13, pp. 2719–2734, 2004.

[38] W. L. Pearn and S. Kotz, *Encyclopedia and Handbook of Process Capability Indices, a Comprehensive Exposition of Quality Control Measures*, George Washington University, Washington, DC, USA, 2006.

[39] H. M. Zhu, "Supplier selection using process capability and price information chart," in *International Conference on Wireless Communications, Networking and Mobile Computing (WiCOM '07)*, pp. 4850–4853, September 2007.

[40] F. H. F. Liu and H. L. Hai, "The voting analytic hierarchy process method for selecting supplier," *International Journal of Production Economics*, vol. 97, no. 3, pp. 308–317, 2005.

[41] J. M. Juran, *A History of Managing for Quality: The Evolution, Trends, and Future Directions of Managing for Quality*, ASQC Quality Press, Milwaukee, Wis, USA, 1995.

[42] G. Taguchi, *introduction to Quality Engineering: Designing Quality in to Products and Processes*, Asian Productivity Organization, Tokyo, Japan, 1986.

[43] B. Ramakrishnan, P. Sandborn, and M. Pecht, "Process capability indices and product reliability," *Microelectronics Reliability*, vol. 41, no. 12, pp. 2067–2070, 2001.

[44] C. W. Reimann and H. S. Hertz, "The Baldrige Award and ISO 9000 registration compared," *Journal for Quality and Participation*, vol. 19, pp. 12–19, 1996.

[45] A. Adamyan and D. He, "Analysis of sequential failures for assessment of reliability and safety of manufacturing systems," *Reliability Engineering and System Safety*, vol. 76, no. 3, pp. 227–236, 2002.

[46] M. Perakis and E. Xekalaki, "A process capability index that is based on the proportion of conformance," *Journal of Statistical Computation and Simulation*, vol. 72, no. 9, pp. 707–718, 2002.

19

Uniqueness of Maximum Likelihood Estimators for a Backup System in a Condition-Based Maintenance

Qihong Duan, Ying Wei, and Xiang Chen

Department of Statistics, School of Mathematics and Statistics, Xi'an Jiaotong University, Shaanxi, Xi'an 710049, China

Correspondence should be addressed to Xiang Chen, xjusgr@126.com

Academic Editor: Tadashi Dohi

A parameter estimation problem for a backup system in a condition-based maintenance is considered. We model a backup system by a hidden, three-state continuous time Markov process. Data are obtained through condition monitoring at discrete time points. Maximum likelihood estimates of the model parameters are obtained using the EM algorithm. We establish conditions under which there is no more than one limitation in the parameter space for any sequence derived by the EM algorithm.

1. Introduction

Suppose a backup system is represented by a continuous time homogeneous Markov chain $X = \{X_t : t \geq 0\}$ with a state space $S = \{0, 1, 2\}$. States 0, 1, and 2 are the healthy state, unhealthy state, and the failure state, respectively. Assume that the system is in a healthy state at time 0, and the transition rate matrix is given by

$$\Lambda = \begin{pmatrix} -1/\theta_1, & 1/\theta_1, & 0 \\ 0, & -1/\theta_2, & 1/\theta_2 \\ 0, & 0, & 0 \end{pmatrix}, \quad (1)$$

where $\theta_i \in (\alpha, +\infty)$ for $i = 1, 2$ are unknown. Here $\alpha > 0$ is a known extreme edge of the parameters. Suppose the system is observed at time points $0, \Delta, 2\Delta, \ldots$, where $[k\Delta, (k + 1)\Delta]$ is a benchmark interval. While the system is failed at an inspection, a new system replaces it. Let two processes $Y = \{Y(k) : k = 1, \ldots, n\}$, and let $R = \{R(k) : k = 1, \ldots, n\}$ be a record of the system. Here $R(k) = 1$ if the system is failed during $(k - 1, k)$, and $R(k) = 0$ otherwise. And $Y(k) = X_{k\Delta}$. As a path of X is a stepped right-continuous function, $Y(k) = 0$ when a replacement occurs at time $k\Delta$. Moreover, we set $Y(0) = 0$ for convenience. The process R represents the replacement of the system, and the process Y represents observable information of the system collected through condition monitoring.

The maximum likelihood estimates (MLE) of the model parameters for such models have been studied by [1, 2]. As the stochastic processes, Y, R are not Markov processes and the sample path of X is not observable, the likelihood function of incomplete data (Y, R) is complex. Hence, it is difficult to obtain directly the MLE $\hat{\theta} = \arg\max_{\theta \in \Theta} \ln L(\theta \mid Y, R)$, where $\theta = (\theta_1, \theta_2)$ and $\Theta = \{(x, y) : x, y > \beta\}$. Here $\beta > 0$ is a prearranged constant. Both [1, 2] suggest the EM algorithm (see e.g., [3, 4]). Let $\theta(0)$ be initial values of the unknown parameters. The EM algorithm works as follows.

The E step. For $n > 0$, compute the following pseudo-likelihood function:

$$Q(\theta' \mid \theta(n)) = E_{\theta(n)}(\ln L(\theta' \mid D) \mid Y, R). \quad (2)$$

Here D is the complete data set of the process X. The forms of the complete data set may be different for different purpose. For example, the forms are different in [1, 2].

The M step. Choose $\theta(n + 1) = \widetilde{M}(\theta(n))$. Here

$$\widetilde{M}(\theta) = \arg\sup_{\theta' \in \Theta} Q(\theta' \mid \theta). \quad (3)$$

The E and M steps are repeated. According to the theory of EM algorithms (Theorem 1 in [3]), $L(\theta(n + 1) \mid Y, R) \geq L(\theta(n) \mid Y, R)$ for any given initial value $\theta(0)$ and $n = 0, 1, \ldots$. It is clear that if an MLE $\hat{\theta}$ in Θ is one of these fixed-points when it exists.

In this paper, we consider the uniqueness of the MLE $\hat{\theta}$. As the likelihood function $\ln L(\theta \mid Y, R)$ of incomplete data is complex, we do not follow the classical method by which the uniqueness of a MLE is demonstrated by establishing the global concavity of the log-likelihood function (see e.g., [5–7].) Alternatively, we investigate conditions under which the operator $\widetilde{M}$ is a contraction. The conditions ensure that the MLE is unique if it exists. Moreover, the conditions implies that there is not more than one limitations in Θ for different sequences derived by the EM algorithm. For the complete data set we present in the next section, we have the following main theorem of this paper.

Theorem 1. *There is not more than one fixed-point of the operator $\widetilde{M}$ in Θ provided that $\triangle < (\sqrt{2}/2)\alpha$ and the record $\{Y, R\}$ has at least one replacement.*

2. Complete Data Set

To establish the expression of operators Q, we present our construction of the path of X and the complete data set D of the process X. Suppose that random variables T_k^i, $i = 1, 2$, $k = 0, 1, \ldots$ are independent, T_k^i has an exponential distribution, and $ET_k^i = \theta_i/\Delta$. As every state of X has an exponential duration distribution, we may construct a path of X through the approach introduced by Theorem 5.4 in [8]. The path of X restricted on $t \in [0, \Delta)$ has the following form:

$$X_t 1_{0 \le t < \Delta} = j \text{ for } \Delta \sum_{i=1}^{j} T_1^i \le t < \Delta \left(\sum_{i=1}^{j+1} T_1^i \wedge 1 \right), \quad j = 0, 1, \tag{4}$$

and $X_t 1_{0 \le t \le \Delta} = 2$ for $\Delta(T_1^1 + T_1^2) \le t < \Delta$. If $m = \lim_{t \uparrow k\Delta} X_t \neq 2$ for $k \ge 1$, we can construct the following path of X restricted on $t \in [k\Delta, (k+1))$. Consider

$$X_t 1_{k \le t < k+1}$$
$$= \begin{cases} j, & \Delta \sum_{i=m}^{j} T_k^i \le t - k\Delta < \Delta \left(\sum_{i=m}^{j+1} T_k^i \wedge 1 \right), \quad j \neq 2, \\ 2, & \Delta \sum_{i=m}^{2} T_k^i \le t - k\Delta < \Delta. \end{cases} \tag{5}$$

If $\lim_{t \uparrow k\Delta} X_t = 2$, that is, the system is failed during $(k\Delta, (k+1)\Delta)$, a new system replaces it at time $(k+1)\Delta$ and a path of X restricted on $t \in [k\Delta, (k+1)\Delta)$ is presented by setting $m = 1$ in (5). In this paper, the complete data set $D = \{T_k^i : i = 1, 2, k = 0, 1, \ldots, n-1\}$.

Our choice of the complete data set D ensures a simple form of the operator Q. The log-likelihood function has the following form:

$$\ln L(\theta' \mid \widetilde{D})$$
$$= 2n \ln \Delta - n(\ln \theta_1' + \ln \theta_2') - \Delta \sum_{k=0}^{n-1} \left(\frac{T_k^1}{\theta_1'} + \frac{T_k^2}{\theta_2'} \right). \tag{6}$$

TABLE 1: The number for values of $(Y(k-1), Y(k), R(k))$.

Values	$(0,0,0)$	$(0,1,0)$	$(1,1,0)$	$(0,0,1)$	$(1,0,1)$
Number	n_1	n_2	n_3	n_4	n_5

And the Markovian property of the process X implies that

$$E_\theta \left(T_k^i \mid Y \right) = E_\theta \left(T_k^i \mid Y(k-1), Y(k), R(k) \right), \tag{7}$$

where $Y(0) = 0$.

In Table 1, we denote by n_i the number of different values of the triple $(Y(k-1), Y(k), R(k))$. For example, n_1 is the number of triple $(Y(k-1), Y(k), R(k)) = (0,0,0)$. It follows from (6) that

$$Q(\theta' \mid \theta)$$
$$= 2n \ln \Delta - n(\ln \theta_1' + \ln \theta_2') - \Delta \sum_{i=1}^{5} n_i \left(\frac{M_i^1}{\theta_1'} + \frac{M_i^2}{\theta_2'} \right). \tag{8}$$

Here the following forms of M_i^m for $m = 1, 2$ follows from (7) and Table 1. Consider

$$M_1^m = E_\theta(T_1^m \mid T_1^1 \ge 1), \tag{9}$$
$$M_2^m = E_\theta(T_1^m \mid T_1^2 \ge 1), \tag{10}$$
$$M_3^m = E_\theta(T_1^m \mid T_1^2 < 1), \tag{11}$$
$$M_4^m = E_\theta(T_1^m \mid T_1^1 < 1, T_1^1 + T_1^2 \ge 1), \tag{12}$$
$$M_5^m = E_\theta(T_1^m \mid T_1^1 + T_1^2 < 1). \tag{13}$$

For any given θ, it is obvious that the function $\widetilde{Q}(\lambda_1, \lambda_2) = Q(1/\lambda_1, 1/\lambda_2 \mid \theta)$ is a concave function for $0 < \lambda_1, \lambda_2 < \infty$. Hence there is a unique vector $\theta' = M(\theta)$ satisfying

$$\frac{\partial Q(\theta' \mid \theta)}{\partial \theta_i'} = 0, \quad i = 1, 2. \tag{14}$$

Moreover, it follows from the definition (3) that if $\widetilde{M}(\theta) \in \Theta$, $M(\theta) = \widetilde{M}(\theta)$. Therefore, every fixed-point of $\widetilde{M}$ in Θ is a fixed-point of M and vice versa. In this paper, we will prove Theorem 1 through studying the number of fixed-points of M in Θ. Here we present the from of $M(\theta) = (M_1(\theta), M_2(\theta))$ derived by (14). Consider

$$M_m(\theta) = \frac{\Delta}{n} \sum_{i=1}^{5} n_i M_i^m, \quad m = 1, 2. \tag{15}$$

3. Two Lemmas

Lemma 2. *Consider the following:*

$$\left| \frac{\partial M_4^m}{\partial \theta_i} \right| \le \frac{\Delta}{\theta_i^2}, \quad i = 1, 2, m = 1, 2. \tag{16}$$

Proof. Writing $\sigma = \{(x_1, x_2) \mid x_1, x_2 \geq 0, x_1 + x_2 \leq 1\}$ and

$$V = \iint_\sigma \frac{\Delta^2 x_m}{\theta_1 \theta_2} e^{(-\Delta/\theta_1)x_1 - (\Delta/\theta_2)x_2} dx_1 dx_2,$$

$$D = \iint_\sigma \frac{\Delta^2}{\theta_1 \theta_2} e^{(-\Delta^2/\theta_1)x_1 - (\Delta/\theta_2)x_2} dx_1 dx_2, \qquad (17)$$

we have $M_4^m = V/D$. Hence,

$$\frac{\partial M_4^m}{\partial \theta_i} = M_4^m \left(V^{-1} \frac{\partial V}{\partial \theta_i} - D^{-1} \frac{\partial D}{\partial \theta_i} \right). \qquad (18)$$

As $0 \leq x_m \leq 1$ on σ, we have that $0 \leq V \leq D$ and $0 \leq M_4^m \leq 1$. Therefore,

$$\left| \frac{\partial M_4^m}{\partial \theta_i} \right| \leq \left| \left(V^{-1} \frac{\partial V}{\partial \theta_i} - D^{-1} \frac{\partial D}{\partial \theta_i} \right) \right|. \qquad (19)$$

Moreover, we have

$$\frac{\partial V}{\partial \theta_i} = -\frac{V}{\theta_i} + \frac{\Delta^3}{\theta_i^2} \iint_\sigma \frac{x_m x_i}{\theta_1 \theta_2} e^{-(\Delta/\theta_1)x_1 - (\Delta/\theta_2)x_2} dx_1 dx_2. \qquad (20)$$

As $0 \leq x_i \leq 1$ on the region σ, it follows from the definition to V that there is $0 < \eta_1 < 1$ such that

$$\frac{\partial V}{\partial \theta_i} = -\frac{V}{\theta_i} + \eta_1 \frac{\Delta V}{\theta_i^2}. \qquad (21)$$

Similarly, there is $\eta_2 \in (0, 1)$ such that

$$\frac{\partial D}{\partial \theta_i} = -\frac{D}{\theta_i} + \eta_2 \frac{\Delta D}{\theta_i^2}. \qquad (22)$$

The result follows from (19), (20), and (22). □

Lemma 3. *Let a function $A : R \longrightarrow R$ be $A(0) = 1/2$ and $A(x) = 1/(1 - e^x) + 1/x$ for $x \neq 0$. We have $0 \leq A'(x) \leq -1/12$ for any $x \in R$.*

Proof. By a routine analysis, we may obtain that $A'(0) = -1/12$ and for $x \neq 0$,

$$A'(x) = \left(e^{-x/2} - e^{x/2} \right)^{-2} - x^{-2}. \qquad (23)$$

Moreover, we may obtain that for $x > 0$,

$$A''(x) = A_2(x) x^{-3} (e^x - 1)^{-3}, \qquad (24)$$

where $A_2(x) = (e^x - 1)^3 - x^3 e^x (1 + e^x)$. We have

$$A_2'(x) = -e^x A_3(x), \qquad (25)$$

where $A_3(x) = 3x^2(1 + e^x) + x^3(1 + 2e^x) - 6(e^x - 1)^2$. A routine analysis may confirm that $A_3^{(4)}(x) = e^x A_4(x)$, where $A_4(x) = 96 + 96x + 27x^2 + 2x^3 - 96e^x$. Then it follows from $e^x \leq 1 + x + x^2/2 + x^3/6$ that $A_4(x) \leq 0$. Moreover, as $A_3(0) = A_3'(0) = A_3''(0) = A_3^{(3)}(0) = 0$, we have that for any $x > 0$, there is $\xi \in [0, x]$ such that

$$A_3(x) = \frac{1}{24} A_3^{(4)}(\xi) x^4 = \frac{1}{24} A_4(\xi) e^\xi x^4 \leq 0. \qquad (26)$$

Then it follows from $A_2(0) = 0$ and (25) that $A_2(x) \geq 0$. Hence, it follows from (24) that $A''(x) \geq 0$ for any $x > 0$. Therefore, for any $x > 0$

$$-\frac{1}{12} = A'(0) \leq A'(x) \leq \lim_{x \to +\infty} A'(x) = 0. \qquad (27)$$

The result follows from the above formula, and the fact that $A'(x)$ is an even function. □

4. Proof of the Main Theorem

We may derive that $M(\theta)$ is a contraction by investigating the Jacobian matrix $\partial(M_1, M_2)/\partial(\theta_1, \theta_2)$.

By a routine analysis, it follows from (12) that

$$M_2^1 = 1 + A\left(\frac{\Delta}{\theta_1} - \frac{\Delta}{\theta_2} \right), \qquad M_2^2 = 1 + \frac{\theta_2}{\Delta} + A\left(\frac{\Delta}{\theta_1} - \frac{\Delta}{\theta_2} \right). \qquad (28)$$

Hence, it follows from Lemma 3 that

$$\left| \frac{\partial M_2^1}{\partial \theta_2} \right| = \frac{\Delta}{\theta_2^2} \left| A'\left(\frac{\Delta}{\theta_1} - \frac{\Delta}{\theta_2} \right) \right| \leq \frac{\Delta}{12\theta_2^2}. \qquad (29)$$

Similarly, we have

$$\left| \frac{\partial M_2^1}{\partial \theta_1} \right| \leq \frac{\Delta}{12\theta_1^2}. \qquad (30)$$

As $\Delta < \alpha < \theta_2$, it follows from (28) and Lemma 3 that

$$\left| \frac{\partial M_2^2}{\partial \theta_1} \right| + \left| \frac{\partial M_2^2}{\partial \theta_2} \right| = \frac{1}{\Delta}. \qquad (31)$$

It follows from (11) that $M_5^2 = A(\theta_2/\Delta)$, and $M_5^1 = \theta_1/\Delta$. Therefore,

$$\frac{\partial M_5^2}{\partial \theta_1} = 0, \qquad \frac{\partial M_5^1}{\partial \theta_1} = 0, \qquad \frac{\partial M_5^1}{\partial \theta_2} = \frac{1}{\Delta}. \qquad (32)$$

From the definition (11), we have that $M_5^2 = A(\Delta/\theta_2)$. Then, it follows from Lemma 3 that

$$\left| \frac{\partial M_5^2}{\partial \theta_2} \right| \leq \frac{\Delta}{12\theta_2^2}. \qquad (33)$$

It follows from (9) and (10) that $M_1^1 = 1 + \theta_1/\Delta$, $M_1^2 = \theta_2/\Delta$, and hence for $m = 1, 2$,

$$\left| \frac{\partial M_1^m}{\partial \theta_1} \right| + \left| \frac{\partial M_1^m}{\partial \theta_2} \right| \leq \frac{1}{\Delta}, \qquad \left| \frac{\partial M_3^m}{\partial \theta_1} \right| + \left| \frac{\partial M_3^m}{\partial \theta_2} \right| \leq \frac{1}{\Delta}. \qquad (34)$$

Write $S = \max\{S_1, S_2\} < 1$, where for $i = 1, 2$,

$$S_i = \sup_{\theta \in \Theta} \left(\left| \frac{\partial M_i}{\partial \theta_1} \right| + \left| \frac{\partial M_i}{\partial \theta_2} \right| \right). \qquad (35)$$

As $\partial M_i/\partial \theta_j$ are continuous on the convex set Θ, it follows from Theorem 5.19 in [9] that for $x = (x_1, x_2), y = (y_1, y_2) \in \Theta$, there exist $z \in \Theta$ such that

$$M(x) - M(y) = \frac{\partial(M_1, M_2)}{\partial(\theta_1, \theta_2)} \bigg|_z (x - y). \qquad (36)$$

Therefore, we have that $\|M(x) - M(y)\| \leq S\|x - y\|$, where $\|x\| = \max\{|x_1|, |x_2|\}$. Hence, there is not more than one fixed-point of M in Θ when $S < 1$. As every fixed-point of $\widetilde{M}$ in Θ is a fixed-point of M, we will prove the theorem by indicating $S < 1$.

It follows from (29) to (35), (15), and Lemma 2 that

$$S_1 \leq \frac{n_1}{n} + \frac{n_3}{n} + \frac{n_5}{n} + \frac{n_4}{n}\frac{2\Delta^2}{\alpha^2} + \frac{n_2}{n}\frac{\Delta^2}{6\alpha^2}, \qquad (37)$$

$$S_2 \leq \frac{n_1}{n} + \frac{n_3}{n} + \frac{n_2}{n} + \frac{n_4}{n}\frac{2\Delta^2}{\alpha^2} + \frac{n_5}{n}\frac{\Delta^2}{12\alpha^2}. \qquad (38)$$

Then it follows from $\Delta < \sqrt{2}\alpha/2$ that $S_1 \leq 1$ and $S_2 \leq 1$.

The record has at least one replacement. That is, there is $k \geq 1$ such that $R(k) = 1$. As a new system replaces the old failed system at time $k\Delta$, we have that $Y(k) = 0$. Now the theorem will be accomplished in two cases.

Case 1. The case $Y(k-1) = 0$. We have that $n_4 \neq 0$. It follows from (37) and (38) that

$$S \leq 1 - \frac{n_4}{n}\left(1 - \frac{2\Delta^2}{\alpha^2}\right) < 1. \qquad (39)$$

Case 2. The case $Y(k-1) \neq 0$. According to the condition-based maintenance policy, a failed system is replaced at an inspection. Hence $Y(k-1) = 1$ follows from $Y(k-1) \neq 0$. As $Y(k-1) = 1$, $Y(k) = 0$, and $R(k) = 1$, we have that $n_5 \neq 0$. Assume that the last replacement occurs at $s\Delta$ if any replacement occurs before $(k-1)\Delta$. And we write $s = 0$ when there is no replacement before $(k-1)\Delta$. Now we study the sequence $Y(s+1), \ldots, Y(k-1)$, which consists of digits 0 and 1, corresponding healthy and unhealthy states of a system without replacement. It is obvious that there is $s+1 \leq t \leq k-1$ such that $Y(t) = 1$ and $Y(t-1) = 0$. Then, $n_2 \neq 0$ follows from $Y(t-1) = 0$, $Y(t) = 1$ and $R(t) = 0$. It follows from (37) and $n_2 \neq 0$ that

$$S_1 \leq 1 - \frac{n_2}{n}\left(1 - \frac{\Delta^2}{6\alpha^2}\right) < 1. \qquad (40)$$

Moreover, it follows from (38) and $n_5 \neq 0$ that

$$S_2 \leq 1 - \frac{n_5}{n}\left(1 - \frac{\Delta^2}{12\alpha^2}\right) < 1. \qquad (41)$$

5. Example and Discussion

We will apply the EM algorithm to a simulation dataset. Based on this example, we will show the efficiency and accuracy of the EM algorithm. Moreover, by this example, we will show some limitations and shortcomings of Theorem 1. In this example, we make ensembles of 10^3 consecutive inspection of a simulating backup system defined by (1). The true parameters are $\theta_1 = 10$ and $\theta_2 = 10/3$, which is adopted from [2].

We describe first the process of iterations described by the EM algorithm (2) and (3). For a given couple of initial value $\theta(0)$ in a given parameter space Θ, we may derive $M(\theta(0))$

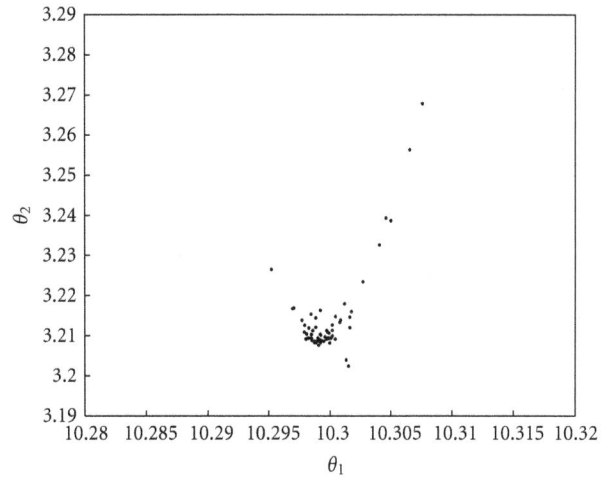

FIGURE 1: Estimated parameters for different initial values.

from (15). If $M(\theta(0)) \in \Theta$, it is the unique solution to (3), and we have that $\theta(1) = M(\theta(0))$. If we are fortunate and can repeat the operation again and again, then we obtain a sequence $\theta(n) \in \Theta, n = 1, 2, \ldots$.

It follows from the expression (8) of $Q(\theta' \mid \theta)$ that Q is continuous with respect to both θ' and θ. Similar to the discussion of the Theorem 1 in [4], we can prove that if the limitation of the sequence $\theta(n) \in \Theta, n = 1, 2, \ldots$ exists and is also in Θ, then the limitation is a fixed-point of the operator $\widetilde{M}$. Theorem 1 shows that the limitation is unique for all such sequences.

In the first experiment, we run the EM algorithm for different initial values. In this experiment, we set $\beta = 1$ and the parameter space $\Theta = \{(x, y) : x, y > \beta\}$. We run the algorithm for 64 couples of initial values which are chosen randomly from 2 to 12. For each couple of initial value, we run the algorithm for 200 iterations. Figure 1 draws the final estimations of the parameters for initial values. We can see that the algorithm converges to the same result for a great range of initial values. As Theorem 1 points out that the number of fixed points is not more than 1. So we can conclude that there is a unique fixed points, and hence it is the MLE of the model parameters on the parameter space $\Theta = \{(x, y) : x, y > 1\}$.

Sometimes, the above procedure of $\theta(n)$, $n = 1, 2, \ldots$ must stop without the output of the estimated parameters. In general, for a $\theta(n) \in \Theta$, if $M(\theta(n))$ derived from (15) is not an element of Θ, then $\widetilde{M}(\theta(n)) \neq M(\theta(n))$. For this case, the solution $\widetilde{M}(\theta(n))$ to (3) is on the boundary of Θ. As we do not obtain the explicit expression $\widetilde{M}(\theta(n))$ for this case, the procedure is aborted.

In the following second experiment, for the same dataset of the first experiment, we run the EM algorithm for another parameter space $\Theta = \{(x, y) : x, y > \beta'\}$ with $\beta' = 12$. We run the algorithm for 32 couples of initial values which are chosen randomly from 20 to 80. As we predict, the procedure is aborted for every couple of initial values. For these initial

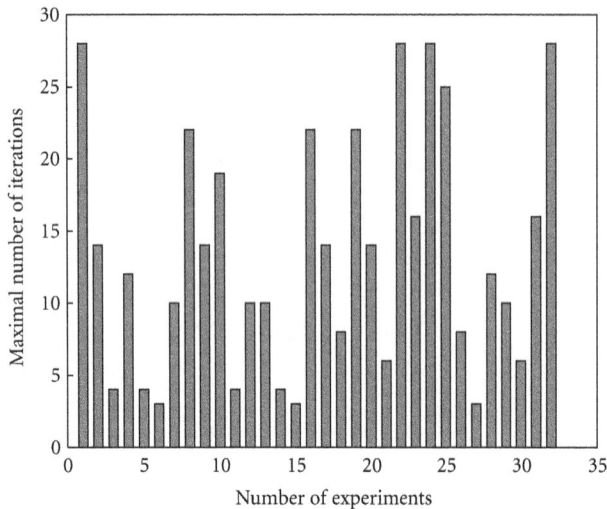

FIGURE 2: Maximal iteration number for different initial values.

values, Figure 2 draws the maximal iteration numbers before the procedure is aborted.

As we know, MLE $\hat{\theta}$ in Θ, when it exists, is one of fixed-points of the operator $\widetilde{M}$. However, there may be other fixed-points of $\widetilde{M}$, such as stationary points of Q. Theorem 1 provides a sufficient condition under which such fixed-point does not exist. Our first experiment and Figure 1 illustrate this fact. However, in some cases, there is not a fixed point of $\widetilde{M}$ on the parameter space Θ. Such a phenomenon occurs when we set a wrong parameter space as in the second experiment.

6. Conclusion

A parameter estimation problem for a backup system has been considered. We established an EM algorithm, which can be used to iteratively determine the maximum likelihood estimators given observations of the system at discrete time points. It has been found that for any initial values, the sequence derived by the EM algorithm converges to a unique point when the limitation belongs to the specified parameter space.

Acknowledgment

This research was supported by the National Natural Science Foundation of China Grant Nos. 50977073 and 70971109.

References

[1] M. J. Kim, V. Makis, and R. Jiang, "Parameter estimation in a condition-based maintenance model," *Statistics and Probability Letters*, vol. 80, no. 21-22, pp. 1633–1639, 2010.

[2] Q. Duan, X. Chen, D. Zhao, and Z. Zhao, "Parameter estimation of a multi-state model for an aging piece of equipment under condition-based maintenance," *Mathematical Problems in Engineering*, vol. 2012, Article ID 347675, 19 pages, 2012.

[3] A. P. Dempster, N. M. Laird, and D. B. Rubin, "Maximum likelihood from incomplete data via the EM algorithm," *Journal of the Royal Statistical Society B*, vol. 39, no. 1, pp. 1–38, 1977.

[4] C. Wu, "On the convergence properties of the EM algorithm," *Annals of Statistics*, vol. 11, pp. 95–103, 1982.

[5] K. Akashi, "On uniqueness of the conditional maximum likelihood estimation for a binary panel model," *Economics Letters*, vol. 112, no. 2, pp. 148–150, 2011.

[6] J. Aragón, D. Eberly, and S. Eberly, "Existence and uniqueness of the maximum likelihood estimator for the two-parameter negative binomial distribution," *Statistics and Probability Letters*, vol. 15, no. 5, pp. 375–379, 1992.

[7] A. Seregin, "Uniqueness of the maximum likelihood estimator for k-Monotone densities," *Proceedings of the American Mathematical Society*, vol. 138, no. 12, pp. 4511–4515, 2010.

[8] R. Bhattacharya and E. Waymire, *Stochastic Processes With Applications*, John Wiley & Sons, New York, NY, USA, 1990.

[9] W. Rudin, *Principles of Mathematical Analysis*, McGraw-Hill, New York, NY, USA, 1976.

20

On the Use of the K-Chart for Phase II Monitoring of Simple Linear Profiles

Walid Gani[1] and Mohamed Limam[2]

[1] LARODEC, ISG, University of Tunis, 41 Avenue de la Liberté, 2000 Le Bardo, Tunisia
[2] Dhofar University, P.O. Box 2509, 211 Salalah, Oman

Correspondence should be addressed to Walid Gani; walid.gani@cct.gov.tn

Academic Editor: Suk joo Bae

Control charts for monitoring linear profiles are used to control quality processes which are characterized by a relationship between a response variable and one or more explanatory variables. In the literature, the majority of control charts deal with phase II analysis of linear profiles, where the objective is to assess the performance of control charts in detecting shifts in the parameters of linear profiles. Recently, the kernel distance-based multivariate control chart, also known as the K-chart, has received much attention as a promising nonparametric control chart with high sensitivity to small shifts in the process. Despite its numerous advantages, no work has proposed the use of the K-chart for monitoring simple linear profiles and that serves the motivation for this paper. This paper proposes the use of the K-chart for monitoring simple linear profiles. A benchmark example is used to show the construction methodology of the K-chart for simultaneously monitoring the slope and intercept of linear profile. In addition, performance of the K-chart in detecting out-of-control profiles is assessed and compared with traditional control charts. Results demonstrate that the K-chart performs better than the T^2 control chart, EWMA control chart, and R-chart under small shift in the slope.

1. Introduction

In the last decade, control charts for monitoring linear profiles have acquired a prominent role in controlling quality processes characterized by a relationship between a response variable and one or more explanatory variables. A control chart for monitoring linear profiles consists of two phases. In phase I, the parameters of the regression line are estimated to determine the stability of the process. In phase II, the goal is to detect shifts in the process from the baseline estimated in phase I. In the literature, the majority of control charts deal with the phase II analysis of linear profiles. Kang and Albin [1] proposed a multivariate T^2 control chart for monitoring both the intercept and the slope, while Kim et al. [2] suggested the use of three univariate exponentially weighted moving average (EWMA$_3$) control charts for simultaneously monitoring the intercept, slope and standard deviation. Zou et al. [3] proposed a multivariate EWMA scheme when the quality process is characterized by a general linear profile. Zhang et al. [4] developed a control chart based on EWMA and

Likelihood ratio test. Zou and Qiu [5] developed the LASSO-based EWMA control chart, for monitoring multiple linear profiles. Li and Wang [6] established an EWMA scheme with variable sampling intervals for monitoring linear profiles.

Recently, the kernel distance-based multivariate control chart, also known as the K-chart, developed by Sun and Tsung [7], has received significant attention as a promising nonparametric control chart with high sensitivity to small shifts in the process mean. According to Gani et al. [8], the K-chart gives the minimum volume closed spherical boundary around the in-control process data. It measures the distance between the kernel center and the incoming new sample to be monitored, which can be calculated using support vectors (SVs). The K-chart relies on support vector data description (SVDD) method, developed by Tax and Duin [9], to determine the shape of the sphere. Any point outside the sphere is considered as out-of-control. When monitoring more than two variables, the K-chart uses kernel methods that provide the advantage of dealing with high-dimensional data. Several works dealt with the K-chart.

Kumar et al. [10] suggested an improvement of the K-chart performance by solving the problem of overfitting due to the existence of outliers in data sets. Camci et al. [11] proposed a robust K-chart that can learn from out-of-control samples and developed an effective heuristic for optimizing the kernel parameters of SVDD. Gani et al. [8] provided an assessment of the K-chart by applying it to a real industrial process and showed that the K-chart is more sensitive to small shifts in mean vector than the T^2 control chart. Furthermore, Gani and Limam [12] provided the MATLAB code for the implementation of the K-chart. Unlike traditional control charts, the K-chart does not require any assumption about the model distribution of quality characteristics and it has the ability to construct flexible control limits based on SVs. All these features serve as incentives to the application of the K-chart for monitoring linear profiles.

In this paper, we propose the use of the K-chart for monitoring simple linear profiles. We show how to construct the K-chart for simultaneously monitoring the slope and intercept of linear profiles. A comparison between the K-chart and traditional control charts, mainly the T^2 control chart, the EWMA control chart, and the R-chart using a benchmark simulated data, is also discussed in this paper.

This paper is organized as follows. Principles of monitoring linear profiles, with a special focus on phase II analysis, are presented in Section 2. Theoretical background of adaptation of the K-chart for monitoring simple linear profiles is presented in Section 3. A benchmark simulated data is used in Section 4 to illustrate the application of K-chart for simultaneously monitoring the slope and intercept of linear profiles, with a comparison with traditional control charts. In Section 5, in order to assess the performance of the K-chart in detecting small shifts in the slope, we compare it to the T^2 and EWMA/R control charts using the average run length criterion. Section 6 provides a conclusion.

2. Monitoring Simple Linear Profiles

2.1. The Linear Profile Model. We consider the following simple linear profile model

$$y_{ij} = \beta_{0j} + \beta_{1j}x_{ij} + \epsilon_{ij}, \quad i = 1, 2, \ldots, n, \ j = 1, 2, \ldots, m, \tag{1}$$

where y_{ij} is the jth measurement, x_{ij} is the value of the explanatory variable corresponding to the jth profile, β_{0j} and β_{1j} are, respectively, the intercept and the slope for profile j, and ϵ_{ij} is the jth random error assumed to be independent and normally distributed with mean zero and variances σ^2.

In phase I, the parameters of the model given in (1) are estimated using the least squares method. The estimated slope for profile j, denoted by b_{1j}, is given by

$$b_{1j} = \frac{\sum_{i=1}^n (y_{ij} - \overline{y}_j)(x_{ij} - \overline{x}_j)}{\sum_{i=1}^n (x_{ij} - \overline{x}_j)^2}, \tag{2}$$

where $\overline{y}_j = (\sum_{i=1}^n y_{ij})/n$ and $\overline{x}_j = (\sum_{i=1}^n x_{ij})/n$, and the estimated intercept for profile j, denoted by b_{0j}, is given by

$$b_{0j} = \overline{y}_j - b_{1j}\overline{x}_j. \tag{3}$$

The variance of residuals, denoted by σ_j^2, is estimated by the jth mean square error (MSE_j) as follows

$$\sigma_j^2 = \text{MSE}_j = \frac{\sum_{i=1}^n (y_{ij} - b_{0j} - b_{1j}x_{ij})^2}{n-2}. \tag{4}$$

The variance of b_{1j}, denoted by $\sigma_{b_{1j}}^2$, is expressed as follows

$$\sigma_{b_{1j}}^2 = \frac{\sigma^2}{\sum_{i=1}^n (x_{ij} - \overline{x}_j)^2}. \tag{5}$$

The variance of b_{0j}, denoted by $\sigma_{b_{0j}}^2$, is expressed as follows

$$\sigma_{b_{0j}}^2 = \sigma^2 \left(\frac{1}{n} + \frac{\overline{x}_j}{\sum_{i=1}^n (x_{ij} - \overline{x}_j)^2} \right). \tag{6}$$

For monitoring simple linear profiles in phase I, several types of control charts have been proposed such as the use of T^2 control chart (Mestek et al. [13], Stover and Brill [14], and Kang and Albin [1]). Besides, Kim et al. [2] proposed the use of three independent Shewhart control charts for monitoring b_{0j}, b_{1j}, and σ^2. However, phase II monitoring of linear profiles remains the most important step since it aims to assess the performance of control charts in detecting shifts in the parameters of linear profiles. In the following, we present the main control charts for phase II analysis of linear profiles.

2.2. Control Charts for Phase II Linear Profile Monitoring. Mahmoud [15] distinguished between two main categories of control charts for phase II. The omnibus control charts category for monitoring simultaneously the intercept and slope and the individual control charts category for monitoring separately individual regression parameters. This paper focuses on the omnibus category, since our objective is to simultaneously monitor the slope and intercept of linear profiles. The most applied traditional control charts in this category are T^2 control chart, EWMA control chart, and R-chart.

For monitoring b_0, b_1, and σ^2 in phase II, Kang and Albin [1] recommended the use of the T^2 control chart for b_{0j} and b_{0j}. The T^2 statistics for monitoring the intercept and the slope are given by

$$T_j^2 = (z_j - \mu)^T \Sigma^{-1} (z_j - \mu), \tag{7}$$

where $z_j = (b_{0j}, b_{1j})$, $\mu = (\overline{b}_0, \overline{b}_1)$, $\overline{b}_0 = (\sum_{j=1}^m b_{0j})/m$, $\overline{b}_1 = (\sum_{j=1}^m b_{1j})/m$, $\Sigma = \begin{pmatrix} \sigma_{b_{0j}}^2 & \sigma_{b_{0j},b_{1j}} \\ \sigma_{b_{0j},b_{1j}} & \sigma_{b_{1j}}^2 \end{pmatrix}$.

The upper control limit (UCL) for the T^2 control chart is given by

$$\text{UCL} = \chi^2_{2,\alpha}, \tag{8}$$

where $\chi^2_{2,\alpha}$ is the $100(1 - \alpha)$ percentile of the chi-squared distribution with 2 degrees of freedom.

In addition, Kang and Albin [1] proposed an EWMA control chart to monitor the average deviation from the in-control line. The EWMA statistics for monitoring σ^2 are given by

$$\text{EWMA}_j = \theta \bar{e}_j + (1 - \theta)\,\text{EWMA}_{j-1}, \tag{9}$$

where $\bar{e}_j = \sum_{i=1}^{n}(e_{ij}/n_j)$ is the average deviation for sample j, $0 < \theta < 1$ is a smoothing constant, and $\text{EWMA}_0 = 0$.

The lower control limit (LCL) and the UCL for this EWMA chart are given by

$$\text{LCL} = -L_1 \sigma \sqrt{\frac{\theta}{n(2 - \theta)}}, \qquad \text{UCL} = +L_1 \sigma \sqrt{\frac{\theta}{n(2 - \theta)}}, \tag{10}$$

where $L_1 > 0$ is a constant chosen to give a specified in-control average run length (ARL).

Also, Kang and Albin [1] suggested the use of an R-chart for monitoring the process variation as follows

$$\text{LCL} = \sigma(d_2 - L_2 d_3), \qquad \text{UCL} = \sigma(d_2 - L_2 d_3), \tag{11}$$

where $L_2 > 0$ is a constant selected to produce a specified in-control ARL and d_2 and d_3 are constants depending on the sample size n.

3. Monitoring Simple Linear Profiles Using the K-Chart

We consider $\beta_j = [b_{0j}, b_{1j}]$, the vector of the intercept and slope for profile j, with $j = 1, \ldots, m$. The construction of the K-chart for simultaneously monitoring the slopes and intercepts of linear profiles requires two steps. In the first step, a sphere around the samples of β_j is constructed using SVDD. The sphere should contain the maximum of β_j with minimum volume. This is equivalent to solving the following quadratic programming

$$\text{Minimize} \quad F(R, a) = R^2 \tag{12}$$

subject to

$$\|\beta_j - a\|^2 \leq R^2, \quad j = 1, 2, \ldots, m, \tag{13}$$

where F, a, and R, are respectively, the cost function to minimize, the center, and the radius of the sphere. Equation (13) shows that samples of β_j having a distance smaller than the radius are considered as targets. To allow the possibility of having outliers in the training set, the distance from β_j to the center a should not be strictly smaller than R^2, and larger

distances should be penalized. Therefore, we introduce slack variables $\xi_j \geq 0$ and the minimization problem becomes

$$\text{Minimize} \quad F(R, a) = R^2 + C\sum_{j=1}^{m} \xi_j, \tag{14}$$

subject to

$$\|\beta_j - a\|^2 \leq R^2 + \xi_j, \tag{15}$$

where $C > 0$ is a parameter introduced for the trade-off between the volume of the sphere and the errors.

Equation (15) can be incorporated into (14) by using Lagrange multipliers

$$L\left(R, a, \alpha_j, \eta_j, \xi_j\right)$$

$$= R^2 + C\sum_{j=1}^{m} \xi_j$$

$$- \sum_{j=1}^{m} \alpha_j \left[R^2 + \xi_j - \left(\|\beta_j\|^2 - 2a \cdot \beta_j + \|a\|^2 \right) \right] - \sum_{j=1}^{m} \eta_j \xi_j, \tag{16}$$

with the Lagrange multipliers $\alpha_j \geq 0$ and $\eta_j \geq 0$; L should be minimized with respect to, R, a, ξ_j and maximized with respect to α_j and η_j. Setting partial derivatives of L, we obtain

$$\sum_{j=1}^{m} \alpha_j = 1, \tag{17}$$

$$a = \sum_{j=1}^{m} \alpha_j \beta_j, \tag{18}$$

$$C - \alpha_j - \eta_j = 0. \tag{19}$$

From (19), $\alpha_j = C - \eta_j$, $\alpha_j \geq 0$, and $\eta_j \geq 0$, then Lagrange multipliers η_j can be removed and we have

$$0 \leq \alpha_j \leq C. \tag{20}$$

By substituting (17) and (19) into (16), we have

$$\text{Maximize} \quad L = \sum_{j=1}^{m} \alpha_j \left(\beta_j \cdot \beta_j \right) - \sum_{j,k=1}^{m} \alpha_j \alpha_k \left(\beta_j \cdot \beta_k \right), \tag{21}$$

subject to

$$0 \leq \alpha_j \leq C. \tag{22}$$

A test sample, denoted by β^{test}, is accepted when its distance is smaller or equal to the radius. This is equivalent to

$$\left(\beta^{\text{test}} - a \right)' \left(\beta^{\text{test}} - a \right)$$

$$= \left(\beta^{\text{test}'} \cdot \beta^{\text{test}} \right) - 2\sum_{j=1}^{m} \alpha_j \left(\beta^{\text{test}} \cdot \beta_j \right) \tag{23}$$

$$+ \sum_{j,k=1}^{m} \alpha_j \alpha_k \left(\beta_j \cdot \beta_k \right) \leq R^2.$$

Generally, data is not spherically distributed. To make the method more flexible, the vectors of β_j are transformed to a higher-dimensional feature space. The inner products in (21) and (23) are substituted by a kernel function $K(\beta_j, \beta_k)$. In a higher dimension, the sphere becomes a complex form called "hypersphere." The problem of finding the optimal hypersphere is given by

$$\text{Maximize} \quad L = \sum_{j=1}^{m} \alpha_j K\left(\beta_j, \beta_k\right) - \sum_{j,k=1}^{m} \alpha_j \alpha_k K\left(\beta_j, \beta_k\right), \tag{24}$$

subject to (23).

A test sample β^{test} is accepted when

$$K\left(\beta^{\text{test}}, \beta^{\text{test}}\right) - 2\sum_{j=1}^{m} \alpha_j K\left(\beta^{\text{test}}, \beta_j\right)$$
$$+ \sum_{j,k=1}^{m} \alpha_j \alpha_k K\left(\beta_j, \beta_k\right) \leq R^2. \tag{25}$$

The second step in the construction of the K-chart consists in determining which samples are SVs by solving the following quadratic programming

$$\text{Maximize} \quad \sum_{j=1}^{m} \alpha_j K\left(\beta_j, \beta_j\right) - \sum_{j,k=1}^{m} \alpha_j \alpha_k K\left(\beta_j, \beta_k\right), \tag{26}$$

subject to

$$0 \leq \alpha_j \leq C. \tag{27}$$

Once the SVs are obtained, the kernel distance (KD) of each sample is computed. For a test sample β^{test}, the KD is computed as follows

$$\text{KD}_{\beta^{\text{test}}} = \left(K\left(\beta^{\text{test}}, \beta^{\text{test}}\right) - 2\sum_{j \in S}^{m} \alpha_j K\left(\beta^{\text{test}}, \beta_j\right) \right.$$
$$\left. + \sum_{j,k \in S}^{m} \alpha_j \alpha_k K\left(\beta_j, \beta_k\right) \right)^{1/2}, \tag{28}$$

where S is the set of SVs.

The KD of SVs, denoted by KD_{SVs}, represents the UCL for the K-chart used to monitor a new sample β^{test}. This can be illustrated by the following hypothesis test

$$H_0 : \text{KD}_{\beta^{\text{test}}} \leq \text{KD}_{\text{SVs}} \quad \text{versus} \quad H_1 : \text{KD}_{\beta^{\text{test}}} > \text{KD}_{\text{SVs}}. \tag{29}$$

Under H_0 the process is considered as in-control and under H_1 the process is considered as out-of-control, when sample β^{test} was taken.

4. Application

This section is devoted to show how to construct the K-chart for simultaneously monitoring the slope and intercept

TABLE 1: Characteristics of SVDD-based one-class when varying the values of γ.

γ	Threshold	SVs	F_{measure}
0.25	0.905	17	0.947
0.50	0.809	9	0.919
0.75	0.705	8	0.947
1.00	0.610	7	0.889
1.25	0.535	4	0.974
1.50	0.469	4	1.000
1.75	0.413	3	0.919
2.00	0.363	2	0.974

of simple linear profiles. In addition, to show the efficacy of our proposed approach, we compare the performance of the K-chart with that of T^2 control chart, EWMA control chart, and R-chart in detecting out-of-control (OOC) profiles. In this application, we use the benchmark simulated data of Mahmoud [15], where we consider the following in-control profile model: $Y_{ij} = 13 + 2X_i + \epsilon_{ij}$ (where the ϵ_{ij} are random errors assumed to be independent and normally distributed with mean zero and variance 1), with fixed X-values of -3, -1, 1, and 3. The simulated data set consists of 29 profiles generated as follows. First, 20 in-control profiles were generated. Then, nine OOC profiles were generated, after shifting the slope from 2.0 to 2.4. Details about the simulated data can be found in Mahmoud [15].

During the training phase, the 20 in-control profiles are used to construct the optimal one class using SVDD algorithm. Then, the nine remainder profiles are used to detect OOC states. For the construction of the optimal SVDD-based one-class, the Gaussian kernel function is used and it is defined as follows

$$K\left(\beta_j, \beta_k\right) = \exp\left(-\frac{\left\| \beta_j - \beta_k \right\|^2}{2\gamma^2} \right), \tag{30}$$

where $\gamma > 0$ is the width of the Gaussian kernel that controls the complexity of the SVDD boundary. For the determination of the optimal value of γ, the F_{measure} criterion is used and it is defined as follows:

$$F_{\text{measure}} = \frac{2 \times \text{precision} \times \text{recall}}{\text{precision} + \text{recall}}, \tag{31}$$

where $0 \leq F_{\text{measure}} \leq 1$, precision = true positive rate/(true positive rate + false positive rate), and recall = $1 -$ precision. The optimal value of γ corresponds to the highest value of F_{measure}.

All calculations were carried out in MATLAB software. The optimal value of the Gaussian kernel width was found to be $\gamma = 1.50$, which corresponds to the highest value of F_{measure}, as shown in Table 1. Some other criteria can be used to verify the construction of the optimal SVDD based one-class such as graphical representations. It can be seen from Figure 1 that, for $\gamma = 0.25$ and 0.5, the construction of SVDD-based one-class is not possible. It is worth noting that the shape of SVDD-based one-class depends on the parameter

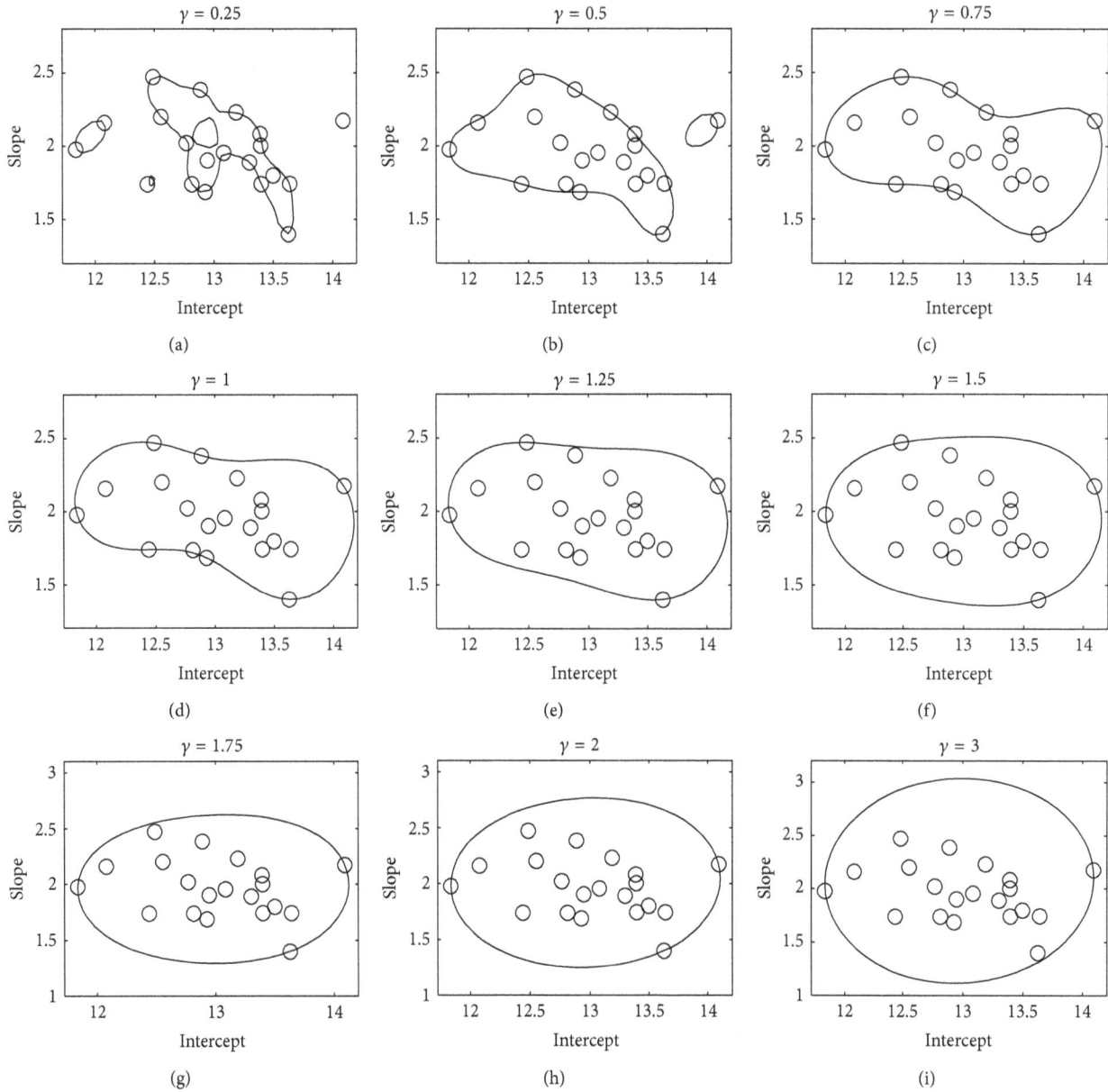

FIGURE 1: Examples of SVDD-based one-classes for simultaneously monitoring the intercept and slope under different values of the parameter γ.

γ. In our application, we stated that the smoothness of the SVDD boundary was enhanced when increasing the value of γ.

Once the optimal SVDD-based one-class is obtained, the construction of the K-chart is done as follows. After obtaining the solutions of the quadratic programming given by (26) and (27), samples with positive α_j are considered as SVs and used to construct the UCL of the K-chart. The latter is based on 4 SVs and it is estimated at 0.469. The T^2 control chart was constructed with an UCL = $\chi^2_{2,0.005}$ = 10.579. For the construction of EWMA control chart, the smoothing parameter θ was set at 0.2 to obtain the charting statistics. Based on (10), the control limits were set at ±0.48 so that they produce an in-control ARL of 200. Using (11), the UCL

of the R-chart was set at 4.94. It is worth noting that there is no LCL for the R-chart since $n < 7$. Figure 2 shows the constructed control charts for phase II.

Regarding the performance of the discussed control charts, the K-chart performed better than the other control charts. In fact, the K-chart detected 6 OOC profiles which are profiles number 23, 25, 26, 27, 28, and 29, while the T^2 chart detected only one OOC state which is profile number 27. The R-chart detected one OOC state which is profile number 26. The EWMA control chart was the weakest control chart since it did not detect any OOC profile. As shown in Table 2, the performance rate of K-chart, estimated at 66.67%, was highly superior to that of traditional control charts. The used performance rate in this application is defined as

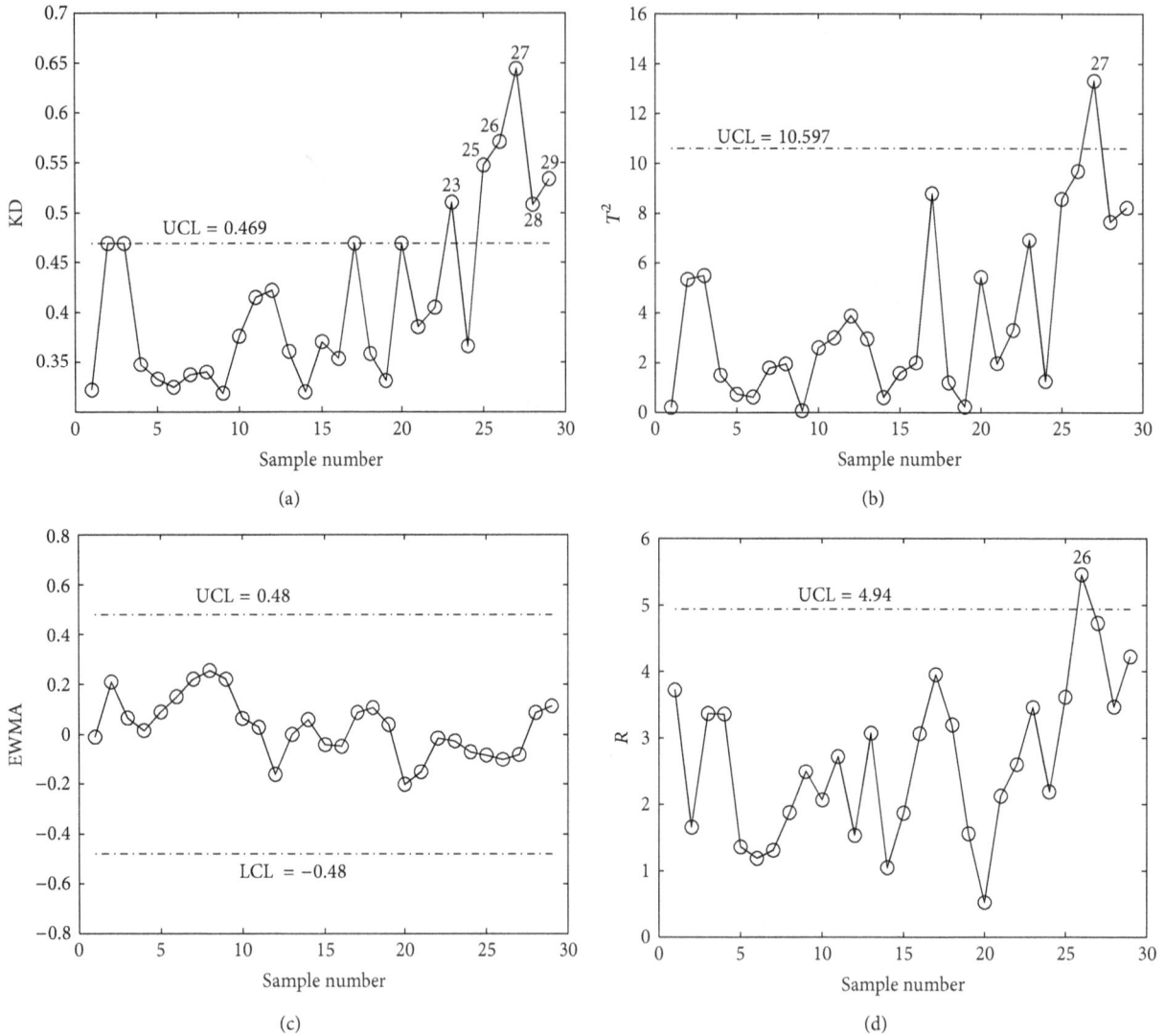

FIGURE 2: Phase II control charts for monitoring linear profiles: (a) the K-chart, (b) the T^2 control chart, (c) the EWMA control chart, and (d) the R-chart.

TABLE 2: Performance evaluation of control charts in detecting OOC profiles.

Control chart	Number of OOC profiles	Performance rate
K-chart	6	66.67%
T^2 chart	1	11.11%
R-chart	1	11.11%
EWMA chart	0	0.00%

the detected number of OOC profiles divided by the generated number of OOC profiles. A key element that can explain the performance of the K-chart is the nature of its control limit based on SVs, which is very sensitive to any change in kernel's width.

5. Performance Assessment

To evaluate the performance of the K-chart in detecting small shifts in the slope, we compare it with the T^2 and EWMA/R

control charts using the ARL criterion. The latter is a common performance measure used to assess the effectiveness of control charts in detecting OOC signal. The ARL is defined as the expected number of samples taken before the shift is detected and it is given by

$$ARL = \frac{1}{p}, \qquad (32)$$

where p is the probability of one point plots OOC.

For the computation of ARL, a simulation study is conducted based on the model discussed in Section 4 and our results are compared with those reported in Kim et al. [2]. To be consistent with the latter, we follow their methodology which consists in introducing shifts in the slope of linear profiles as follows: β_1 shifts to $\beta_1 + \delta\sigma$, where $\delta = [-0.2; -0.3; -0.4; -0.5; -0.6; -0.7; -0.8; -0.9; -1]$.

Table 3: ARL values when β_1 shifts to $\beta_1 + \delta\sigma$.

δ	T^2 chart	EWMA/R chart	K-chart
0.00	200.000	200.000	200.000
−0.20	52.20	76.70	14.30
−0.30	21.20	33.70	11.80
−0.40	9.60	15.30	10.50
−0.50	4.90	7.50	9.50
−0.60	2.90	4.20	9.10
−0.70	1.90	2.60	7.70
−0.80	1.50	1.80	7.40
−0.90	1.20	1.40	6.70
−1.00	1.10	1.20	5.40

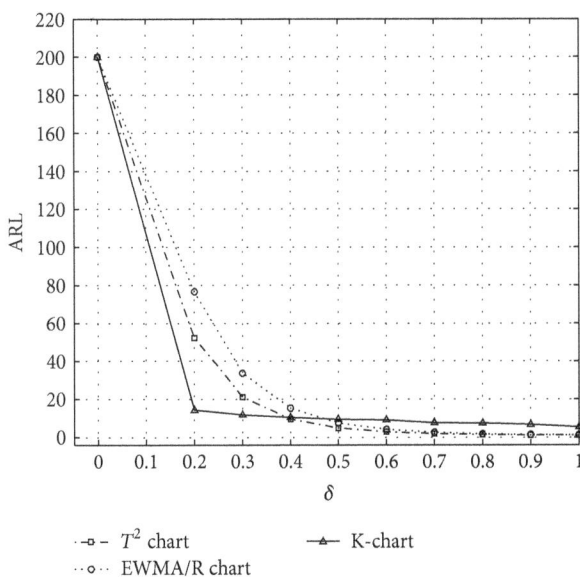

Figure 3: ARL comparisons under different values of the shift δ in the slope.

The K-chart is designed to achieve an overall in-control ARL of 200. The ARL value is estimated by averaging the run lengths obtained by running 10000 simulated charts.

It is clear from Table 3 that the K-chart has better ARL performance under small shifts in the slope. In fact, for $\delta = -0.20$, the K-chart gives an ARL of 14.30, while the T^2 and EWMA/R control charts give an ARL of 52.20 and 76.70, respectively. This means that the K-chart requires only 14.30 samples to detect the process shift, while the T^2 and EWMA/R control charts need 52.20 and 76.70 samples, respectively, to detect the process shift. Figure 3 shows the ARL behavior of the three control charts under different shifts in the slope. It is worth noticing that, for a moderate shift ($\delta = -0.40$), the T^2 control chart was slightly better than the K-chart since its ARL was estimated at 9.60 against an ARL of 10.50 for the K-chart. Broadly speaking, from the results of the simulation study one can draw the conclusion that the K-chart is more sensitive to small shifts in the slope of linear profiles than the T^2 and EWMA/R charts.

6. Conclusion

In this paper, we have adapted the K-chart for monitoring simple linear profiles. We show how to construct the K-chart for simultaneously monitoring the slope and intercept. Based on the ARL criterion, the simulation study shows that the K-chart performs better in detecting small shifts in the slope in comparison with T^2 and EWMA/R control charts. In addition, our application demonstrated that the K-chart is an effective tool for detecting OCC profiles in comparison with traditional control charts. The high sensitivity level of the K-chart is explained by its flexible control limit based on SVs, making it adaptive to any shift in the process. Many interesting extensions are possible for the use of the K-chart for monitoring simple linear profiles. One possible extension is to apply the K-chart for monitoring multivariate linear profiles and compare it with the multivariate EWMA control charts. This extension could constitute a promising research field in the future.

Conflict of Interests

The authors declare that they do not have a direct financial relation with the software mentioned in this paper and no competing interests.

Acknowledgments

The authors appreciate the valuable comments of the reviewer which led to a significant improvement of this paper. The authors express their appreciation to LARODEC of ISG, University of Tunis for the support of this work.

References

[1] L. Kang and S. L. Albin, "On-line monitoring when the process yields a linear profile," *Journal of Quality Technology*, vol. 32, no. 4, pp. 418–426, 2000.

[2] K. Kim, M. A. Mahmoud, and W. H. Woodall, "On the monitoring of linear profiles," *Journal of Quality Technology*, vol. 35, no. 3, pp. 317–328, 2003.

[3] C. Zou, F. Tsung, and Z. Wang, "Monitoring general linear profiles using multivariate EWMA schemes," *Technometrics*, vol. 49, no. 4, pp. 395–408, 2007.

[4] J. Zhang, Z. Li, and Z. Wang, "Control chart based on likelihood ratio for monitoring linear profiles," *Computational Statistics and Data Analysis*, vol. 53, no. 4, pp. 1440–1448, 2009.

[5] C. Zou and P. Qiu, "Multivariate statistical process control using LASSO," *Journal of the American Statistical Association*, vol. 104, no. 488, pp. 1586–1596, 2009.

[6] Z. Li and Z. Wang, "An exponentially weighted moving average scheme with variable sampling intervals for monitoring linear profiles," *Computers and Industrial Engineering*, vol. 59, no. 4, pp. 630–637, 2010.

[7] R. Sun and F. Tsung, "A kernel-distance-based multivariate control chart using support vector methods," *International Journal of Production Research*, vol. 41, no. 13, pp. 2975–2989, 2003.

[8] W. Gani, H. Taleb, and M. Limam, "An assessment of the kernel-distance-based multivariate control chart through an industrial

application," *Quality and Reliability Engineering International*, vol. 27, no. 4, pp. 391–401, 2011.

[9] D. M. J. Tax and R. P. W. Duin, "Support vector data description," *Machine Learning*, vol. 54, no. 1, pp. 45–66, 2004.

[10] S. Kumar, A. K. Choudhary, M. Kumar, R. Shankar, and M. K. Tiwari, "Kernel distance-based robust support vector methods and its application in developing a robust K-chart," *International Journal of Production Research*, vol. 44, no. 1, pp. 77–96, 2006.

[11] F. Camci, R. B. Chinnam, and R. D. Ellis, "Robust kernel distance multivariate control chart using support vector principles," *International Journal of Production Research*, vol. 46, no. 18, pp. 5075–5095, 2008.

[12] W. Gani and M. Limam, "Performance evaluation of one-class classification-based control charts through an industrial application," *Quality and Reliability Engineering International*, 2012.

[13] O. Mestek, J. Pavlík, and M. Suchánek, "Multivariate control charts: control charts for calibration curves," *Fresenius' Journal of Analytical Chemistry*, vol. 350, no. 6, pp. 344–351, 1994.

[14] F. S. Stover and R. V. Brill, "Statistical quality control applied to ion chromatography calibrations," *Journal of Chromatography A*, vol. 804, no. 1-2, pp. 37–43, 1998.

[15] M. A. Mahmoud, "Simple linear profiles," in *Statistical Analysis of Profile Monitoring*, R. Noorossana, A. Saghaei, and A. Amiri, Eds., pp. 21–92, John Wiley & Sons, New York, NY, USA, 2011.

21

Productivity Improvement of a Special Purpose Machine Using DMAIC Principles: A Case Study

Sunil Dambhare,[1] Siddhant Aphale,[2] Kiran Kakade,[2] Tejas Thote,[2] and Atul Borade[3]

[1] *Department of Mechanical Engineering, PVPIT, Bavdhan, Pune, Maharashtra 411021, India*
[2] *Mechanical Engineering, PVPIT, Bavdhan, Pune, Maharashtra 411021, India*
[3] *Department of Mechanical Engineering, JDIET, Yavatmal, Maharashtra 445001, India*

Correspondence should be addressed to Atul Borade; atulborade@rediffmail.com

Academic Editor: Shey-Huei Sheu

Six Sigma is one of the popular methodologies used by the companies to improve the quality and productivity. It uses a detailed analysis of the process to determine the causes of the problem and proposes a successful improvement. Various approaches are adopted while following Six Sigma methodologies and one of them is DMAIC. The successful implementation of DMAIC and FTA is discussed in this paper. In this study, the major problem was of continuous rework up to 16%, which was leading to wastage of man hours and labor cost. Initially, fault tree analysis (FTA) was used to detect the key process input variables (KPIVs) affecting the output. Multivariable regression analysis was performed to know the possible relationship between the KPIVs and the output. The DMAIC methodology was successfully implemented to reduce the rework from 16% bores per month to 2.20% bores per month. The other problem of nonuniform step bores was also reduced significantly.

1. Introduction

Diesel engines have very wide applications in this modern technological era. They are designed to cater for the need of construction, mining, power generation, locomotives, marine transport, compressors, and so forth market segments. These are heavy power requirement applications. For such applications, the output requirement may vary from 10 horsepower to 3500 horsepower. The engine with this huge power output is also huge. The firm in which this study was conducted was involved in manufacturing 1 V- and 16 V-cylinder diesel engines. The sand casted engine block went through various operations and then was assembled. Operations like undercutting, water clearance chamfer, boring, step-boring, and so forth were performed on each cylinder. The step-boring operation was the most difficult operation. It needs special attention as the liner is resting in it. A cylinder liner is pressed into an engine block and houses the piston. The cylinder liner is much harder than the engine block and prevents the piston from wearing out through the cylinder bore. Typically used in aluminium engine blocks and diesel

engines, the cylinder liner is either pressed into position or held in place by the cylinder head. In large engines such as the engines found in diesel locomotives, the cylinder liner is part of an assembly containing a new piston, piston rings, and a connecting rod. During scheduled maintenance or repair, the liner is changed as a complete unit. In aluminium engine blocks, the block material is too soft to house a piston. The friction of a piston moving up and down inside the alloy block would make the piston wear out, resulting in a loss of compression and severe oil consumption [1, 2]. In such cases, the steel cylinder liner is pressed into the engine block and then the engine block is machined to assure that the cylinder head mating surface is smooth and flat. With this modification, the engine is able to operate for many years without failure. The flat surface resulting from the machining of the engine block assures a proper seal of the head between the cylinder head and the engine block. An improperly sealed head gasket will result in overheating of the engine, loss of power, and the potential to ruin the block and cylinder head. It is clear that the depth of a V-cylinder engine is one of the most crucial parameters for the efficient working of an engine. A slight variation in depth of the step can damage the cylinder and piston as well as hamper the engine operation. A cylinder's head and liner rest on the bored depth. Thus maintaining the depth is the necessity for any engine manufacturing firm.

Six Sigma is a business management strategy, first developed by Motorola in 1986, which seeks to improve the quality of process outputs [3]. It identifies and removes the cause of defects and minimizes variability in manufacturing and business processes [4, 5]. A set of quality management methods and statistical methods is used. Each Six Sigma project carried out within an organization follows a defined sequence of steps and quantified financial targets. According to the definition, a Six Sigma process is one in which 99.99966% of the products manufactured are statistically expected to be free of defects (3.4 defects per million) [6–8]. Six Sigma uses a group of improvement specialists for problem solving and improving the process continuously [9, 10]. Six Sigma techniques have two main methodologies DMAIC and DMADV [11]. Define, Measure, Analyse, Improve, and Control (DMAIC) methodology was followed for reducing the rework [12, 13]. The reasons for the main problem can be detected using again two methodologies FTA and failure mode and effects analysis. FTA deals with identifying all possible causes related to a particular problem. It is a stepwise approach to identify causes and all the parameters related to every cause. It can be used in almost every application which involves cause-effect analysis. Once the problem is defined, it is to be measured so as to collect statistical data about the problem. Once the statistical data is collected, it is analysed using various analysis techniques like chi-square test, regression analysis, ANOVA, and so forth. Minitab statistical software was used for the analysis of various phases of the project.

2. Review of the Literature

Six Sigma methodologies have become a top agenda for many companies which are continuously trying to improve productivity at lesser costs. It was shown previously that Six Sigma continued to be a predominate target to try and obtain a competitive advantage [11]. Many Fortune 500 companies have adopted Six Sigma to improve the productivity and reduce cost. Six Sigma has been described as a data driven approach for problem solving, business process, disciplined statistical approach, and a management strategy [14–18]. Six Sigma methods can prove to be beneficial when applied to labour-intensive, repeatable processes according to Swink and Jacobs [14]. Six Sigma benefits are significantly correlated with intensity in manufacturing and with financial performance before adoption in services [14]. The Six Sigma improvement method is problem-focused and its main objectives are decreasing scrap, earning income, and creating value. Previous researches have shown the effective implementation of Six Sigma techniques in problem solving and process improvements. FTA was successfully utilised to analyse the bridge erecting tripping problem [19]. FTA can be used to analyse hazards and calculate system reliability for simple as well as complex systems [20]. Six Sigma methodologies have been previously used for reducing the rejection level [21].

FTA can be effectively used for finding the effective causes from accident cases to scrap reduction [22]. According to Wang et al. [23], FTA is a simple, effective, reliable method which is recognised internationally and is used on guiding system optimization and analysing and repairing the system of weak links. FTA helps deepen the research to find out the possible causes for the fault [24]. FTA has previously been used successfully to establish priorities for the manufacturing plant for future projects aimed at improving the manufacturing plant [25].

According to Büyüközkan and Öztürkcan [4] and de Koning and de Mast [26], Six Sigma program offers a wide range of tools and techniques, which might be statistical and nonstatistical, that are intended to assist the project leader. Swink and Jacobs [14] showed solid support for the hypothesis that Six Sigma adoption tends to produce significant benefits for firm's profitability. Positive return on asset (ROA) changes were frequently observed in latter periods (years +3 and +4), and moreover these benefits also appear to be persistent. These findings hint at potential differences in how Six Sigma programs are possibly being applied in front-office versus back-office contexts. Findings suggest that Six Sigma methods may be most beneficial when applied to labour-intensive, repeatable processes. However, less labour-intensive, quality experienced, manufacturing firms will not experience the profit impact from Six Sigma adoption the same way that others will. Their outcomes also revealed marginally significant positive effects on sales growth Six Sigma projects can accomplish successfully using FTA. FTA has been used extensively by the military, the space program, and nuclear industry. FTA is a very structured, systematic, and rigorous that lends itself well to quantification [25]. FTA can be used to express the logical relationship between the possibility of certain accident and causes of undesired events or accidents in fault tree diagram [19]. It could be successfully implemented to find out all the possible causes for the problem or fault that has occurred. According to Shalev and Tiran [20], FTA method analysts apply top-down logic in building their models. The problem of reducing process variation and the associated defect rate can be solved using the DMAIC methodology of Six Sigma. Six Sigma is a useful problem-solving methodology and provides a valuable measurement approach. Six Sigma focuses on some vital dimension of business processes, reducing the variation around the mean value of the process [20]. The original task of Six Sigma's DMAIC Methodology is variation reduction. As stated by de Mast and Lokkerbol [27], Six Sigma and its DMAIC are built on insights from the quality engineering field, incorporating ideas from statistical quality control, total quality management, and Taguchi's offline quality control. It has also been used for general tasks like quality improvement, efficiency improvement, cost reduction, and other pursuits in operations. Thus Six Sigma is a generic method and its original task domain was variation reduction, typically in manufacturing processes. Li et al. [28] successfully implemented DMAIC approach to improve the capability of the solder paste printing process by reducing thickness variations from a nominal value. The DMAIC approach has shown a wider application and how the engineering organisation can achieve competitive

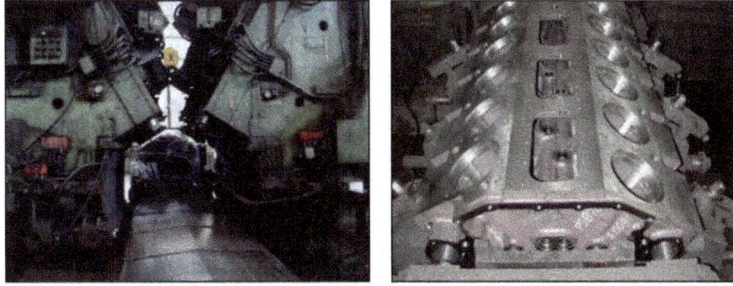

FIGURE 1: Ingersoll special purpose machine and engine block.

advantages, efficient decision making, and problem-solving capabilities within a business context. According to Li and Al-Refaie [29], adopting the DMAIC procedure including GR&R study turns out to be an effective method in improving the quality system including measurements.

3. Background for the Study

The case study was conducted at a leading manufacturer of 12 V- and 16 V-cylinder diesel engines. The sand casted engine block is processed with operations like rough boring, water clearance chamfer, surface milling, finish boring, and so forth before accomplishing the engine assembly. Critical operations which demand precise dimension control are performed on special purpose machines. Engine block boring is one of the critical operations performed on special purpose machines under study.

The finished boring operation is completed in 3 stages. Initially, the V-surface of the engine is milled using milling head. Considering the milled surface as datum, step-boring operation and water clearance chamfering operation are performed. As the engine is a V-cylinder engine, performing machining operations on the block is a tedious task. If these operations are performed on a CNC machine, it requires nearly 5 hours and tiresome programming. Thus there was a need of special purpose machine.

Ingersoll machine performs the mentioned operations in 17 minutes approximately. The machine uses hydraulic circuits for performing operations which were designed around 40 years back. The machine performs three operations on each bore, namely, undercutting, chamfering, and step-boring, after performing milling on the V-surface. Operations are carried out in a sequence as undercut/chamfering/step-boring. Machine starts its forward stroke at 800 psi pressure acting in forward direction. In this stroke it performs two operations, undercutting and chamfering. Both operations are carried out at 800 psi pressure and flow control valves are used to control speed of sliding tool during operations. After completion of chamfering, tool slide completes the forward stoke and at the same time a lever attached to the tool post touches the inclined milling surface, actuating the boring tool slide mechanism. This provides forward motion to the boring tool. Simultaneously, the lever actuates return pressure valve and 1000 psi pressure acts to carry out backward stroke of the tool post slide. Step-boring finishing operation is

completed in return stroke of tool post. Once tool post returns to its original position, another engine bore gets lined for the same set of operations. Figure 1 shows the Ingersoll special purpose machine and the engine block machined using this machine.

4. Case Study

The machine under consideration is a special purpose machine which was specially developed for performing the specified operations on the V-engines manufactured by the firm. The operations performed are milling, of the V-surface where the material on the surface is removed. After milling, crevice chamfering is completed. Step-boring is performed at last in which the existing hole is enlarged to the designed depth. The step bore is a critical dimension as the liner of the cylinder rests on it. The boring tool performs the operation, and after it reaches the required depth, a lever senses its position. As soon as the lever touches the surface, the boring tool retracts and the operation is completed.

During the study, readings for the step bore depth were noted for every bore of each engine block. The allowable tolerance for the step bore was $0.7190'' \pm 0.0013''$. From the data, it was observed that there was a variation in the readings and many of the readings were outside the tolerance limit of $0.7190'' \pm 0.0013''$. This was a serious problem. Bringing the dimension within the tolerance limits means rework on the bores was necessary. The rework data was gathered and it was found that average rework per month was 16% of the bores machined. For performing rework operations, 369 man-hours per month and approximately INR 60400 were spent. This hampered the productivity of the firm. The study showed that rework was needed mostly on the left bank of the engine block indicating need of the improvement actions on this section of engine boring operation. Therefore the objective of the study was set to minimize the rework percentage per month close to zero without affecting the operations and cycle time. To improve the productivity and reduce the rework expenses, the Six Sigma technique was selected.

5. Methodology

As the study aimed at improving the existing business process, DMAIC methodology was considered [11, 12, 21, 30].

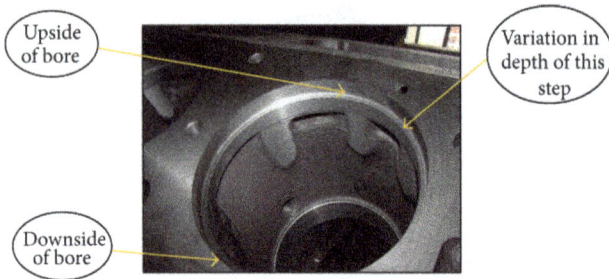

FIGURE 2: Step bore and depth variation.

0.7177″ to 0.7203″. Based on the lower limit and upper limit of the bore, they are categorised as undersize, in size, or oversize bores. Each bore failing to achieve these tolerances was subjected to rework. The depth of the bore was measured using a depth gauge which was calibrated before data was collected. The depth gauge indicated the measured depth about the mean depth, that is, 0.7190″. The depth was measured at two points of the bore, upside and downside of the bore on both banks, as shown in Figure 2.

The machine behaviour was unpredictable because of the following reasons.

(1) Dimensional variation was observed mostly on the left bank of the block even under the same machining conditions on both sides.

(2) A fix pattern in dimensional variation was not observed in the finished bores. Some blocks were oversize while some were undersize. Some cases were reported where all the bore categories were involved.

(3) There was no pattern repetition in dimensional variation of the bores. If on a particular block all the bores went out of tolerance, then for the immediate next block it could be a block without any fault.

The following objectives were set to achieve the target:

(1) to reduce rework of bores from 16% bores per month close to zero without adversely affecting the cycle time,

(2) to improve the overall quality of the process,

(3) to reduce the energy consumption involved in the process by reducing the rework,

(4) to reduce the cost of rework.

It consists of phases, namely, Define, Measure, Analyse, Improve, and Control [31, 32]. The whole Six Sigma project starts with Define phase and is defined based on the customer requirement and company strategy and mission [33, 34]. Measure phase helps the project team to refine the problem and begin the search for various causes of the failure. In Analyse phase, the causes found are analysed using various data analysis tools and the data is validated for Improvement phase. Improvement phase helps in finding solutions and implementing them so that the problems can be eliminated. In Control phase, the performance of the process after Improvement is measured routinely and accordingly adjustments are made in operations. If the Control phase is not implemented, it may revert the project to its previous state.

In the case study presented, the DMAIC methodology was applied to identify the probable sources of deviation in machined surface and successfully reduced the rework to 2.20% from an initial 16% per month. The following sections explain the methodology applied for the purpose.

5.1. Step I—Define

Problem Statement. Reduce engine block liner bore counter depth rework close to zero from 16% bores per month without adversely affecting the cycle time.

The special purpose machine was in regular use with heavy production for a long time. Due to the continuous course of action and heavy load, parts of the machine are worn out. Thus a variation in the depth of step bore was observed as shown in Figure 1. This variation occurred on a number of blocks leading to increased rework. The major concern was the unpredictable behaviour of the machine. Each V-block has two sides: left bank and right bank when looked to from the rear end of the engine. The data showed that the majority of the rework was required on the left bank of the block. If there is a variation in the depth of the bore either if it can be reworked or if the depth is out of rework range, then the entire block is scrapped. Reworking of the cylinder bore is possible at the expense of 369 man-hours per month and approximately INR 60400.

The rework demands skilled manpower due to precise tolerances, man-/machine-hours, and other considerable resources. Thus to increase productivity and reduce the rework cost, there was a need to reduce this rework. It is expected that the depth of each bore must lie within

5.2. Step II—Measure.
In the proposed study, a variation in the depth of step bore was observed on the blocks used for V 12- and V 16-engines. These variations were not uniform and of same pattern.

The detailed data for total number of bores produced from the month of April 2012 to December 2012 was collected. The up and down measurements of both banks were recorded for six months. The measurements for the engine bore which were not in the specified tolerances were also counted in all. The total number of bores produced per month was counted and accordingly the rework percentage was found out by plotting the I-MR chart as shown in Figure 3. The bores which required rework were classified into two major categories: oversize bores and undersize bores. These two categories were again divided into two subcategories depending on the banks of the block where variation was recorded, that is, left bank and right bank. For the collected data, individual value and moving range chart (I-MR) was plotted using Minitab 16 software. I-MR chart plots individual observations on one chart accompanied with another chart of the range of the individual observations, normally from each consecutive data point. Figure 3 is the I-MR chart of the rework data collected during April–December 2012.

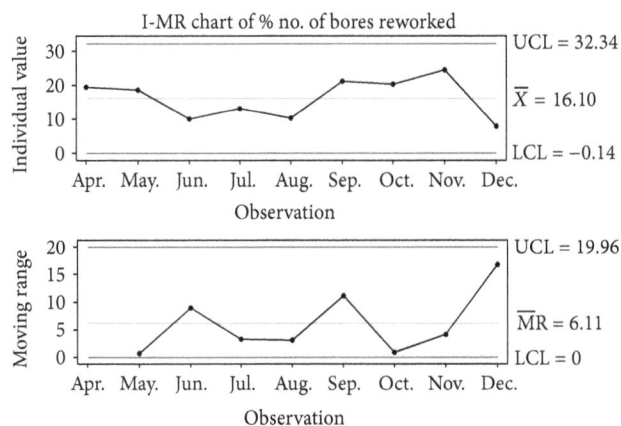

FIGURE 3: I-MR chart of rework data.

It can be inferred from Figure 3 that the average monthly block liner bores reworked for DC are 16% of total production. Each undersize bore required 10 minutes of rework time and each oversize bore required 60 minutes of rework time. The manual rework cost incurred per bore whether oversize or undersize was INR 2.33. Accordingly, eliminating rework would save monthly 369 man-hours and INR 60400. It also saved average sleeve rework cost of INR 30000 per month. Hence the total average monthly cost saving could be INR 90400. The projected annual cost saving could be INR 1,084,800 or USD 19,3700 approximately.

5.3. Step III—Analyse. The Analyse phase is the third and usually the longest phase in the Six Sigma methodology. Most of the crucial data analysis is performed in this phase. This eventually leads you to isolate the root causes of the problem and provides insight into how to eliminate them.

The operational working of the machine was considered for the FTA. FTA is not a cause and effect diagram. FTA can be used when the problem has already occurred in the current business process. As the case of the project was of current business process, FTA was used instead of FMEA. FMEA or failure mode and effect analysis is used for "what can happen," whereas FTA is used for "what has happened." FTA is a method to analyse a failure mode in order to identify possible assignable causes and find the failure mechanism [25]. FTA connects failure mode to assignable causes.

In this case study the fault tree was started from the definition of problem and then it was directed to primary causes and secondary causes. This procedure was followed till all possible causes were listed. FTA provided all areas to be improved in single view and helped in stepwise analysis. The critical parameters were segregated from experience of the persons using the machine and further analysis was carried out on these key input parameters. Figure 4 represents the fault tree drawn for the case study.

Factors that were considered the most influential key inputs are shown in Figure 5.

Once the key inputs were obtained from the FTA, there was a need to check the reliability of all the readings taken

by the operators. This was done by performing measurement system analysis. Three inspectors measured two blocks separately once in a serial order and then in a random order. These readings were analysed using Minitab software to check the gage reproducibility and repeatability [6]. Figure 6 was the outcome of the measurement system analysis.

From the above results around 90% confidence level was obtained. Thus there was no error in measurement system and now all the readings can be called Data.

The third objective of this phase was to find out how they are related. The continuous key inputs, namely, slide pressure, lever pressure, ambient temperature, and oil temperature, were analysed using one of the multivariable regression analyses. Tests that can be used in this phase are regression, correlation, analysis of variance, hypothesis testing, t-tests, chi-squared tests, graphical analyses, GLM, logistic regression, and so forth. These tests come under Multi-Vari Studies. Before proceeding, to select the test, type of data was analysed.

Multi-Vari analysis is a graphical tool, which, through logical subgrouping, analyzes the effects of categorical X's on continuous Y's. The graphical results of Multi-Vari analysis can be quantified using nested analysis of variance.

Multi-Vari was chosen because of the following reasons:

(1) to determine with high statistical confidence the capability of the KPOVs of a process,

(2) to identify assignable causes of variability,

(3) to obtain initial components of variability (shift-to-shift, run-to-run, and operator-to-operator),

(4) to get a first look at process stability over time,

(5) to provide direction and input for design of experiments (DOE) activities.

Selection of Test to Be Performed. The selection of test depends on the type of data whether it is continuous or discrete, single inputs or multiple inputs, single outputs or multiple outputs, and so forth. In this study, depth variation, that is, Y, and all the X's were continuous, so multiple regression analysis was performed as seen from Figure 7.

The general equation of approach was

$$y = f(x_1, x_2, x_3, \ldots, x_k). \tag{1}$$

Depending on the key inputs obtained from the FTA, the data was sorted into continuous and discrete data. Multi-Vari regression analysis was performed for the continuous data. The key inputs varying continuously with time included slide pressure, lever pressure, ambient temperature, and the oil temperature. As the data was categorized as continuous, the data collection was done depending on time. Data was collected during all three shifts. The slide pressure, lever pressure, oil temperature, and ambient temperature were noted for every bore. Per shift 2 engine blocks were considered for this data collection. The data was collected at the start of each shift and at the end of each shift.

Before performing the regression analysis, null hypothesis (H_0) was set. Null hypothesis (H_0) is equal to the specified

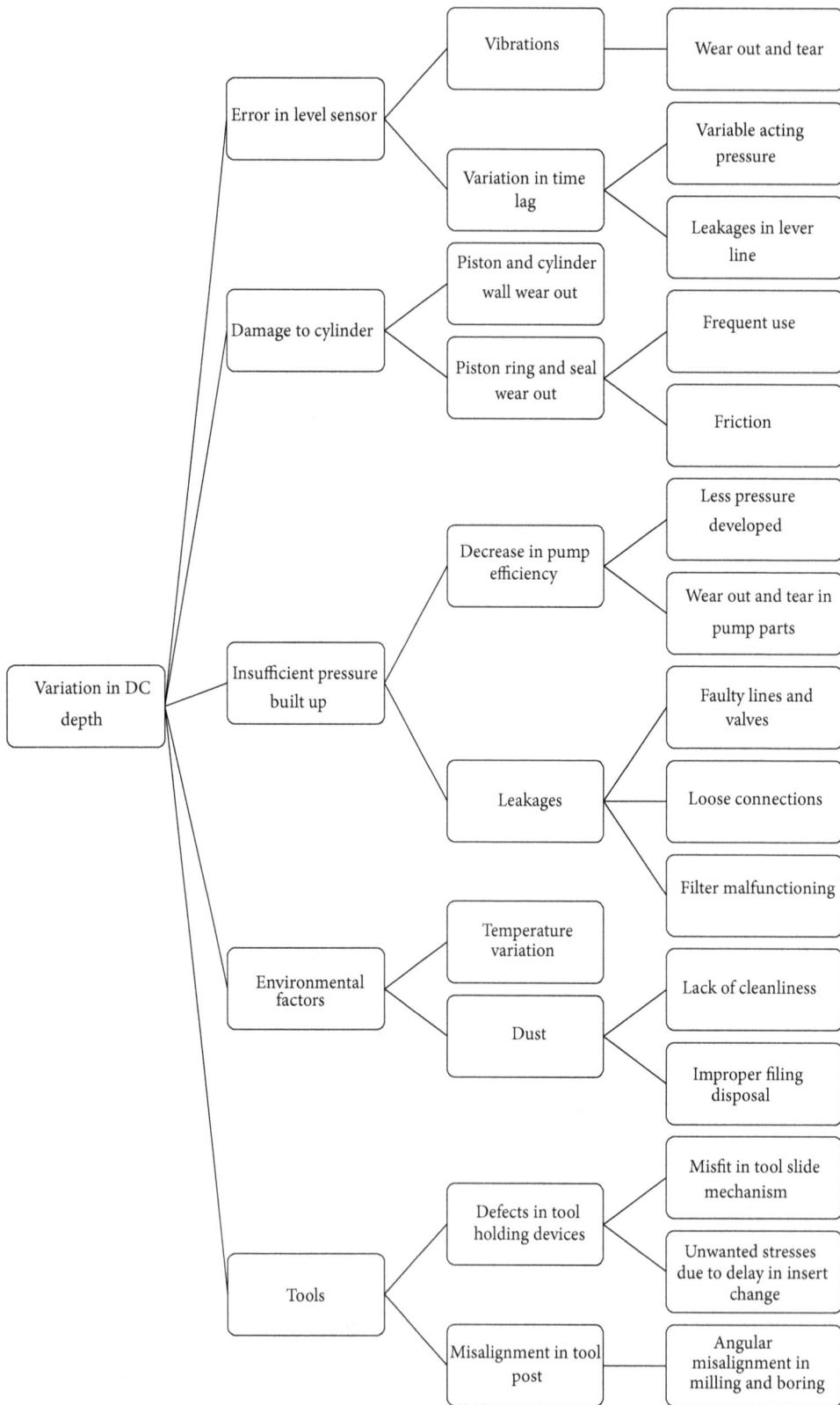

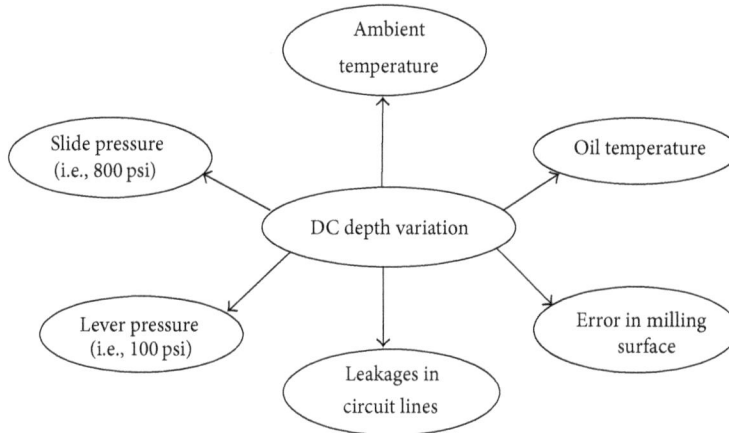

FIGURE 5: FTA key inputs.

value or parameter from another population. Alternative hypothesis (H_a) is not equal to the specified value or parameter from another population. P value is the value used to reject or fail to reject the null hypothesis. A is the probability that true null hypothesis is rejected:

$$\text{If } P \leq \alpha - \text{Reject } H_0,$$
$$\text{If } P > \alpha - \text{Fail to reject } H_0. \tag{2}$$

The statistical analysis is done with the development of a theory, null hypothesis. The analysis will "fail to reject" or "reject" the theory.

Null Hypothesis (H_0): data are independent (not related).

Alternative Hypothesis (H_a): data are dependent (related).

If the P value is ≥ 0.05, then accept the H_0 (no statistical relationship).

If the P value is < 0.05, then reject H_0 (a statistical relationship exists).

According to this theory it was assumed that the inputs ambient temperature, oil temperature, slide pressure, and lever pressure were not affecting the process; that is, the X and Y are not related. Thus it was called null hypothesis (H_0). Once the null hypothesis was set, the very first step was to find the correlations between each of the four inputs, that is, ambient temperature, oil temperature, slide pressure, and lever pressure. Figure 7 shows the correlation results provided by Minitab software. From Table 1 it can be easily seen that correlation exists only between ambient temperature and oil temperature. Thus, for regression, slide pressure and lever pressure can be neglected.

Once the correlation test was done, the next step was to perform multiple regression using the terms obtained from correlation test, that is, oil temperature and ambient temperature.

For the left bank the multiple regression failed. The results and residual plots obtained are shown in Figure 8 and Table 2.

TABLE 1: Correlation results from Minitab.

Correlations: amb temp., oil temp., slide pr., lever pr.		
	Amb. temp.	Oil temp.	Slide pr.
Correlation for left bank		
Oil temp.	0.790		
Slide pr.	*	*	
Lever pr.	*	*	*
Correlation for right bank		
Oil temp.	0.091		
Slide pr.	*	*	
Lever pr.	*	*	*

Cell Contents: Pearson correlation.
*All values in column are identical.

TABLE 2: Minitab results for regression LB.

Regression analysis: down versus amb temp., oil temp.				
The regression equation is				
down = 0.712 + 0.000077 amb. temp. + 0.000126 oil temp.				
21 cases used, 1 case contain missing values				
Predictor	Coef.	SE Coef.	T	P
Constant	0.712296	0.002422	294.07	0.000
Amb. temp.	0.00007696	0.00003924	1.96	0.066
Oil temp.	0.00012577	0.00009384	1.34	0.197
$S = 0.000293334$, R-Sq $= 60.0\%$, R-Sq (adj) $= 55.5\%$				

It can be observed from Table 2 that the P value is zero, that is, ≤ 0.05. Thus, null hypothesis (H_0) is accepted. But the variance, that is, R-Sq, value is just 60%. The R-Sq value must be at least 80% for multiple regressions to be successful.

Figure 9 and Table 3 show the residual plots and results for Right Bank of the engine block. It can be seen that the R-Sq value is very low for right bank, which is just 39%. Thus these results are strictly rejected. The multiple regression analysis for left as well as right banks is not successful.

The null hypothesis (H_0) set that there is no relation between the X's and the Y was true. Thus we fail to reject

MSA study liner bore counter depth -KVBLOCK -INGERSOLL-AUG'10

Gage name: depth dial
Date of study: 24/09/10

Reported by: Anil
Tolerance: 0.0001"
Misc:

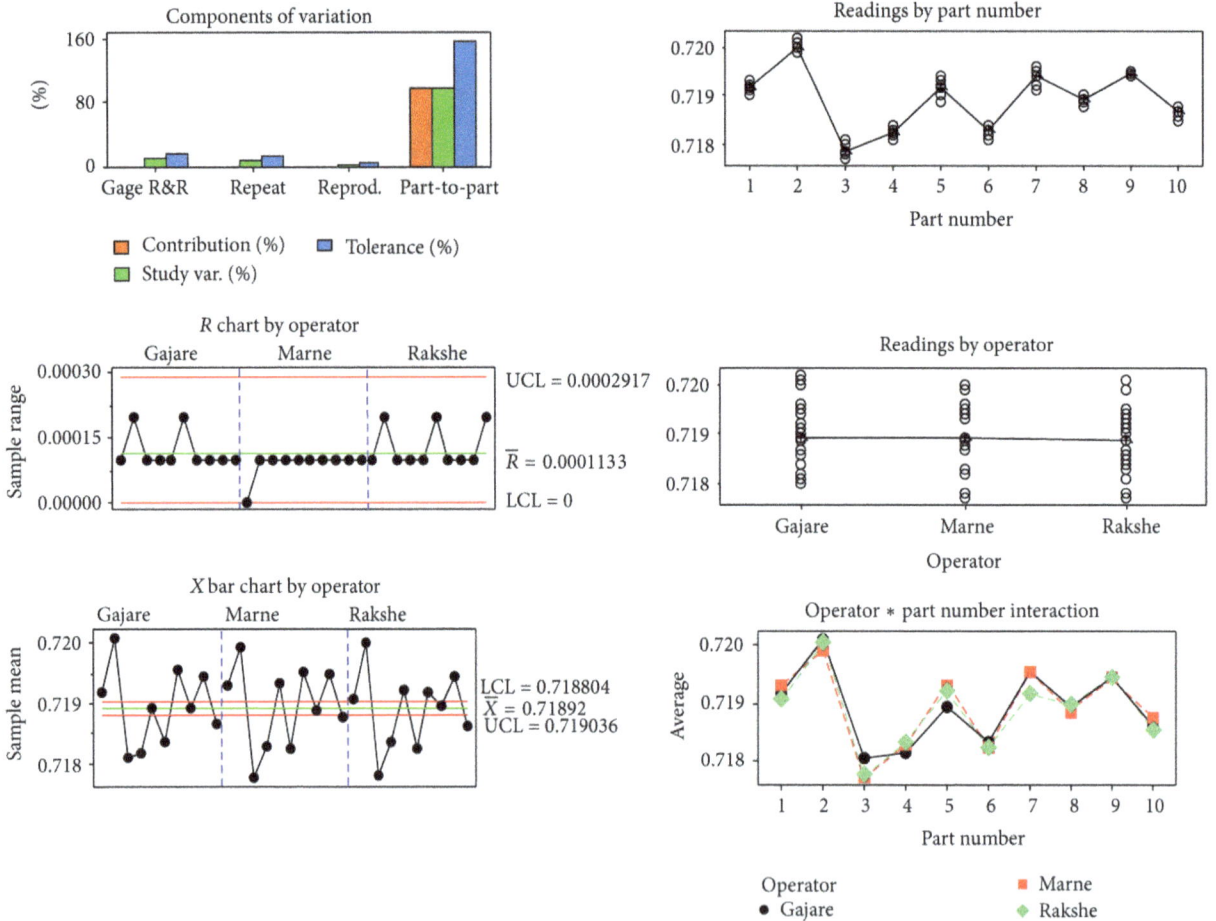

FIGURE 6: MSA plots.

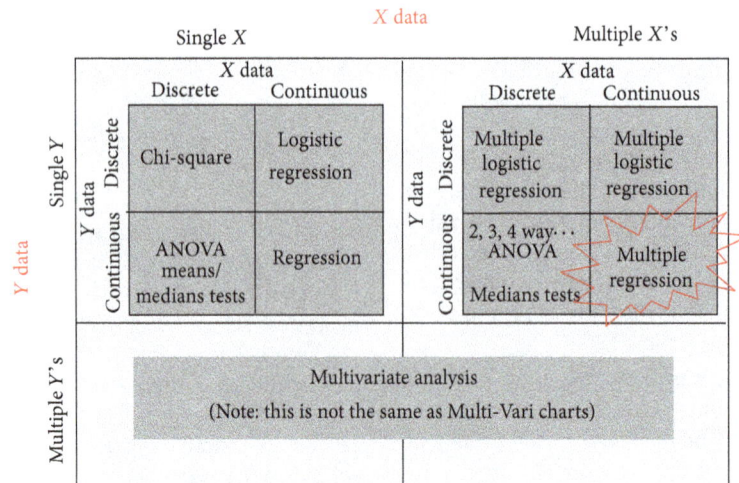

FIGURE 7: Selection of test for analysis [35].

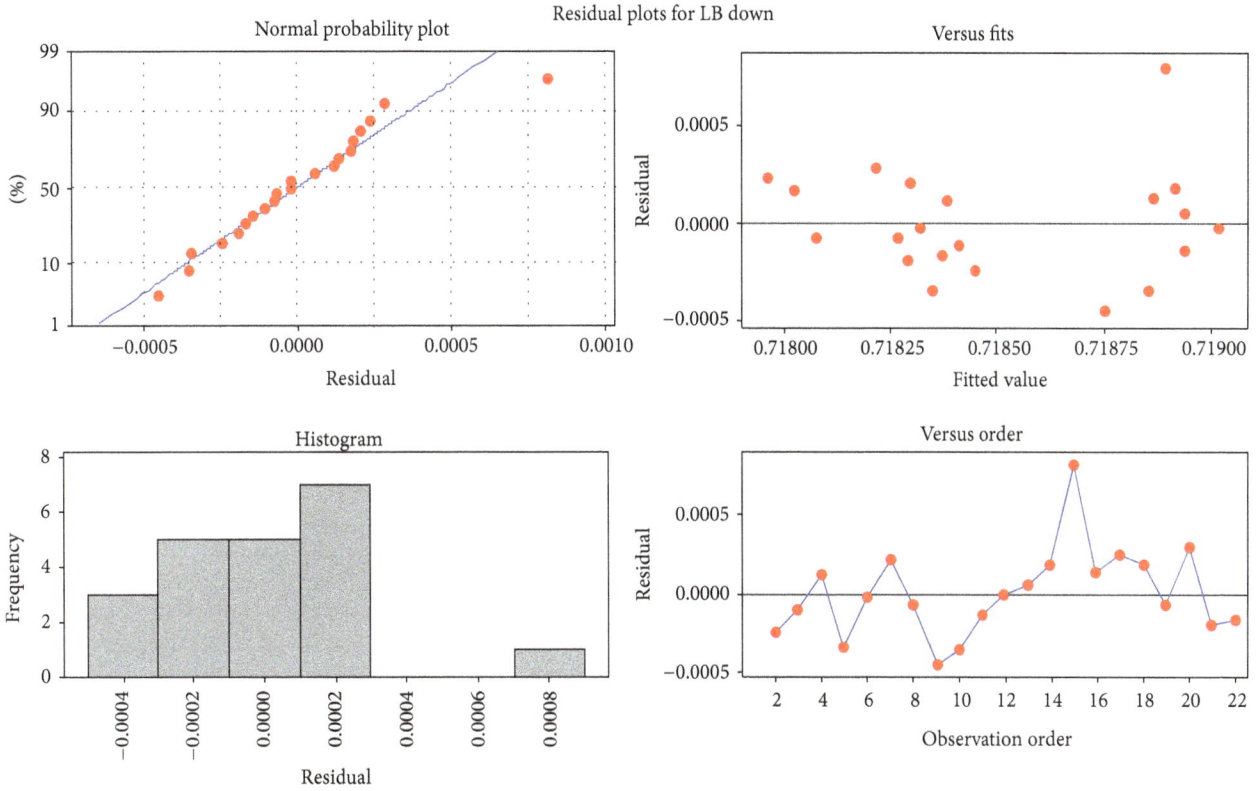

FIGURE 8: Residual plots of LB.

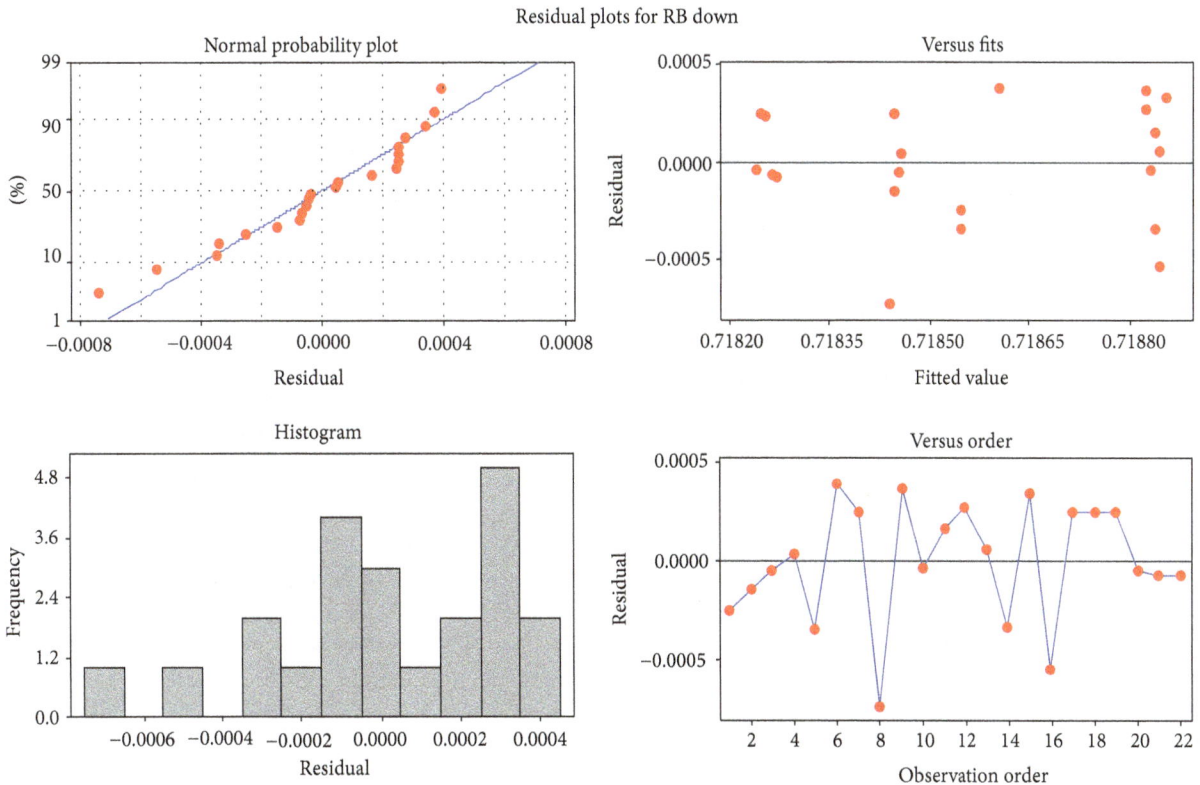

FIGURE 9: Residual plots for right bank.

Table 3: Minitab results for multiple regression of RB.

Regression analysis: down versus amb. temp., oil temp.				
The regression equation is				
down = 0.718 + 0.000097 amb. temp. − 0.000060 oil temp.				
Predictor	Coef.	SE Coef.	T	P
Constant	0.718100	0.003310	216.97	0.000
Amb. temp.	0.00009655	0.00002780	3.47	0.003
Oil temp.	−0.00005955	0.00009589	−0.62	0.542
$S = 0.000320196$, R-Sq = 39.0%, R-Sq (adj) = 32.6%				

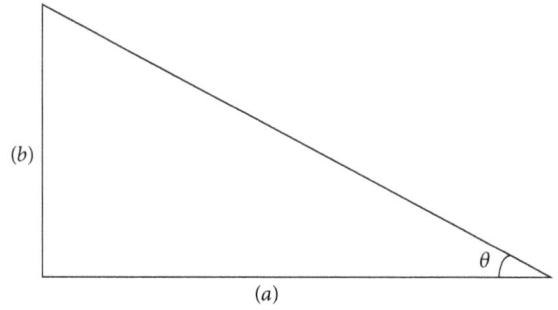

Figure 10: Milling head slide sleeve front view.

the null hypothesis (H_0). It was concluded that the four key process input variables do not really dominate in contribution to key process output variable. Hence some other parameters were dominating the output. This could be easily seen from the regression equation as well as the residual plot. Two rifle shot inputs were obtained from the FTA, namely, leakages in the hydraulic circuit and errors in milling surface.

5.4. Step IV—Improvement. As the hydraulic circuit worked with high pressure of 800–1000 psi, leakage was the factor responsible for pressure drop in the circuit. Hydraulic circuit was of very complex nature, involving many directional control valves and many pressure switches. There were many pressure switching actions causing stress on various joints. These leakages were creating problem to maintain pressure. Leakages were observed on both banks of the machine. As pressure maintaining was critical, all leakages were removed. These leakages were removed by cleaning all the pipes and valves in the circuit and changing pipes which were cut. After removing the leakages the rework percentage dropped but the change was not significant. Thus it was decided to check for milling surface error and remove if any.

Boring operation needs to be performed precisely as it can go wrong very easily. It requires precise alignment with drilled hole as well as surface on which hole is drilled. Alignment with drilled hole never creates a problem. Alignment with surface was another critical issue involved, especially when the surface was inclined. Surface alignment with boring tool was perfectly perpendicular when the machine was manufactured. But in course of time due to vibrations and other undesirable actions, misalignment was produced in the milling head, that is, milling surface, and boring tool, which created undesirable difference between up and down depths of step bore. On performing the analysis it was observed that the difference between up and down readings was $0.0040''$ which was almost 40% of total allowable tolerance. Thus a sleeve of $0.0040''$ was manufactured and inserted behind the milling head. The front view of sleeve is shown in Figure 10 where "b" = $0.0040''$.

This solved the problem significantly and the results were proven by plotting I-MR charts for rework.

After making the suggested improvements, the rework data was collected similarly as collected in Define phase. The I-MR chart was plotted and both charts were compared to study the results obtained before and after making improvements.

The I-MR chart of the revised rework for 3 months is shown below. It can be seen that the rework has been reduced to approximately 0% in March 2013. The value of mean rework for the months of April 2012–December 2012 was 16%. After doing the improvements to the machine the rework reduced continuously from January 2013. In January the rework was 4.22%. For the month of February 2013 the rework was further reduced to 2.33%. The main objective was achieved in March 2013. The percentage rework dropped drastically to 0.33%. Thus the target to make rework close to zero was successfully achieved.

At the beginning there was a lot of variation in the depths of bores of a single block. With reduction in rework, the other aim was to reduce this variation in the depth of the bores. This variation in depths was nonuniform. As discussed earlier the variation for a bore may go oversize and the very next bore would be undersize. On completion of the Improvement Phase, the box plot was plotted to compare the moving rages of the depths of the blocks which is shown in Figure 12. It could be observed that the moving range in March 2012 varied from $0.7170''$ to $0.7210''$ for a block. In November 2012, this moving range was decreased. It varied from $0.7183''$ to $0.7205''$. Thus the improvement could be seen. In March 2013 this range was drastically decreased and the new range variation was between $0.7187''$ and $0.7195''$ thus making the variation uniform within less moving range.

5.5. Step V—Control. After completing the Improve phase, factors affecting the depth variation of the step bore were proposed. The actions proposed were implemented in the manufacturing process. The results of these improvements were monitored in Control phase. A control plan was prepared which is the major action of this phase. This control plan consisted of all the actions that were proposed for decreasing the rework of the blocks. It included training and certifying the operators, employees, maintenance plan preparation, regular inspection, and preparation of control charts. And thus from Figure 11, it can be observed that the goal set of reducing the rework to zero percent was achieved.

6. Results

The case study was carried out on a special purpose machine developed by Ingersoll. The machine was in continuous

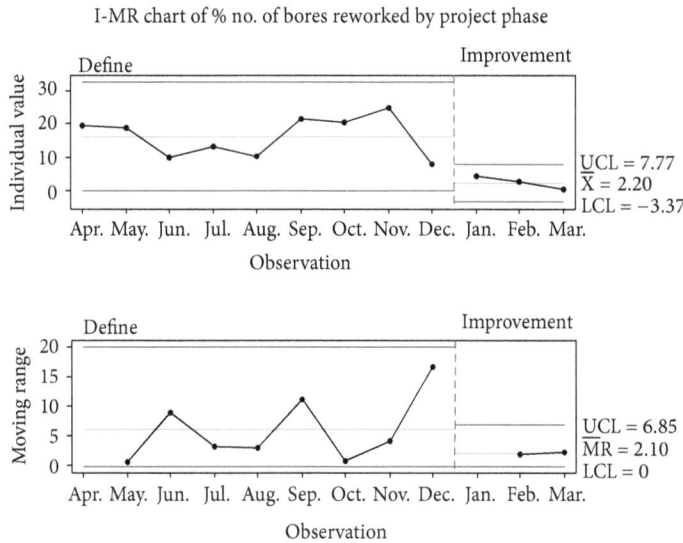

FIGURE 11: I-MR chart after improvement.

production for last 40 years. The hydraulic circuit components, seals, hoses were worn out resulting in the inefficient working of the machine. This was leading to high percentage rejection of the engine blocks. The study was carried out in phases and the principles of DMAIC were proved to be useful for reducing the rework rate and hence improving the productivity of the machine. As the machine was in operation and there were many factors contributing to deformation of the surface, it was difficult to carry out the experimentation for finding the reasons for the rework. Hence the FTA was selected for the purpose.

At the first stage the goals were set to reduce the rework from 16% to the minimum possible value as the cost of rework was very high and small deviation in work could reject the entire engine block. Later at Measurement phase, the actual measurement of deviation was carried out. It was found that the left bank of the engine was prone to deviation as compared to the right bank. Hence concentration was focused on this part of the engine block. The analysis of the deviation was an important issue. Probable reasons for the deviation were listed and categorized and the FTA was performed. After discussion with the experienced staff actually working on the machine, the principle factors contributing more to the deviation were identified for the study. The key inputs thus obtained from the FTA were needed to be checked for the reliability of all the readings taken by the operators. This was done by performing measurement system analysis. The results are shown in Figure 6. The multivariable regression analysis was performed to understand the relationship between the parameters. Figure 8 shows the residual plots for the left bank; from Table 2 it was observed that the R-Sq value is just 60% for the left bank. Figure 9 shows residual plots for right bank, and from Table 3 it was observed that the R-Sq value is just 39%. For the multivariable regression test, to be successful, the R-Sq value must be at least 80%. Thus it was found that the four key process input variables were not

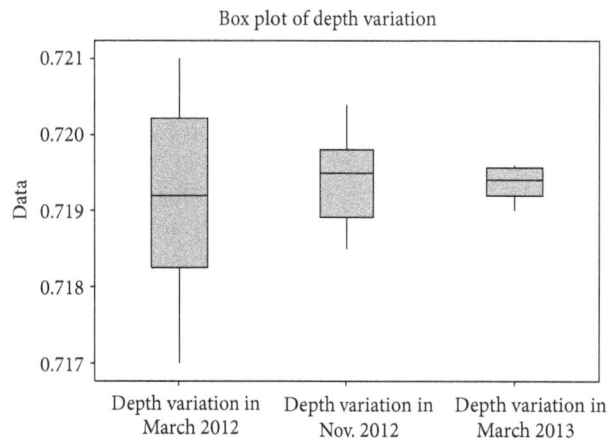

FIGURE 12: Box plot for depth variation.

dominating the key process output variable on both banks. Thus some other parameters were dominating the output. FTA provided the inputs, namely, leakages in the hydraulic circuit and errors in locating milling surface which were then studied for the performance.

The entire machine was operating on a complex hydraulic circuit with an oil pressure in the range of 800–1000 psi. The leakages in the hydraulic circuit were traced and removed. The results of this step showed improvement in the process but were not significant. Another parameter was the milling head location error. It was removed by inserting a sleeve as shown in Figure 11. This time the efforts worked and the rework was reduced drastically close to 2.20% on average for 3 months. A control plan for all the contributing factors was prepared for reducing the rework. It included training and certifying the operators, employees, maintenance plan preparation, regular inspection, and preparation of control charts for further reduction in the rework. Figure 12 shows the improvement in the process after the DMAIC was successfully implemented for the machine.

7. Conclusion

Industries have to deal with a host of problems related to quality control. Substandard quality hampers the productivity of the plant which directly affects the company targets. Organizations have to suffer huge losses which are not easy to cope up with. Thus there is a need to improve the productivity simultaneously keeping in mind the quality of the product. Six Sigma can be effectively applied and the existing business processes can be improved and made error free. Six Sigma provides statistical proof to each and every action, thus helping making decisions more efficient. It can work even with less number of readings in the database. Thus Six Sigma is completely an industry oriented methodology of quality and productivity improvement.

In the presented case study the rework percentage was much higher, that is, 16%. The firm had to sustain the rework cost and the man-hours required in the reworking decreased the productivity. Establishing the relationship between the input parameters and the output parameter is a challenge in a complex system like the one discussed above. The decision of using Six Sigma methodology proved to be facile. FTA was implemented to find all the key inputs that are affecting the output. The KPIVs were categorised as continuous and discrete depending on their property whether they vary with time or not and were analysed using multivariable regression analysis. The Multi-vari regression analysis proved that the selected continuous parameters were not dominating the output, failing to reject the null hypothesis (H_0); that is, these input variables did not affect the output. Hence, the rifle shot input parameters leakages in hydraulic circuit and milling surface errors were checked. Leakages were observed in the circuit for the left bank. These leakages were removed. The milling surface error was removed by inserting a sleeve as shown in Figure 10. After these errors were removed, the rework reduced to 2.20% per month thus achieving the set goal. The rework time of 369 hours per month was reduced to 42 man-hours per month. The cost of rework was reduced to USD 3500 per year. Thus there was significant improvement in the productivity and losses the firm incurred.

It is thus concluded that Six Sigma methodologies could be applied successfully in small firms. Practitioners could refer this case study and implement it in a similar kind of study. With the help of case study we try to prove that Six Sigma tools help to reduce the wastage and help improve quality of product. In this study we had considered limited parameters that affected the output. Depending upon the expert's experience, the study and improvements were done on milling surface errors and leakages. By performing experimentation on other parameters from the KPIVs, further improvements could be done. This would reduce the rework closer to zero in the future. DOE can be planned for the remaining KPIVs which can provide detailed effects on the output.

Abbreviations

DMAIC: Define, Measure, Analyse, Improve, and Control
DMADV: Define, Measure, Analyse, Design, and Verify
ROA: Return on asset
FTA: Fault tree analysis
FMEA: Failure mode and effect analysis
MSA: Measurement system analysis
DOE: Design of experiments
H_0: Null hypothesis
H_a: Alternative hypothesis
R-Sq: R-squared value
KPIV: Key process input variables
KPOV: Key process output variable.

References

[1] Andrew Spencer, *Optimising Surface Textures for Combustion Engine Cylinder Liners*, Luleå University of Technology, 2010.

[2] "What is Cylinder Liner?" http://www.wisegeek.com/what-is-a-cylinder-liner.htm.

[3] J. Antony and R. Banuelas, "A strategy for survival," *Manufacturing Engineer*, vol. 80, no. 3, pp. 119–121, 2001.

[4] G. Büyüközkan and D. Öztürkcan, "An integrated analytic approach for six sigma project selection," *Expert Systems with Applications*, vol. 37, no. 8, pp. 5835–5847, 2010.

[5] A. Y. T. Szeto and A. H. C. Tsang, "Antecedent to successful implementation of six sigma," *Journal of Six Sigma and Competitive Advantage*, vol. 1, no. 3, pp. 307–322, 2005.

[6] H. C. Hung and M. H. Sung, "Applying six sigma to manufacturing processes in the food industry to reduce quality cost," *Scientific Research and Essays*, vol. 6, no. 3, pp. 580–591, 2011.

[7] A. Saghaei, H. Najafi, and R. Noorossana, "Enhanced rolled throughput yield: a new six sigma-based performance measure," *International Journal of Production Economics*, vol. 140, no. 1, pp. 368–373, 2012.

[8] F. W. Breyfogle, *Implementing Six Sigma: Smarter Solutions Using Statistical Methods*, John Wiley & Sons, New York, NY, USA, 1999.

[9] X. Zu, L. D. Fredendall, and T. J. Douglas, "The evolving theory of quality management: the role of six sigma," *Journal of Operations Management*, vol. 26, no. 5, pp. 630–650, 2008.

[10] C. T. Su and C. J. Chou, "A systematic methodology for the creation of six sigma projects: a case study of semiconductor foundry," *Expert Systems with Applications*, vol. 34, no. 4, pp. 2693–2703, 2008.

[11] U. D. Kumar, D. Nowicki, J. E. Ramírez-Márquez, and D. Verma, "On the optimal selection of process alternatives in a six sigma implementation," *International Journal of Production Economics*, vol. 111, no. 2, pp. 456–467, 2008.

[12] A. A. Junankar and P. N. Shende, "Minimization of rework in belt industry using dmaic," *International Journal of Applied Research in Mechanical Engineering*, vol. 1, no. 1, 2011.

[13] Y. H. Kwak and F. T. Anbari, "Benefits, obstacles, and future of six sigma approach," *Technovation*, vol. 26, no. 5-6, pp. 708–715, 2006.

[14] M. Swink and B. W. Jacobs, "Six sigma adoption: operating performance Impacts and contextual drivers of success," *Journal of Operations Management*, vol. 30, no. 6, pp. 437–453, 2012.

[15] J. A. Blakeslee Jr., "Implementing the six sigma solution," *Quality Progress*, vol. 32, no. 7, pp. 77–85, 1999.

[16] G. J. Hahn, W. J. Hill, R. W. Hoerl, and S. A. Zinkgraf, "The impact of six sigma improvement—a glimpse into the future of statistics," *The American Statistician*, vol. 53, no. 3, pp. 208–215, 1999.

[17] M. Harry and R. Schroeder, *Six Sigma: The Breakthrough Management Strategy Revolutionising the World's Top Corporations*, Doubleday, New York, NY, USA, 2000.

[18] M. J. Braunscheidel, J. W. Hamister, N. C. Suresh, and H. Star, "An institutional theory perspective on six sigma adoption," *International Journal of Operations and Production Management*, vol. 31, no. 4, pp. 423–451, 2011.

[19] X. Li, J. Zhan, F. Jiang, and S. Wang, "Cause analysis of bridge erecting machine tipping accident based on fault tree and corresponding countermeasures," *Procedia Engineering*, vol. 45, pp. 43–46, 2012.

[20] D. M. Shalev and J. Tiran, "Condition-based fault tree analysis (CBFTA): a new method for improved fault tree analysis (FTA), reliability and safety calculations," *Reliability Engineering and System Safety*, vol. 92, no. 9, pp. 1231–1241, 2007.

[21] E. V. Gijo, J. Scaria, and J. Antony, "Application of six sigma methodology to reduce defects of a grinding process," *Quality and Reliability Engineering International*, vol. 27, no. 8, pp. 1221–1234, 2011.

[22] M. Xia, X. Li, F. Jiang, and S. Wang, "Cause analysis and countermeasures of locomotive runway accident based on fault tree nalysis method," *Procedia Engineering*, vol. 45, pp. 38–42, 2012.

[23] Y. Wang, Q. Li, M. Chang, H. Chen, and G. Zang, "Research on fault diagnosis expert system based on the neural network and the fault tree technology," *Procedia Engineering*, vol. 31, pp. 1206–1210, 2012.

[24] R. McClusky, "The rise, fall and revival of six sigma," *Measuring Business Excellence*, vol. 4, no. 2, pp. 6–17, 2000.

[25] A. E. Summers, "Achieving six sigma through FTA," in *Proceedings of the Process Plant Reliability Symposium*, Houston, Tex, USA, October 1997.

[26] H. de Koning and J. de Mast, "A rational reconstruction of six-sigma's breakthrough cookbook," *International Journal of Quality and Reliability Management*, vol. 23, no. 7, pp. 766–787, 2006.

[27] J. de Mast and J. Lokkerbol, "An analysis of the six sigma DMAIC method from the perspective of problem solving," *International Journal of Production Economics*, vol. 139, no. 2, pp. 604–614, 2012.

[28] M. H. C. Li, A. A. Al-Refaie, and C. Y. Yang, "DMAIC approach to improve the capability of SMT solder printing process," *IEEE Transactions on Electronics Packaging Manufacturing*, vol. 31, no. 2, pp. 126–133, 2008.

[29] M. H. C. Li and A. Al-Refaie, "Improving wooden parts' quality by adopting DMAIC procedure," *Quality and Reliability Engineering International*, vol. 24, no. 3, pp. 351–360, 2008.

[30] D. Starbird, "Business excellence: Six Sigma as a management system," in *Proceedings of the Annual Quality Congress*, pp. 47–55, Milwaukee, Wis, USA, May 2002.

[31] http://en.wikipedia.org/wiki/Six_Sigma.

[32] "What is Six Sigma," http://www.isixsigma.com/new-to-six-sigma/getting-started/what-six-sigma/.

[33] G. W. Frings and L. Grant, "Who moved my sigma ...effective implementation of the six sigma methodology to hospitals," *Quality and Reliability Engineering International*, vol. 21, no. 3, pp. 311–328, 2005.

[34] R. McAdam and A. Evans, "Challenges to six sigma in a high technology mass-manufacturing environments," *Total Quality Management and Business Excellence*, vol. 15, no. 5-6, pp. 699–706, 2004.

[35] http://www.six-sigma-material.com/Hypothesis-Testing.html.

22

A Family of Lifetime Distributions

Vasileios Pappas,[1] Konstantinos Adamidis,[2] and Sotirios Loukas[1]

[1] Department of Mathematics, University of Ioannina, 45110 Ioannina, Greece
[2] Department of Business Administration of Food and Agricultural Enterprises, University of Ioannina, 30100 Agrinio, Greece

Correspondence should be addressed to Konstantinos Adamidis, cadamid@cc.uoi.gr

Academic Editor: Suk joo Bae

A four-parameter family of Weibull distributions is introduced, as an example of a more general class created along the lines of Marshall and Olkin, 1997. Various properties of the distribution are explored and its usefulness in modelling real data is demonstrated using maximum likelihood estimates.

1. Introduction

Probability distributions are often used in survival analysis for modeling data, because they offer insight into the nature of various parameters and functions, particularly the failure rate (or hazard) function. Throughout the last decades, a considerable amount of research was devoted to the creation of lifetime models with more than the classical increasing and decreasing hazard rates; apparently, the motivation for this trend was to provide with more freedom of choice in the description of complex practical situations (see e.g., [1–9], and the references therein). In this paper a general class of models is introduced, by adding an extra parameter to a distribution in the sense of Marshall and Olkin [10], and subsequently used in developing a four-parameter modified Weibull extension distribution, with various failure rate curves that compete well with other alternatives in fitting real data. Specifically, Xie et al. [11] generalized the Chen [12] distribution by adding the lacking scale parameter, thus creating a three-parameter Weibull distribution; although the variety of shapes of the reliability curves was not enriched, the resulting model provided better fit to real data. The proposed distribution extends the Xie et al. [11] distribution by adding a shape parameter; it will be seen that compared to the previous and other models, the cost of the addition is balanced by the improvement in fitting real data.

The paper is organized as follows. Section 2 includes the general class of models and some properties. The proposed four-parameter Weibull model is introduced in Section 3 and some properties and reliability aspects are studied. The parameters are estimated by the method of maximum likelihood and the observed information matrix is obtained; the fit of the proposed distribution to two sets of real data is examined against three and two parameter competitors.

2. The Class of Distributions

It is possible to generalize a distribution by adding a shape parameter, in the sense of Marshall and Olkin [10]. Thus, starting with a distribution with survival function s_0, the survival function of the proposed family with the additional parameter p is given by

$$s(x) = \frac{\ln\{1 - (1 - p)s_0(x)\}}{\ln p}, \quad x \in \mathbb{R}, \ p \in \mathbb{R}_+ - \{0\}, \tag{1}$$

and when $p \to 1$, then $s \to s_0$. The probability density and hazard functions are readily found to be

$$f(x) = \frac{(p - 1)f_0(x)}{\{1 - (1 - p)s_0(x)\}\ln p},$$

$$h(x) = \frac{(p - 1)s_0(x)h_0(x)}{\{1 - (1 - p)s_0(x)\}\ln\{1 - (1 - p)s_0(x)\}}, \tag{2}$$

where f_0 and h_0 are the probability density and hazard functions corresponding to the distribution with survival function s_0, and $x \in \mathbb{R}$, $p \in \mathbb{R}_+ - \{0\}$. Since,

$$\lim_{x \to -\infty} h(x) = \frac{p-1}{p \ln p} \lim_{x \to -\infty} h_0(x), \qquad \lim_{x \to +\infty} h(x) = \lim_{x \to +\infty} h_0(x) \tag{3}$$

it follows from (2) that

$$\frac{p-1}{p \ln p} h_0(x) \le h(x) \le h_0(x), \quad x \in \mathbb{R}, \ p \ge 1,$$

$$h_0(x) \le h(x) \le \frac{p-1}{p \ln p} h_0(x), \quad x \in \mathbb{R}, \ p \in (0,1]. \tag{4}$$

Therefore, $h(x)/h_0(x)$ with $x \in \mathbb{R}$ is increasing for $p \ge 1$ and decreasing for $p \in (0,1]$. When $s_0(0) = 1$, the hazard function at the origin, $h(0)$, behaves quite differently than the corresponding functions for the Weibull and gamma distributions; for both these families, the distribution can be exponential, or $h(0) = 0$, or $h(0) = \infty$, so that $h(0)$ is discontinuous in the shape parameter. This is not the case for the hazard functions in (2), and therefore the proposed family may be useful in fine-tuning the distribution with survival function s_0.

3. A Weibull Extension Model

The survival function of the modified Weibull extension distribution, introduced by Xie et al. [11] and studied further by Tang et al. [13], is given by

$$s_0(x) = e^{-\alpha \lambda (e^{(x/\alpha)^\beta} - 1)}, \tag{5}$$

for $x, \alpha, \beta, \lambda \in \mathbb{R}_+ - \{0\}$; hereinafter we shall be referring to the distribution with survival function given by (5) as the XTG distribution for brevity. By substituting (5) in (1), the survival function of the proposed distribution is obtained in the form

$$s(x; \theta) = \frac{\ln\left\{1 - (1-p)e^{-\alpha\lambda(e^{(x/\alpha)^\beta} - 1)}\right\}}{\ln p}, \tag{6}$$

where $\theta = (\alpha, \beta, \lambda, p)$ and $x, \alpha, \beta, \lambda, p \in \mathbb{R}_+ - \{0\}$. From (2) or (6), the pdf is readily obtained as

$$f(x; \theta) = \frac{\beta\lambda(p-1)(x/\alpha)^{\beta-1}e^{(x/\alpha)^\beta - \alpha\lambda(e^{(x/\alpha)^\beta} - 1)}}{\left\{1 - (1-p)e^{-\alpha\lambda(e^{(x/\alpha)^\beta} - 1)}\right\}\ln p}, \tag{7}$$

where α is a scale parameter and β, λ, p are shape parameters. It can be shown that the pdf is monotone decreasing, unimodal or even roller-coaster type; the different shapes of the pdf are illustrated in Figure 1 for selected values of the parameters.

Clearly, for $p \to 1$ the proposed distribution reduces to the XTG distribution therefore the proposed model can be viewed as an extension of the XTG model (which is asymptotically related to the usual two-parameter Weibull

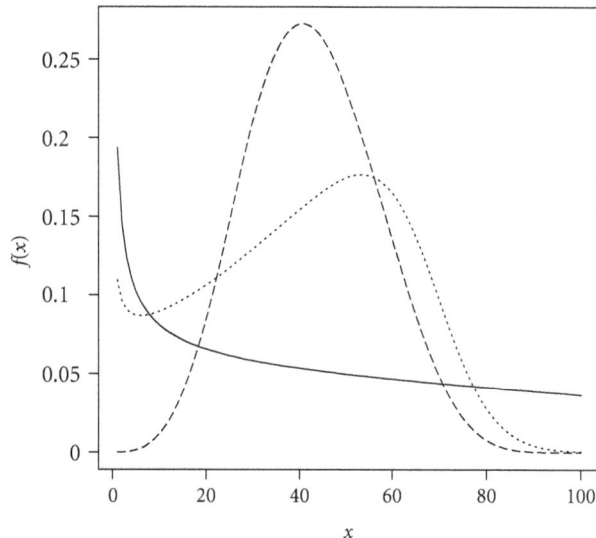

FIGURE 1: Probability density functions of the EXTG distribution for $\alpha = 2$, $\beta = 0.5$, $\lambda = 0.03$, $p = 0.2$ (solid line), $\alpha = 12$, $\beta = 4$, $\lambda = 2$, $p = 0.2$ (dashed line), and $\alpha = 2$, $\beta = 0.7$, $\lambda = 0.3$, $p = 0.2$ (dotted line).

distribution) and if, in addition, $\alpha = 1$, then (7) defines the Chen [12] distribution; hereinafter we shall be referring to this distribution as the extended XTG distribution (EXTG distribution for brevity). Furthermore, it can be shown that for $p \in (0, 1)$ (7) is a compound of the logarithmic and the XTG distributions. Indeed, by incorporating the results of Barlow and Proschan [14] and Arnold et al. [15], consider the lifetime $X = \min(X_1, X_2, \ldots, X_Z)$ of a "series-system" of Z identical components, where failure occurs if at least one component ceases to function. If the lifetimes of the components are iid random variables with survivals given by (5) and the distribution of their number Z is logarithmic, independently of the X's, with pmf

$$p(z; p) = -\frac{(1-p)^z}{z \ln p}, \tag{8}$$

for $z \in \mathbb{N} - \{0\}$, $p \in (0, 1)$, then the distribution of $X \mid Z$ has pdf

$$f(x \mid z; \beta, \lambda) = \beta\lambda z \left(\frac{x}{\alpha}\right)^{\beta-1} e^{(x/\alpha)^\beta - \alpha\lambda(e^{(x/\alpha)^\beta} - 1)}, \tag{9}$$

for $x, \alpha, \beta, \lambda \in \mathbb{R}_+ - \{0\}$, and the distribution of X is the EXTG with pdf given by (7).

The calculations of the rth raw moments of the EXTG distribution involve the use of standard numerical integration procedures available in every mathematical package; for $p \in (0, 1)$ they can be expressed in the form,

$$E(X^r) = -\frac{1}{\ln p} \sum_{k=1}^{\infty} \frac{(1-p)^k}{k} \int_0^\infty \alpha^r \left\{\ln\left(\frac{x}{\alpha\kappa\lambda} + 1\right)\right\}^{r/\beta} e^{-x} dx. \tag{10}$$

By straightforward reversal of the cdf, obtained from (6) using that $F(x; \theta) = 1 - S(x)$, the quantile function is calculated to be

$$F^{-1}(q) = \alpha \left[\ln \left\{ 1 - \frac{1}{\alpha\lambda} \ln \left(\frac{1 - p^{1-q}}{1 - p} \right) \right\} \right]^{1/\beta}, \quad (11)$$

for $p \in (0, 1)$; hence the median is $M = \alpha[\ln\{1 + (1/\alpha\lambda)\ln(1 + \sqrt{p})\}]^{1/\beta}$.

3.1. Failure Rate and Mean Residual Life Functions.

From (6) and (7) the failure rate (also known as hazard rate) function of the EXTG distribution is

$$h(x; \theta)$$

$$= \frac{\beta\lambda(p-1)(x/\alpha)^{\beta-1}e^{(x/\alpha)^{\beta}-\alpha\lambda(e^{(x/\alpha)^{\beta}}-1)}}{\left\{1 - (1-p)e^{-\alpha\lambda(e^{(x/\alpha)^{\beta}}-1)}\right\}\ln\left\{1 - (1-p)e^{-\alpha\lambda(e^{(x/\alpha)^{\beta}}-1)}\right\}}.$$

$$(12)$$

It can be shown that for $\beta \geq 1$, the EXTG is an IFR distribution [16]. However, for $\beta < 1$ it can be IFR, DFR, and BTFR distribution, although it is not easy to determine analytically the ranges of the parameter values; the IFR, DFR, and BTFR characteristics are depicted in Figure 2 for selected values of the parameters. Given that there is no failure prior to x_0, the residual life is the period from time x_0 until the time of failure. The mean residual lifetime, for $p \in (0, 1)$, is

$$m(x_0; \theta) = E(X - x_0 \mid X \geq x_0)$$

$$= \int_{x_0}^{\infty} \frac{(x - x_0)f(x; \theta)}{1 - F(x_0; \theta)} dx$$

$$= -\frac{1}{\ln\left\{1 - (1-p)e^{-\alpha\lambda(e^{(x_0/\alpha)^{\beta}}-1)}\right\}} \sum_{k=1}^{\infty} \frac{(1-p)^k}{k}$$

$$\times \int_{\alpha k\lambda(e^{(x_0/\alpha)^{\beta}}-1)}^{\infty} \left[\alpha \left\{ \ln\left(\frac{x}{\alpha k\lambda} + 1 \right) \right\}^{1/\beta} - x_0 \right] e^{-x} dx.$$

$$(13)$$

Other reliability aspects of the distribution can be obtained numerically. For example, the renewal function, which is important for maintenance, can be calculated approximately either by the well-known method of considering its limit at infinity and the first and second raw moments given by (10), or by applying the method of the linear combination of the cdf and the hazard function; see Cui and Xie [17], Jiang [18] and the references therein.

3.2. Inference.

Assuming a random sample of n observations, $y_{\text{obs}} = (x_i; i = 1, \ldots, n)$, from (7), the log-likelihood is given by

$$\ell(\theta; y_{\text{obs}}) = n[\ln\{(p-1)\beta\lambda\} - \ln(\ln p) + \alpha\lambda + (1-\beta)\ln\alpha]$$

$$+ (\beta - 1) \sum_{i=1}^{n} \ln x_i + \sum_{i=1}^{n} \left(\frac{x_i}{\alpha} \right)^{\beta} - \alpha\lambda \sum_{i=1}^{n} e^{(x_i/\alpha)^{\beta}}$$

$$- \sum_{i=1}^{n} \ln\left\{ 1 - (1-p)e^{-\alpha\lambda(e^{(x_i/\alpha)^{\beta}}-1)} \right\},$$

$$(14)$$

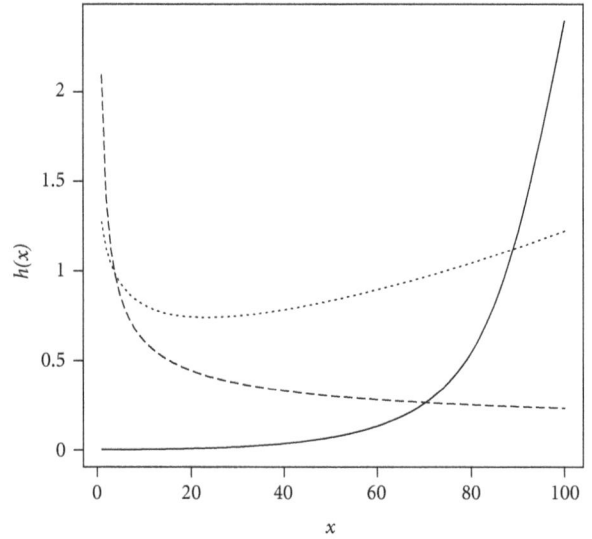

FIGURE 2: Hazard rate functions of the EXTG distribution for $\alpha = 0.05$, $\beta = 0.45$, $\lambda = 0.002$, $p = 20$ (solid line), $\alpha = 0.2$, $\beta = 0.2$, $\lambda = 3$, $p = 2$ (dashed line), and $\alpha = 10$, $\beta = 0.9$, $\lambda = 0.5$, $p = 200$ (dotted line).

and by differentiating with respect to θ the gradients are

$$\frac{\partial \ell}{\partial \alpha} = n\lambda - n\alpha^{-1}(\beta - 1) - \alpha^{-1}\beta \sum_{i=1}^{n} \left(\frac{x_i}{\alpha} \right)^{\beta}$$

$$- \lambda \sum_{i=1}^{n} e^{(x_i/\alpha)^{\beta}} \left\{ 1 + \beta \left(\frac{x_i}{\alpha} \right)^{\beta} \right\} - \lambda(1-p)$$

$$\times \sum_{i=1}^{n} \frac{\left[e^{(x_i/\alpha)^{\beta}} \left\{ 1 + \beta(x_i/\alpha)^{\beta} \right\} - 1 \right] e^{-\alpha\lambda(e^{(x_i/\alpha)^{\beta}}-1)}}{1 - (1-p)e^{-\alpha\lambda(e^{(x_i/\alpha)^{\beta}}-1)}},$$

$$\frac{\partial \ell}{\partial \beta} = n\beta^{-1} + \sum_{i=1}^{n} \left\{ 1 + \left(\frac{x_i}{\alpha} \right)^{\beta} - \alpha\lambda \left(\frac{x_i}{\alpha} \right)^{\beta} e^{(x_i/\alpha)^{\beta}} \right\} \ln\left(\frac{x_i}{\alpha} \right)$$

$$- \alpha\lambda(1-p) \sum_{i=1}^{n} \frac{(x_i/\alpha)^{\beta} \ln(x_i/\alpha) e^{(x_i/\alpha)^{\beta}-\alpha\lambda(e^{(x_i/\alpha)^{\beta}}-1)}}{1 - (1-p)e^{-\alpha\lambda(e^{(x_i/\alpha)^{\beta}}-1)}},$$

$$\frac{\partial \ell}{\partial \lambda} = n(\alpha + \lambda^{-1}) - \alpha \sum_{i=1}^{n} e^{(x_i/\alpha)^{\beta}} - \alpha(1-p)$$

$$\times \sum_{i=1}^{n} \frac{\left(e^{(x_i/\alpha)^{\beta}} - 1 \right) e^{-\alpha\lambda(e^{(x_i/\alpha)^{\beta}}-1)}}{1 - (1-p)e^{-\alpha\lambda(e^{(x_i/\alpha)^{\beta}}-1)}},$$

$$\frac{\partial \ell}{\partial p} = n\left\{ (p-1)^{-1} + (p\ln p)^{-1} \right\}$$

$$- \sum_{i=1}^{n} \frac{e^{-\alpha\lambda(e^{(x_i/\alpha)^{\beta}}-1)}}{1 - (1-p)e^{-\alpha\lambda(e^{(x_i/\alpha)^{\beta}}-1)}}.$$

$$(15)$$

The mle of θ, $\hat{\theta}$, is obtained by solving simultaneously the four nonlinear normal equations, $\partial\ell/\partial\alpha = 0$, $\partial\ell/\partial\beta = 0$,

$\partial \ell / \partial \lambda = 0$ and $\partial \ell / \partial p = 0$ by any iterative numerical method such as the Newton-Raphson, quasi-Newton, or Nelder-Mead procedures.

If the true parameter vector is an interior point of the parameter space, then $\hat{\theta}$ can be treated as being approximately trivariate normal with mean θ and covariance matrix the inverse of the Fisher's expected information matrix $J(\theta) = E(I; \theta)$, where $I(\theta)$ is the observed information matrix with elements $I_{ij} = -\partial^2 \ell / \partial \theta_i \partial \theta_j$, $i, j = 1, \ldots, 4$ and the expectation is to be taken with respect to the distribution of X. By differentiating the normal equations in (15) the elements in the upper triangular part of $I(\theta)$ are found to be

$$I_{11} = -n\alpha^{-2}(\beta - 1)$$

$$- \alpha^{-1}\beta \sum_{i=1}^{n} \left(\frac{x_i}{\alpha}\right)^{\beta} \left[\alpha^{-1}(\beta + 1) - \lambda e^{(x_i/\alpha)^{\beta}} \right.$$

$$\left. \times \left\{ \beta \left(\frac{x_i}{\alpha}\right)^{\beta} + \beta + 1 \right\} \right]$$

$$+ \lambda(1-p) \sum_{i=1}^{n} \frac{e^{-\alpha\lambda(e^{(x_i/\alpha)^{\beta}}-1)}}{1 - (1-p)e^{-\alpha\lambda(e^{(x_i/\alpha)^{\beta}}-1)}}$$

$$\times \left\{ e^{(x_i/\alpha)^{\beta}} \left(\lambda \left[e^{(x_i/\alpha)^{\beta}} \left\{ \beta^2 \left(\frac{x_i}{\alpha}\right)^{2\beta} - 1 \right\} + 2 \right] \right. \right.$$

$$\left. - \alpha^{-1}\beta \left(\frac{x_i}{\alpha}\right)^{\beta} \left\{ \beta \left(\frac{x_i}{\alpha}\right)^{\beta} + \beta + 1 \right\} \right)$$

$$\left. - \frac{\lambda(1-p)e^{-\alpha\lambda(e^{(x_i/\alpha)^{\beta}}-1)} \left[e^{(x_i/\alpha)^{\beta}} \left\{ 1 + \beta \left(\frac{x_i}{\alpha}\right)^{\beta} \right\} - 1 \right]^2}{1 - (1-p)e^{-\alpha\lambda(e^{(x_i/\alpha)^{\beta}}-1)}} \right\},$$

$$I_{12} = n\alpha^{-1}$$

$$+ \alpha^{-1} \sum_{i=1}^{n} \left(\frac{x_i}{\alpha}\right)^{\beta} \left[\left\{ 1 + \beta \ln\left(\frac{x_i}{\alpha}\right) \right\} + \lambda e^{(x_i/\alpha)^{\beta}} \right.$$

$$\left. \times \left\{ 1 + (1+\beta) \ln\left(\frac{x_i}{\alpha}\right) + \beta \left(\frac{x_i}{\alpha}\right)^{\beta} \right\} \right]$$

$$+ \lambda(1-p) \sum_{i=1}^{n} \frac{(x_i/\alpha)^{\beta} e^{(x_i/\alpha)^{\beta}} e^{-\alpha\lambda(e^{(x_i/\alpha)^{\beta}}-1)}}{1 - (1-p)e^{-\alpha\lambda(e^{(x_i/\alpha)^{\beta}}-1)}}$$

$$\times \left\{ 1 + (1+\beta) \ln\left(\frac{x_i}{\alpha}\right) + \beta \left(\frac{x_i}{\alpha}\right)^{\beta} \right.$$

$$\left. - \alpha\lambda \ln\left(\frac{x_i}{\alpha}\right) \left[e^{(x_i/\alpha)^{\beta}} \left\{ 1 + \beta \left(\frac{x_i}{\alpha}\right)^{\beta} \right\} - 1 \right] \right.$$

$$\left. \times \left(1 + \frac{(1-p)e^{-\alpha\lambda(e^{(x_i/\alpha)^{\beta}}-1)}}{1 - (1-p)e^{-\alpha\lambda(e^{(x_i/\alpha)^{\beta}}-1)}} \right) \right\},$$

$$I_{13} = -n + \sum_{i=1}^{n} e^{(x_i/\alpha)^{\beta}} \left\{ 1 + \beta \left(\frac{x_i}{\alpha}\right)^{\beta} \right\}$$

$$+ (1-p) \sum_{i=1}^{n} \frac{e^{-\alpha\lambda(e^{(x_i/\alpha)^{\beta}}-1)} \left[e^{(x_i/\alpha)^{\beta}} \left\{ 1 + \beta(x_i/\alpha)^{\beta} \right\} - 1 \right]}{\left\{ 1 - (1-p)e^{-\alpha\lambda(e^{(x_i/\alpha)^{\beta}}-1)} \right\}^2}$$

$$\times \left\{ 1 - (1-p)e^{-\alpha\lambda(e^{(x_i/\alpha)^{\beta}}-1)} - \alpha\lambda \left(e^{(x_i/\alpha)^{\beta}} - 1 \right) \right\},$$

$$I_{14} = - \sum_{i=1}^{n} \frac{e^{-\alpha\lambda(e^{(x_i/\alpha)^{\beta}}-1)}}{\left\{ 1 - (1-p)e^{-\alpha\lambda(e^{(x_i/\alpha)^{\beta}}-1)} \right\}^2},$$

$$I_{22} = n\beta^{-2} - \sum_{i=1}^{n} \left(\frac{x_i}{\alpha}\right)^{\beta} \left\{ 1 - \alpha\lambda \left(1 + \left(\frac{x_i}{\alpha}\right)^{\beta} \right) e^{(x_i/\alpha)^{\beta}} \right\} \ln^2\left(\frac{x_i}{\alpha}\right)$$

$$+ \alpha\lambda(1-p) \sum_{i=1}^{n} \frac{(x_i/\alpha)^{\beta} \ln^2(x_i/\alpha) e^{(x_i/\alpha)^{\beta} - \alpha\lambda(e^{(x_i/\alpha)^{\beta}}-1)}}{\left\{ 1 - (1-p)e^{-\alpha\lambda(e^{(x_i/\alpha)^{\beta}}-1)} \right\}^2}$$

$$\times \left[1 + \left(\frac{x_i}{\alpha}\right)^{\beta} - \alpha\lambda \left(\frac{x_i}{\alpha}\right)^{\beta} e^{(x_i/\alpha)^{\beta}} \right.$$

$$\left. - (1-p) \left\{ 1 + \left(\frac{x_i}{\alpha}\right)^{\beta} \right\} e^{-\alpha\lambda(e^{(x_i/\alpha)^{\beta}}-1)} \right],$$

$$I_{23} = \alpha \sum_{i=1}^{n} \left(\frac{x_i}{\alpha}\right)^{\beta} e^{(x_i/\alpha)^{\beta}} \ln\left(\frac{x_i}{\alpha}\right)$$

$$+ \alpha(1-p) \sum_{i=1}^{n} \frac{(x_i/\alpha)^{\beta} \ln(x_i/\alpha) e^{(x_i/\alpha)^{\beta} - \alpha\lambda(e^{(x_i/\alpha)^{\beta}}-1)}}{\left\{ 1 - (1-p)e^{-\alpha\lambda(e^{(x_i/\alpha)^{\beta}}-1)} \right\}^2}$$

$$\times \left\{ 1 - \alpha\lambda \left(e^{(x_i/\alpha)^{\beta}} - 1 \right) - (1-p)e^{-\alpha\lambda(e^{(x_i/\alpha)^{\beta}}-1)} \right\},$$

$$I_{24} = -\alpha\lambda \sum_{i=1}^{n} \frac{(x_i/\alpha)^{\beta} \ln(x_i/\alpha) e^{(x_i/\alpha)^{\beta} - \alpha\lambda(e^{(x_i/\alpha)^{\beta}}-1)}}{\left\{ 1 - (1-p)e^{-\alpha\lambda(e^{(x_i/\alpha)^{\beta}}-1)} \right\}^2},$$

$$I_{33} = n\lambda^{-2} - \alpha^2(1-p) \sum_{i=1}^{n} \frac{\left(e^{(x_i/\alpha)^{\beta}} - 1 \right)^2 e^{-\alpha\lambda(e^{(x_i/\alpha)^{\beta}}-1)}}{\left\{ 1 - (1-p)e^{-\alpha\lambda(e^{(x_i/\alpha)^{\beta}}-1)} \right\}^2},$$

$$I_{34} = -\alpha \sum_{i=1}^{n} \frac{\left(e^{(x_i/\alpha)^{\beta}} - 1 \right) e^{-\alpha\lambda(e^{(x_i/\alpha)^{\beta}}-1)}}{\left\{ 1 - (1-p)e^{-\alpha\lambda(e^{(x_i/\alpha)^{\beta}}-1)} \right\}^2},$$

$$I_{44} = n\left\{ (p-1)^{-2} - (1 + \ln p)(p \ln p)^{-2} \right\}$$

$$- \sum_{i=1}^{n} \frac{e^{-2\alpha\lambda(e^{(x_i/\alpha)^{\beta}}-1)}}{\left\{ 1 - (1-p)e^{-\alpha\lambda(e^{(x_i/\alpha)^{\beta}}-1)} \right\}^2}.$$

$$(16)$$

The latter is a consistent estimator of $J(\theta)$ and can be used for constructing asymptotic confidence intervals for the parameters. However, if any of the true parameter values is zero then the asymptotic distribution of the maximum likelihood estimators is a mixture distribution [19]; in this

TABLE 1: Parameter estimates, values of the log-likelihood (LL) and Akaike information criterion (AIC), and Kolmogorov-Smirnov (K-S) statistic obtained from the fit of each of the four distributions, to the times of first failure of devices (data set 1) and the machine's subsystems lifetimes (data set 2).

Data set	Distribution	$\hat{\theta} = (\hat{\alpha}, \hat{\beta}, \hat{\lambda}, \hat{p})$	LL	AIC	K-S (P value)
1 ($n = 50$)	EXTG	$(14.06, .4838, .042, 313.57)$	227.25	462.50	.1470 (.2300)
	XTG	$(13.747, .588, .00876, —)$	231.65	469.29	.1597 (.1562)
	Chen	$(—, .3444, .0205, —)$	233.17	470.34	.1676 (.1204)
	DAL	$(6.2975, .8074, .0053, —)$	236.04	478.08	.1981 (.03945)
2 ($n = 44$)	EXTG	$(.0831, .2007, .0211, 20.4087)$	348.01	704.02	.1397 (.3262)
	XTG	$(.0806, .2086, .0057, —)$	351.43	708.86	.1667 (.1544)
	Chen	$(—, .2473, .0024, —)$	351.25	706.50	.1737 (.1244)
	DAL	$(1.752, .7771, .0021, —)$	359.44	724.89	.2875 (.0001)

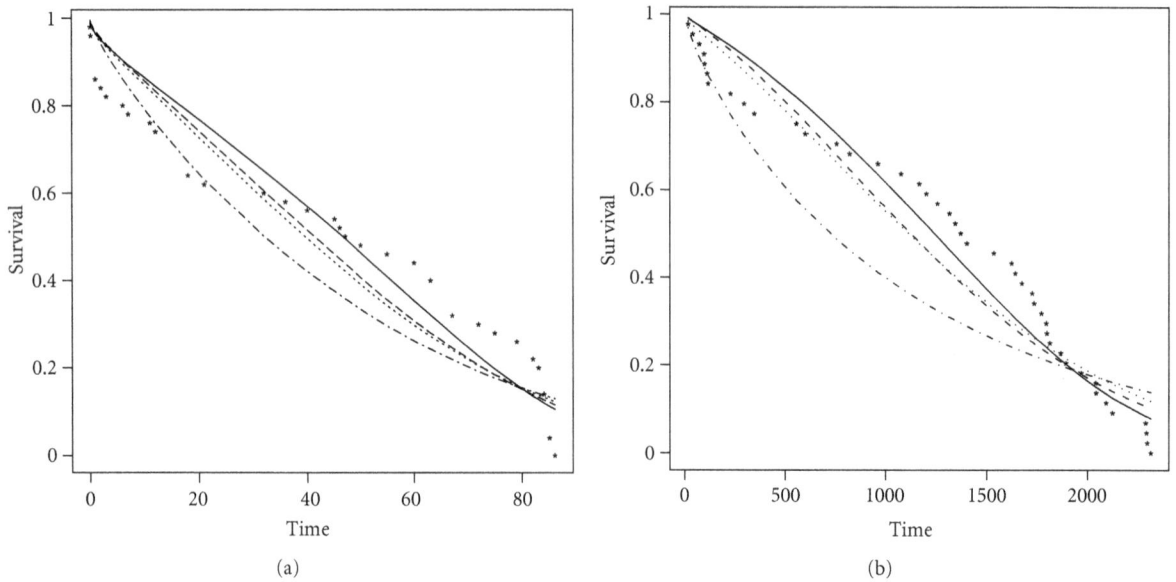

(a)   (b)

FIGURE 3: Reliability curves of the empirical distribution (starred line), the EXTG distribution (solid line), the XTG distribution (dashed line), the Chen distribution (dotted line), and the DAL distribution (dot-dashed line) for the times of first failure of fifty devices (a) and the forty-four failure times of a machine's subsystem (b).

case obtaining the asymptotic confidence intervals becomes quite difficult and shall not be pursued here.

3.3. Examples. In this section two sets of real data are considered in order to test the goodness of fit of the proposed model. The first set of data consists of times to first failure of fifty devices [20]. The second set of data involves forty-four observations obtained from a life test concerning failure times (in hours) of all subsystems of a machine, that is, engine, hydraulic and air-conditioning subsystems, brakes, transmissions, tyres and wheels, body and chassis [21, 22]; in both cases, the data were grouped and the empirical hazard rate was estimated by many methods to indicate a BT shape. In addition to the EXTG, the XTG distribution, the two-parameter Chen [12] distribution, and the three-parameter model introduced by Dimitrakopoulou et al. [3], were fitted to the datasets; for brevity, hereinafter we shall be referring to the latter two models as the Chen and DAL distributions respectively. The fit of each distribution was

examined by the Akaike information criterion (AIC) and the Kolmogorov-Smirnov (K-S) goodness-of-fit test using maximum likelihood estimates; the estimates, the maximized log-likelihoods, the values of the AIC, and the values of the K-S statistic with the associated P-values are presented in Table 1. Furthermore, the values of the likelihood ratio test statistic for testing $H_0 : p = 1$, calculated from the first and the second set of data, were 8.7939 ($P = 0.003$) and 6.8366 ($P = 0.0089$), respectively; the analogous computations for testing $H_0 : p = \alpha = 1$ were 11.8371 ($P = 0.0027$) and 6.479 ($P = 0.0392$). All the results indicate that the EXTG distribution describes these data better than the other models; these findings are also supported by the empirical and fitted survivor functions, plotted in Figure 3.

Acknowledgment

The authors would like to thank a referee for useful comments and suggestions.

References

[1] K. Adamidis and S. Loukas, "A lifetime distribution with decreasing failure rate," *Statistics & Probability Letters*, vol. 39, no. 1, pp. 35–42, 1998.

[2] K. Adamidis, T. Dimitrakopoulou, and S. Loukas, "On an extension of the exponential-geometric distribution," *Statistics & Probability Letters*, vol. 73, no. 3, pp. 259–269, 2005.

[3] T. Dimitrakopoulou, K. Adamidis, and S. Loukas, "A lifetime distribution with an upside-down bathtub-shaped hazard function," *IEEE Transactions on Reliability*, vol. 56, no. 2, pp. 308–311, 2007.

[4] C. Kus, "A new lifetime distribution," *Computational Statistics and Data Analysis*, vol. 51, no. 9, pp. 4497–4509, 2007.

[5] A. W. Marshall and I. Olkin, *Life Distributions*, Springer, New York, NY, USA, 2007.

[6] W. Barreto-Souza, A. L. de Morais, and G. M. Cordeiro, "The weibull-geometric distribution," *Journal of Statistical Computation and Simulation*, vol. 81, no. 5, pp. 645–657, 2011.

[7] R. Tahmasbi and S. Rezaei, "A two-parameter lifetime distribution with decreasing failure rate," *Computational Statistics and Data Analysis*, vol. 52, no. 8, pp. 3889–3901, 2008.

[8] W. Barreto-Souza and F. Cribari-Neto, "A generalization of the exponential-Poisson distribution," *Statistics & Probability Letters*, vol. 79, no. 24, pp. 2493–2500, 2009.

[9] M. Chahkandi and M. Ganjali, "On some lifetime distributions with decreasing failure rate," *Computational Statistics and Data Analysis*, vol. 53, no. 12, pp. 4433–4440, 2009.

[10] A. W. Marshall and I. Olkin, "A new method for adding a parameter to a family of distributions with application to the exponential and Weibull families," *Biometrika*, vol. 84, no. 3, pp. 641–652, 1997.

[11] M. Xie, Y. Tang, and T. N. Goh, "A modified Weibull extension with bathtub-shaped failure rate function," *Reliability Engineering & System Safety*, vol. 76, no. 3, pp. 279–285, 2002.

[12] Z. Chen, "A new two-parameter lifetime distribution with bathtub shape or increasing failure rate function," *Statistics & Probability Letters*, vol. 49, no. 2, pp. 155–161, 2000.

[13] Y. Tang, M. Xie, and T. N. Goh, "Statistical analysis of a Weibull extension model," *Communications in Statistics—Theory and Methods*, vol. 32, no. 5, pp. 913–928, 2003.

[14] R. E. Barlow and F. Proschan, *Statistical Theory of Reliability and Life Testing:*, vol. Probability Models, To Begin With, Silver Spring, Md, USA.

[15] B. C. Arnold, N. Balakrishnan, and H. N. Nataraja, *First Course in Order Statistics*, John Wiley, New York, NY, USA, 1992.

[16] R. E. Glaser, "Bathtub and related failure rate characterizations," *Journal of the American Statistical Association*, vol. 75, pp. 667–672, 1980.

[17] L. Cui and M. Xie, "Some normal approximations for renewal function of large Weibull shape parameter," *Communications in Statistics B*, vol. 32, no. 1, pp. 1–16, 2003.

[18] R. Jiang, "A simple approximation for the renewal function with an increasing failure rate," *Reliability Engineering & System Safety*, vol. 95, pp. 963–969, 2010.

[19] S. G. Self and K.-L. Liang, "Asymptotic properties of maximum likelihood estimators and likelihood ratio tests under nonstandard conditions," *Journal of the American Statistical Association*, vol. 82, pp. 605–610, 1987.

[20] M. V. Aarset, "How to identify a bathtub hazard rate," *IEEE Transactions on Reliability*, vol. 36, no. 1, pp. 106–108, 1987.

[21] U. Kumar, B. Klefsjö, and S. Granholm, "Reliability investigation for a fleet of load haul dump machines in a Swedish mine," *Reliability Engineering & System Safety*, vol. 26, no. 4, pp. 341–361, 1989.

[22] G. Pulcini, "Modeling the failure data of a repairable equipment with bathtub type failure intensity," *Reliability Engineering & System Safety*, vol. 71, no. 2, pp. 209–218, 2001.

Permissions

The contributors of this book come from diverse backgrounds, making this book a truly international effort. This book will bring forth new frontiers with its revolutionizing research information and detailed analysis of the nascent developments around the world.

We would like to thank all the contributing authors for lending their expertise to make the book truly unique. They have played a crucial role in the development of this book. Without their invaluable contributions this book wouldn't have been possible. They have made vital efforts to compile up to date information on the varied aspects of this subject to make this book a valuable addition to the collection of many professionals and students.

This book was conceptualized with the vision of imparting up-to-date information and advanced data in this field. To ensure the same, a matchless editorial board was set up. Every individual on the board went through rigorous rounds of assessment to prove their worth. After which they invested a large part of their time researching and compiling the most relevant data for our readers. Conferences and sessions were held from time to time between the editorial board and the contributing authors to present the data in the most comprehensible form. The editorial team has worked tirelessly to provide valuable and valid information to help people across the globe.

Every chapter published in this book has been scrutinized by our experts. Their significance has been extensively debated. The topics covered herein carry significant findings which will fuel the growth of the discipline. They may even be implemented as practical applications or may be referred to as a beginning point for another development. Chapters in this book were first published by Hindawi Publishing Corporation; hereby published with permission under the Creative Commons Attribution License or equivalent.

The editorial board has been involved in producing this book since its inception. They have spent rigorous hours researching and exploring the diverse topics which have resulted in the successful publishing of this book. They have passed on their knowledge of decades through this book. To expedite this challenging task, the publisher supported the team at every step. A small team of assistant editors was also appointed to further simplify the editing procedure and attain best results for the readers.

Our editorial team has been hand-picked from every corner of the world. Their multi-ethnicity adds dynamic inputs to the discussions which result in innovative outcomes. These outcomes are then further discussed with the researchers and contributors who give their valuable feedback and opinion regarding the same. The feedback is then collaborated with the researches and they are edited in a comprehensive manner to aid the understanding of the subject.

Apart from the editorial board, the designing team has also invested a significant amount of their time in understanding the subject and creating the most relevant covers. They scrutinized every image to scout for the most suitable representation of the subject and create an appropriate cover for the book.

The publishing team has been involved in this book since its early stages. They were actively engaged in every process, be it collecting the data, connecting with the contributors or procuring relevant information. The team has been an ardent support to the editorial, designing and production team. Their endless efforts to recruit the best for this project, has resulted in the accomplishment of this book. They are a veteran in the field of academics and their pool of knowledge is as vast as their experience in printing. Their expertise and guidance has proved useful at every step. Their uncompromising quality standards have made this book an exceptional effort. Their encouragement from time to time has been an inspiration for everyone.

The publisher and the editorial board hope that this book will prove to be a valuable piece of knowledge for researchers, students, practitioners and scholars across the globe.

List of Contributors

Saad T. Bakir
College of Business Administration, Alabama State University, P.O. Box 271, Montgomery, AL 36101, USA

Sonia M. Orlando Gibelli and Sérgio Q. Bogado Leite
Comiss~ao Nacional de Energia Nuclear, DRS/CGRC, 22294-900 Rio de Janeiro, RJ, Brazil

P. F. Frutuoso e Melo
Programa de Engenharia Nuclear, COPPE/UFRJ, 21941-972 Rio de Janeiro, RJ, Brazil

Sajid Ali
Department of Decision Sciences, Bocconi University, via Roenthen 1, 20136 Milan, Italy

Harish Garg, Monica Rani and S. P. Sharma
Department of Mathematics, Indian Institute of Technology, Roorkee 247667, Uttarakhand, India

Wenhao Gui
Department of Mathematics and Statistics, University of Minnesota Duluth, Duluth, MN 55812, USA

Mohamed Mubarak
Mathematics Department, Faculty of Science, Minia University, El-Minia 61519, Egypt
Mathematics Department, University College in Lieth, Umm Al-Qura University, Makkah 311, Saudi Arabia

Navid Feroze
Department of Mathematics and Statistics, AIOU, Islamabad 44000, Pakistan

Muhammad Aslam
Department of Statistics, Quaid-i-Azam University, Islamabad 44000, Pakistan

Saad T. Bakir
College of Business Administration, Alabama State University, 915 South Jackson Street, Montgomery, AL 36104, USA

Bander Al-Zahrani
Department of Statistics, Faculty of Sciences, King Abdulaziz University, Jeddah 21589, Saudi Arabia

Mashail Al-Sobhi
Department of Mathematics, Umm Al-Qura University, Makkah, Saudi Arabia

Faisal G. Khamis
Faculty of Economics and Administrative Sciences, AL-Zaytoonah University of Jordan, P.O. Box 130, Amman 11733, Jordan

Ali Salmasnia
Department of Industrial Engineering, Faculty of Engineering, Tarbiat Modares University, Tehran, Iran

Mahdi Bastan
Department of Industrial Engineering, Eyvanekey University, Semnan, Iran

Asghar Moeini
Department of Industrial Engineering, Faculty of Engineering, Shahed University, Tehran, Iran

Kanwar Sen
Department of Statistics, University of Delhi, Delhi 7, India

Pooja Mohan
RMS India, A-7, Sector 16, Noida 201 301, India

Manju Lata Agarwal
The Institute for Innovation and Inventions with Mathematics and IT (IIIMIT), Shiv Nadar University, Greater Noida 203207, India

Alessandro Barbiero
Department of Economics, Management and Quantitative Methods, University of Milan, 20122 Milan, Italy

Eftychia C. Marcoulaki
System Reliability and Industrial Safety Laboratory, National Centre for Scientific Research, "Demokritos", P.O. Box 60228, Agia Paraskevi, 15310 Athens, Greece

Asokan Mulayath Variyath and Jayasankar Vattathoor
Department of Mathematics and Statistics, Memorial University of Newfoundland, St. John's, NL, Canada A1C 5S7

Patrick Hester
Engineering Management and Systems Engineering Department, Old Dominion University, Kaufman Hall, Room 241, Norfolk, VA 23529, USA

Mohammad Abdolshah
Engineering Faculty, Islamic Azad University, Semnan Branch, Semnan 35136-93688, Iran

Qihong Duan, Ying Wei and Xiang Chen
Department of Statistics, School of Mathematics and Statistics, Xi'an Jiaotong University, Shaanxi, Xi'an 710049, China

Walid Gani
LARODEC, ISG, University of Tunis, 41 Avenue de la Libert'e, 2000 Le Bardo, Tunisia

Mohamed Limam
Dhofar University, P.O. Box 2509, 211 Salalah, Oman

Sunil Dambhare
Department of Mechanical Engineering, PVPIT, Bavdhan, Pune, Maharashtra 411021, India

Siddhant Aphale, Kiran Kakade and Tejas Thote
Mechanical Engineering, PVPIT, Bavdhan, Pune, Maharashtra 411021, India

Atul Borade
Department of Mechanical Engineering, JDIET, Yavatmal, Maharashtra 445001, India

Vasileios Pappas and Sotirios Loukas
Department of Mathematics, University of Ioannina, 45110 Ioannina, Greece

Konstantinos Adamidis
Department of Business Administration of Food and Agricultural Enterprises, University of Ioannina, 30100 Agrinio, Greece

www.ingramcontent.com/pod-product-compliance
Lightning Source LLC
Chambersburg PA
CBHW050441200326
41458CB00014B/5028